CYBERSECURITY AND COGNITIVE SCIENCE

CYBERSECURITY AND COGNITIVE SCIENCE

Edited by

AHMED A. MOUSTAFA

ELSEVIER

ACADEMIC PRESS

An imprint of Elsevier

Academic Press is an imprint of Elsevier
125 London Wall, London EC2Y 5AS, United Kingdom
525 B Street, Suite 1650, San Diego, CA 92101, United States
50 Hampshire Street, 5th Floor, Cambridge, MA 02139, United States
The Boulevard, Langford Lane, Kidlington, Oxford OX5 1GB, United Kingdom

ISBN 978-0-323-90570-1

For information on all Academic Press publications
visit our website at https://www.elsevier.com/books-and-journals

Publisher: Nikki P. Levy
Acquisitions Editor: Joslyn Chaiprasert-Paguio
Editorial Project Manager: Barbara L. Makinster
Production Project Manager: Omer Mukthar
Cover Designer: Christian J. Bilbow

Typeset by STRAIVE, India

Contents

14. Modeling the effects of network size in a deception game involving honeypots

Harsh Katakwar, Shashank Uttrani, Palvi Aggarwal, and Varun Dutt

15. Computational modeling of decisions in cyber-security games in the presence or absence of interdependence information

Zahid Maqbool, V.S. Chandrasekhar Pammi, and Varun Dutt

Contributors

Palvi Aggarwal Department of Computer Science, The University of Texas at El Paso, El Paso, TX, United States

Micheline Al Harrack Faculty of Computing and Information Technology, King Abdulaziz University, Jeddah, Saudi Arabia

Hany Alashwal College of Information Technology, United Arab Emirates University, Abu Dhabi, United Arab Emirates

Prashant Anantharaman Department of Computer Science, Dartmouth College, Hanover, NH, United States

Annika Andreasson KTH Royal Institute of Technology, Stockholm, Sweden

Henrik Artman KTH Royal Institute of Technology; FOI Swedish Defence Research Agency, Stockholm, Sweden

Adham Atyabi Department of Computer Science, University of Colorado Colorado Springs, Colorado Springs, CO, United States

Abubakar Bello Cyber Security and Behaviour, School of Social Sciences, Western Sydney University, Sydney, NSW, Australia

Shada Bennbaia Department of Mechanical Engineering, College of Engineering, Qatar University, Doha, Qatar

Lucy R. Betts Department of Psychology, Nottingham Trent University, Nottingham, United Kingdom

Samuela Bolgan Department of General Psychology, University of Padova, Padova, Italy

Joel Brynielsson KTH Royal Institute of Technology; FOI Swedish Defence Research Agency, Stockholm, Sweden

Ferhat Ozgur Catak Department of Electrical Engineering and Computer Science, University of Stavanger, Stavanger, Norway

Anthony An Duong School of Social Sciences, Western Sydney University, Sydney, NSW, Australia

Varun Dutt Applied Cognitive Science Laboratory, Indian Institute of Technology Mandi, Kamand, India

Mark Evans School of Computer Science and Informatics, De Montfort University, Leicester, United Kingdom

Ulrik Franke KTH Royal Institute of Technology, Stockholm; RISE Research Institutes of Sweden, Kista, Sweden

Ying He School of Computer Science, University of Nottingham, Nottingham, United Kingdom

Helge Janicke Cyber Security Cooperative Research Centre, Joondalup, WA, Australia

Harsh Katakwar Applied Cognitive Science Laboratory, Indian Institute of Technology Mandi, Kamand, India

Blerina Kellezi Department of Psychology, Nottingham Trent University, Nottingham, United Kingdom

Vijay H. Kothari Department of Computer Science, Dartmouth College, Hanover, NH, United States

Peter J.R. Macaulay School of Psychology, University of Derby, Derby, United Kingdom

Leandros Maglaras School of Computer Science and Informatics, De Montfort University, Leicester, United Kingdom

Zahid Maqbool Applied Cognitive Science Laboratory, Indian Institute of Technology Mandi, Kamand, India

Alana Maurushat School of Social Sciences, Western Sydney University, Sydney, NSW, Australia

Alex V. Mbaziira College of Business, Innovation, Leadership and Technology, Marymount University, Arlington, VA, United States

John McAlaney Department of Psychology, Bournemouth University, Poole, United Kingdom

Rosana Montañez Department of Computer Science, University of Texas at San Antonio, San Antonio, TX, United States

Ahmed A. Moustafa School of Psychology, Faculty of Society and Design, Bond University, Gold Coast, Queensland, Australia; Department of Human Anatomy and Physiology, the Faculty of Health Sciences, University of Johannesburg, South Africa

V.S. Chandrasekhar Pammi Centre of Behavioral and Cognitive Sciences, University of Allahabad, Allahabad, India

Veronica Power School of Social Sciences, Western Sydney University, Sydney, NSW, Australia

Elena Rusconi Department of Psychology and Cognitive Science, University of Trento, Rovereto; Centre of Security and Crime Sciences, University of Trento and University of Verona, Trento, Italy

Maha F. Sabir Faculty of Computing and Information Technology, King Abdulaziz University, Jeddah, Saudi Arabia

Sean W. Smith Department of Computer Science, Dartmouth College, Hanover, NH, United States

James Stiller Department of Psychology, University of Chichester, Chichester, United Kingdom

Devrim Unal KINDI Center for Computing Research; Department of Mechanical Engineering, College of Engineering, Qatar University, Doha, Qatar

Shashank Uttrani Applied Cognitive Science Laboratory, Indian Institute of Technology Mandi, Kamand, India

Stefan Varga KTH Royal Institute of Technology; Swedish Armed Forces Headquarters, Stockholm, Sweden

Niklas Vilhelm KTH Royal Institute of Technology, Stockholm, Sweden; Norwegian National Security Authority, Sandvika, Norway

Jennifer Wagner School of Applied Sciences, Abertay University, Dundee, United Kingdom

Shouhuai Xu Department of Computer Science, University of Colorado Colorado Springs, Colorado Springs, CO, United States

PART I

Social engineering, security, and cyber attacks

CHAPTER

1

Social engineering attacks and defenses in the physical world vs. cyberspace: A contrast study

Rosana Montañez[a], *Adham Atyabi*[b], *and Shouhuai Xu*[b]

[a]Department of Computer Science, University of Texas at San Antonio, San Antonio, TX, United States, [b]Department of Computer Science, University of Colorado Colorado Springs, Colorado Springs, CO, United States,

1 Introduction

Social engineering attacks are prevalent in both the physical world and cyberspace. Intuitively, these attacks attempt to cause an error, or failure, in a target or victim's decision-making process to benefit the attacker. The prevalence of these attacks can be attributed to their low cost and effectiveness. In the physical world, social engineering attacks share many similarities with scams and fraud. In cyberspace, social engineering attacks are often the first step of sophisticated attacks that can cause substantial damages.

1.1 Our contributions

In this chapter, we make four contributions. First, we propose a methodology to unify social engineering attacks in the physical world and their counterpart in cyberspace into a single framework. The methodology is novel because it takes a unique perspective based on the following observation. In principle, a social engineering attack attempts to manipulate a victim into complying with a request from the attacker by leveraging

Cybersecurity and Cognitive Science
https://doi.org/10.1016/B978-0-323-90570-1.00012-7

aspects of *individual and social cognition*, which provides a compelling perspective for the study. Individual cognition examines the internal processes that lead to decision making and behavior in an individual, whereas social cognition explores the external social aspects that affect these internal processes. These two complementary aspects of cognition provide a basis for interpreting information: one based on a self-centered perspective influenced by mental processing of sensory input and the other influenced by the interaction with other humans. This perspective guides us to propose a model for describing attacker-victim interactions in both the physical world and cyberspace.

Second, we systematize social engineering attack techniques in the physical world and cyberspace. In total, we systematize 7 techniques (belonging to five categories) used in social engineering attacks in the physical world and 13 techniques (belonging to four categories) used in social engineering attacks in cyberspace. To the best of our knowledge, this is the first systematization of social engineering attack techniques in the physical world and cyberspace.

Third, we systematize defenses against social engineering attacks (i.e., social engineering defenses) in the physical world and cyberspace. For social engineering defenses in the physical world, we systematize six defenses (belonging to two categories, namely preventive defenses and proactive defenses). For social engineering defenses in cyberspace, we systematize 11 defenses (belonging to three categories, namely preventive defenses, proactive defenses, and reactive defenses). To the best of our knowledge, this is the first systematization of social engineering defenses in the physical world and cyberspace.

Fourth, we conduct a contrast analysis of the social engineering attacks and defenses in the physical world and cyberspace. The analysis draws a number of insights, such as (i) we should strive to achieve social engineering resistance by design and (ii) there are no "silver bullet" defenses that can work against social engineering attacks in both the physical world and cyberspace because these two worlds exhibit different features and demand tailored solutions. The analysis sheds light on the future research.

1.2 Related work

We reviewed prior studies on social engineering attacks through the lens of cognition, leading to the distinction of *individual* versus *social cognition*. The former studies the mental processes that affect attention, memory, perception, decision making, and information processing; the latter studies how social interactions affect those mental processes. These prior studies are relevant to the present one because they provide insight into the factors and interactions that facilitate the exploitation of human psychological weaknesses in social engineering attacks. Since most studies on social engineering attacks in the physical world are based on anecdotal accounts (Dove, 2020; Stajano & Wilson, 2009), meaning there is a lack of peer-reviewed, quantitative experiments, we mitigate this limitation by including studies in scams, because scams share many characteristics with social engineering attacks.

1.2.1 *Prior studies related to individual cognition*

We adopt the distinction between cognitive *attributes* and *attitudes* described in the literature, namely that an *attribute* is "an emergent property of an individual embodied in social practices" (Guyon, Falissard, & Kop, 2017), whereas an *attitude* is the "relatively enduring predisposition to respond favorably or unfavorably toward something" (Simon, 1976).

Prior studies related to individual cognitive attributes

We consider the following attributes that are related to social engineering attacks: *personality, expertise, individual difference, culture, workload, stress,* and *vigilance*. These attributes affect one's susceptibility to social engineering attacks.

For *personality*, there is no consensus on its relationship with one's susceptibility to social engineering attacks in cyberspace. Personality can be characterized by five main domains, namely *neuroticism, openness, extroversion, consciousness,* and *agreeableness*. We revise the state-of-the-art understanding about their relationships to social engineering attacks. (i) In terms of neuroticism, two studies suggest that a high neuroticism is associated with lower self-efficacy (i.e., user confidence to manage a cyber risk) (Halevi et al., 2016) and increase one's susceptibility to phishing (Halevi, Lewis, & Memon, 2013), but a study on phishing (Cho, Cam, & Oltramari, 2016) suggests that a high neuroticism decreases one's susceptibility to phishing attacks. (ii) In terms of *openness*, one study (Halevi et al., 2013) suggests that a high openness increases one's susceptibility to privacy attacks, but other studies (Halevi et al., 2016; Pattinson, Jerram, Parsons, McCormac, & Butavicius, 2012) suggest a high openness reduces one's susceptibility to phishing attacks. (iii) In terms of *extroversion*, one study (Lawson, Crowson, & Mayhorn, 2018) suggests that a high extroversion increases one's susceptibility to phishing attacks, but another study (Pattinson et al., 2012) suggests that a high extroversion decreases one's susceptibility to phishing attacks. (iv) In terms of *consciousness*, two studies (Halevi et al., 2016; Lawson et al., 2018) suggest that a high consciousness reduces one's susceptibility to phishing, but another study (Halevi,

Memon, & Nov, 2015) suggests that a high consciousness increases one's susceptibility to targeted social engineering attacks. (v) In terms of *agreeableness*, two studies (Cho et al., 2016; Darwish, El Zarka, & Aloul, 2012) show that a high agreeableness increases one's susceptibility to phishing attacks.

The impact of the rest of the factors is briefly reviewed as follows. For *expertise*, cybersecurity expertise does reduce one's susceptibility to phishing (Kumaraguru, Acquisti, & Cranor, 2006). In the physical world, it improves threat appraisal and risk perception (Klein & Calderwood, 1991). For nonexperts, experience in combination with knowledge reduces one's susceptibility to phishing (Abbasi, Zahedi, & Chen, 2016; Downs, Holbrook, & Cranor, 2006; Gavett et al., 2017; Harrison, Svetieva, &Vishwanath, 2016; Pattinson et al., 2012; Wright & Marett, 2010), and malicious social media messages (Redmiles, Chachra, & Waismeyer, 2018). But pure awareness of cyber threats does not appear to reduce one's susceptibility to social engineering attacks (Downs et al., 2006; Junger, Montoya, & Overink, 2017; Sheng, Holbrook, Kumaraguru, Cranor, & Downs, 2010). However, in scams, knowledge of the threat reduces vulnerabilities (Langenderfer & Shimp, 2001), but domain-specific knowledge may increase susceptibility to scams (Lea, Fischer, & Evans, 2009). For *individual differences*, age increases both young (18–25) (Howe, Ray, Roberts, Urbanska, & Byrne, 2012; Sheng et al., 2010) and old (65+) (Gavett et al., 2017; Howe et al., 2012) people's susceptibility to phishing and online threats. In addition, older people are generally more susceptible to spear phishing (Lin et al., 2019) and scams both in the physical world (Langenderfer & Shimp, 2001) and cyberspace (Grimes, Hough, & Signorella, 2007). Most studies have found no relationship between one's gender and susceptibility to social engineering attacks in cyberspace (Bullee, Montoya, Junger, & Hartel, 2017; Purkait, Kumar De, & Suar, 2014; Rocha Flores, Holm, Svensson, & Ericsson, 2014; Sawyer & Hancock, 2018) or in the physical world (Lea et al., 2009). For *culture*, studies show that individuals are more susceptible to social engineering messages (Al-Hamar, Dawson, & Guan, 2010; Bohm, 2011; Redmiles et al., 2018; Sharevski et al., 2019; Tembe et al., 2014) and scams (Dove, 2020) that align with their cultural norms. For *workload*, it is known that a high email load (Vishwanath, Herath, Chen, Wang, & Rao, 2011), work overload (Jalali, Bruckes, Westmattelmann, & Schewe, 2020), and inattentional blindness as a result of workload (Pfleeger & Caputo, 2012) might increase one's susceptibility to social engineering attacks. For *stress*, one study (Stajano & Wilson, 2009) shows that stress increases one's susceptibility to scams attacks. For *vigilance*, one study (Purkait et al., 2014) shows that a high attentional vigilance reduces one's susceptibility to phishing websites. In the physical world, vigilance reduces scam victimization (Dove, 2020).

Prior work related to individual cognitive attitudes

We consider the following attitudes that are related to social engineering attacks: *trust attitude*, *suspicion attitude*, and *risk attitude*. These attitudes affect one's susceptibility to social engineering attacks. For *trust attitude*, a high trust attitude incurs a high susceptibility to social engineering attacks (Abbasi et al., 2016; Halevi et al., 2013; Rocha Flores et al., 2014; Workman, 2008) and scams (Langenderfer & Shimp, 2001). For *suspicion attitude*, an individual with a high suspicion is less susceptible to social engineering attacks (Tembe et al., 2014; Vishwanath, Harrison, & Ng, 2018; Wright & Marett, 2010). For *risk attitude*, a high-risk perception reduces one's susceptibility to social engineering attacks (Halevi et al., 2016, 2015; Howe et al., 2012; Rocha Flores et al., 2014; Sheng et al., 2010) and low-risk perception increase susceptibility to scams (Dove, 2020).

1.2.2 *Prior work related to social cognition*

Two factors have been investigated in the literature. One is *persuasion*, where social engineering attacks (Rajivan & Gonzalez, 2018; Van Der Heijden & Allodi, 2019) and scams (Langenderfer & Shimp, 2001; Lea et al., 2009) with a high persuasive capability are more successful. The other is *scam compliance* where using persuasion along with emotional and visceral triggers increases scam successes (Lea et al., 2009; Stajano & Wilson, 2009).

1.3 Outline

The rest of the chapter is organized as follows. Section 2 presents the terminology and methodology used in this study. Section 3 characterizes social engineering attack model, techniques, and defenses in the physical world. Section 4 characterizes social engineering attack model, techniques, and defenses in the physical world. Section 5 describes our contrast analysis between the physical world and cyberspace and future research directions. Section 6 concludes this chapter.

2 Terminology and methodology

2.1 Terminology

We propose using the following terminology to describe social engineering attacks in the physical world and cyberspace, while noting that many of these terms are adapted from cyberattack terminology. The term *target*, which is often used in cyber social engineering literature, describes a human being who may have some exploitable psychological weakness which can be leveraged by the attacker in question. Since the term *victim* is often used in other social engineering literature, we will use these two terms interchangeably. A *target* (i.e., victim) can be characterized by an *attack surface*, which is defined as the set of vulnerabilities (i.e., attack vectors) that can be exploited by the attacker to victimize the target. We use the term *social engineering attacks* to describe the attacks that exploit psychological weaknesses or vulnerabilities of humans to achieve malicious goals, such as acquiring information, access, or assets.

2.2 Methodology

Our methodology is inspired by the following observations. Information processing is central to understanding social engineering attacks because (i) it describes the process by which external sensory inputs are processed by internal cognitive units to interpret information and (ii) its output shapes one's decision making, judgment, and behavior. Social engineering attacks involve the processing of persuasive messaging to influence a desired behavior. The elaboration likelihood model (ELM) has been used to study information processing of persuasive messages (Cacioppo & Petty, 1984). ELM is a dual information processing system in which the processing of information can take one of two routes: *central* versus *peripheral* route, despite that the information processing path can alter between these two routes in the course of information processing. Peripheral processing is fast, based on heuristics and requires less cognitive effort; whereas central processing is slow, analytical, and requires cognitive effort. Studies have demonstrated how the processing route affects the outcome of a social engineering attack. A social engineering attack often succeeds by making a target trap into the peripheral processing route (Dove, 2020; Lea et al., 2009). That is, a sophisticated social engineering attack often induces peripheral information processing and an effective defense should trigger central information processing. Although factors beyond the attacker's control may also affect the information processing route (e.g., the environment where the interaction occurs), in this chapter, we focus on the factors that are under the attacker's control. Two key factors that affect the selection of routes are *trust* and *suspicion*, where trust encourages peripheral processing and suspicion encourages central processing.

Although social engineering attacks in cyberspace exploit the same psychological principles that are exploited by social engineering attacks in the physical world, there is a significant difference between social engineering attacks in these two worlds, namely how communication is mediated. This difference may affect victims' performance and provide different opportunities to the attackers. The proceeding observations prompt us to characterize social engineering attacks in the two worlds via: *model*, *attacks*, and *defenses*.

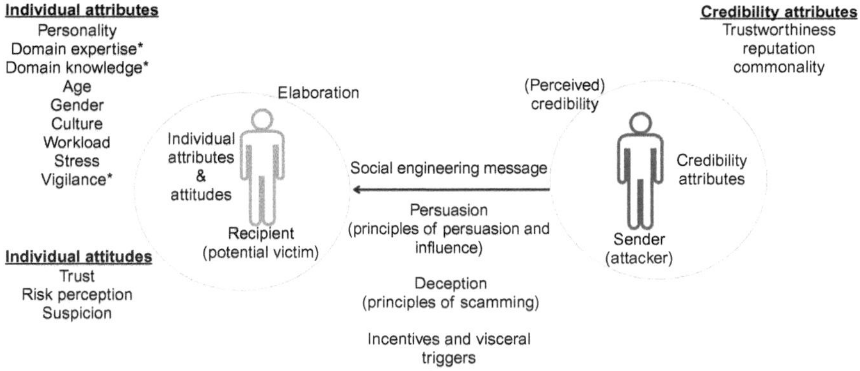

FIG. 1 Our social engineering attack model, where asterisk (*) indicates that the attribute in question can reduce one's susceptibility to social engineering attacks.

2.2.1 *Social engineering attack model*

Fig. 1 highlights our social engineering model, which illustrates the elements and interactions pertinent to social engineering attacks. The model describes the three main components of a social engineering attack: *the attacker, the message,* and *the victim*. The model is equally applicable to social engineering attacks in the physical world and in cyberspace.

2.2.2 *Social engineering attack techniques*

In order to describe social engineering attacks, we use the term *attack kill chain* to describe the sequence of orchestrated attack *phases*, where each *attack phase* can be described by a high-level abstraction that provides a logical grouping of distinct actions conducted by the attacker. From the planning to the termination of an attack, there are multiple phases, such as those specified by the Lockheed Martin kill chain (Hutchins, Cloppert, & Amin, 2011) and the Mandiant kill chain (Mandiant, 2013). *Phases* are divided into *tactics* and *techniques*, which are adapted from the MITRE ATT&CK framework (Strom, 2018). The term *tactic* describes a short-term objective of performing an attack action and the term *technique* describes the actions performed in support of a tactic (Strom, 2018). Although techniques are independent activities, they often supplement or assist one another. For example, during the *planning phase*, an attacker would execute a *reconnaissance tactic* to gather information about a target and identify vulnerabilities and methods for exploiting these vulnerabilities. In support of this tactic, the attacker can use some of the following techniques: *passive surveillance, dumpster diving,* and *open-source reconnaissance*.

2.2.3 *Social engineering defenses*

In order to defend against social engineering attacks, multiple kinds of defenses can be employed. The following terms are adapted from their counterparts in cyber defense context (Han, Lu, & Xu, 2021; Lin, Lu, & Xu, 2019; Pendleton, Garcia-Lebron, Cho, & Xu, 2016; Xu, 2014, 2019, 2020; Xu, Lu, & Xu, 2012; Xu, Lu, Xu, & Zhan, 2014; Xu, Yung, & Wang, 2021; Zheng, Lu, & Xu, 2018): *preventive defenses* aim to prevent a target from falling victim to social engineering attacks; *proactive defenses* aim to mitigate the attacks that may have been successful but are not detected by (or known to) the target and/or the defender; and *reactive defenses* aim to detect successful attacks and recover from the compromised state to the secure state. Note that reactive defenses do not apply to the physical world because the targets are humans who would be punished once detected; whereas in cyberspace, the victims are often innocent and their compromised computers need to recover from a compromised state to a secure state.

3 Characterizing social engineering attack model, techniques, and defenses in the physical world

In the physical world, social engineering attacks are characterized by face-to-face interactions between an attacker and a victim. On one hand, face-to-face interactions increase the victim's trust in the attacker (Riegelsberger, Sasse, & McCarthy, 2003) and allow the attacker to tailor attacks against the victim (Alexander, 2016). On the other hand, face-to-face interactions often expose attacker identities, increasing the risk of exposure (Dimkov, Van Cleeff, Pieters, & Hartel, 2010).

3.1 Social engineering attack model of the physical world

As highlighted in Fig. 1, we consider the following social engineering model of attacks against victims.

3.1.1 *Attacker*

An attacker's goal is to manipulate a victim into performing an action that would grant the attacker access to the intended information or asset. An attacker's weapon to manipulate the victim into compliance is through message exchange. To be effective, an attacker must keep victims' risk perceptions low by projecting qualities associated with credibility and approaching a victim with a message or offer which induces peripheral processing. Credibility is a victim's perception that the attacker will deliver on the offer. Credibility increases trust and lowers risk perceptions (Stajano & Wilson, 2009). An attacker's credibility is characterized by the following attributes: *commonality*, *reputation*, and *trustworthiness*.

Commonality is the perceived common ground between a victim and an attacker. By establishing commonality, an attacker can inherit the trust extended to members of the group. Commonality can be established by providing the details that are only known to group members (contextualization), demonstrating familiarity with the victim (personalization), or sharing common biases and beliefs.

Reputation is a property describing the assessment of others about an individual or source. Reputation is often extrapolated based on characteristics like associates (or social network) and affiliation to institutions. Reputation increases cooperation, explaining why social engineering attacks often exploit both social networks and affiliations to reputable institutions. For example, an attacker might assume the persona of an authority (e.g., government agencies like Internal Revenue Services, or law enforcement) or imply common social network connections with a victim to elicit cooperation.

Trustworthiness is related to trust, which is an individual's "willingness to be vulnerable based on positive expectations about the actions of others" (Adams & Sasse, 1999). Projecting trustworthiness requires an attacker to be perceived as providing services in good faith. Trust can also be developed through continuous interactions between a victim and an attacker. This kind of trust, known as *affection trust* (McAllister, 1995), is based on an affection bond built over time. Under this condition, an individual might willingly take risks based on the relationship and disregard their risk perceptions.

3.1.2 Message

In order to be successful, an attacker's message would intend to induce peripheral processing. For messages involving a high-risk request, an attacker might wage multiple interactions with a victim to achieve compliance because multiple message exchanges might allow an attacker to develop a familiar relationship with a victim, increasing the levels of trust (McAllister, 1995) and making the risk more acceptable for the victim. To increase the chance of success, the following psychological techniques could be leverage to craft messages: *persuasion, scamming, incentive and motivator*, and *visceral trigger*. (i) Persuasion is the act of presenting an argument that encourages an individual to behave in a desired manner (Cialdini & Cialdini, 2007). (ii) Scamming (or deception) is the act of presenting an argument with the intention to create a false belief (Buller & Burgoon, 1996; Stajano & Wilson, 2009). (iii) Incentive and motivators encourage cooperation (Dove, 2020; Lea et al., 2009), where incentives leverage external rewards and motivators leverage internal psychological attributes. (iv) Visceral triggers are motivational manipulations, which trigger emotional response by exploiting needs and desires (Lea et al., 2009; Stajano & Wilson, 2009).

3.1.3 *Victim*

A victim's goal is to identify social engineering attacks while avoiding a high false-positive rate. A social engineering attack succeeds when the victim complies with the attacker's request. In the physical world, the attacker might have to ensure that the victim feels positive about their interaction after that the victim complies, which reduce the regret that discourages reporting of the incident to authorities.

To prevent victimization, the recipient of a message must process it through the central route. Activating central processing requires that the victim detects inconsistencies and deception cues in the message. Attributes facilitating this include *domain expertise, domain knowledge,* and *vigilance.* (i) Domain expertise can reduce victimization to social engineering attacks in several ways: individuals with expertise have more accurate threat mental models which improve threat appraisal and risk perceptions (Klein & Calderwood, 1991), while having better strategies to cope with threat when their risk assessment is erroneous; domain expertise also facilitates the detection of deceptive cues. (ii) Domain knowledge can be developed through training or previous negative experience. It helps with pattern recognition and deceptive cue detection (Langenderfer & Shimp, 2001), while noting that deceptive cue detection is a precursor to suspicion. (iii) Vigilance is the process of dedicating cognitive resources to perform a demanding task, such as detecting cues which can indicate deceptive intent in a message (Dove, 2020; Duffield & Grabosky, 2001). Note that vigilance is affected by suspicion.

3.2 Social engineering attack techniques in the physical world

As highlighted in Fig. 2, we classify social engineering attack techniques in the physical world into five categories: *information gathering, pretexting, impersonation, physical reverse engineering* (physical RE), and *tailgating,* which are elaborated below.

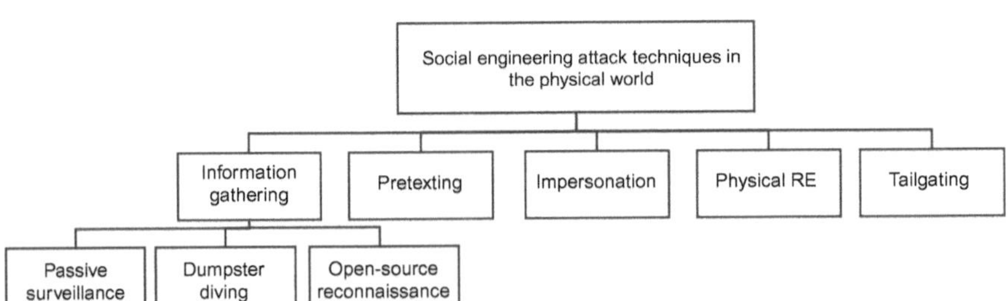

FIG. 2 Taxonomy of social engineering attack techniques in the physical world.

3.2.1 *Information gathering*

This category of attack techniques focuses on acquiring information about a target. These techniques can support social engineering in the physical world. This category has three specific attack techniques: *passive surveillance*, *dumpster diving*, and *open-source reconnaissance*. First, the passive surveillance attack technique (Greenlees, 2009) attempts to collect information about a victim and the environment. The information is used in later phases of an attack to develop a cover story and artifacts supporting an objective without raising the victim's suspicion. Second, the dumpster diving attack technique is the act of searching through the trash for information (Mitnick & Simon, 2003). It is effective because all the content in the trash is specific to the victim. The information can be used along with other social engineering attack techniques (Redmon, 2005). Third, the open-source reconnaissance attack technique is the gathering of information that is publicly accessible (Mitnick & Simon, 2003). In the past, libraries were one of the main sources of gathering open-source information (Thompson, 2006). Nowadays, this attack often involves the mining of information available online and in social media (Ariu, Frumento, & Fumera, 2017). For example, a simple Google search on the name of a person of interest would result in a set of links, images, and videos that either directly involves the person or have been searched/accessed by the person. These types of information can be used to generate a psycho-behavioral profile for the target person to generate a personalized social engineering attack.

3.2.2 *Pretexting*

The pretexting attack technique attempts to obtain information by using false pretenses (Baer, 2008). It requires that the attacker invents a background story to create a scenario that is relevant to the victim and persuade the victim to perform an action or release information (Indrajit, 2017). For example, in the HP pretexting scandal (Baer, 2008), the HP security department hired a third-party investigator to identify the source of private HP Board conversations disclosure to the press. Using pretexting, the third-party investigator was able to acquire records from phone service providers by impersonating HP Board members and journalists that were suspected to be involved in the leak (Workman, 2008).

3.2.3 *Impersonation*

The impersonation attack technique uses a persona (e.g., that can increase compliance of a victim). Personas allow the attacker to keep a low profile and blend into the targeted environment. Examples of personas are authority personas (Greenlees, 2009) (e.g., manager or IT auditor), or a layperson persona (Redmon, 2005) (e.g., custodian or delivery person). Persona can facilitate access to information, assets, or places. Personas also allow the attacker to leverage different persuasion techniques. Authority persona allows an attacker to leverage the persona perceived position and potential consequences if the victim does not comply. Persona selection can be based on the security of the environment where the asset is maintained, or based on the individual that is responsible for safeguarding the asset (Dimkov et al., 2010). For example, an attacker might impersonate a company employee to convince a cleaning staff employee to give them access to an asset. An attacker that focuses on exploitation of the custodian of an asset might choose to impersonate service desk staff, a coordinator representative, or an individual that needed urgent access to the asset.

3.2.4 *Physical reverse engineering (physical RE)*

The physical reverse engineering attack technique requires that an attacker creates problem that gives them access to the victim and then offers assistance to fix it (Gragg, 2003). This attack technique may proceed in the following three steps (Nelson, 2001). (i) Sabotage: the attacker introduces a fault that causes a problem for the victim. (ii) Advertising: once the victim recognizes the problem, the attacker makes it known that the attacker can provide assistance which gives the attacker access to the target. (iii) Assisting: with the victim's consent, the attacker fixes the problem, while using the opportunity as a mechanism to launch an attack. An example of this attack technique in the physical world is an attacker modifying a system to give the appearance that the system is corrupted by displaying an error message. When the user notices that the system is "corrupted," they reach out to the attacker for help, because the attacker had previously advertised its expertise by leaving behind business cards or providing their contact information in the error message (Nelson, 2001).

3.2.5 *Tailgating*

The tailgating attack technique involves gaining access into a control access facility or a restricted area (Alexander & Wanner, 2016), by following an individual with access into the facility. This technique is often combined with impersonation. In a typical scenario, the attacker impersonates a delivery service individual. When an authorized individual opens the door, the attacker asks the individual to hold the door open or simply follows the individual without their notice.

3.3 Social engineering defenses in the physical world

It would be ideal that humans could identify or recognize social engineering attacks in the physical world. However, human cognitive resources are limited, while other tasks are also competing for the limited cognitive resources. For example, vigilance is cognitively costly and unsustainable over extended periods of time (Warm, Matthews, & Finomore, 2018). In order to address these limitations, social engineering defenses should follow a multilayered defense approach (i.e., defense-in-depth). Corresponding to the methodology described earlier, we propose classifying social engineering defenses in the physical world into two categories: *preventive* defenses, which aim to prevent social engineering attacks from succeeding and *proactive* defenses, which aim to mitigate the attacks that may have been successful but are not detected by the target. *Reactive* defenses do not appear to be relevant here because humans are the target in the physical world, rather than computers.

3.3.1 *Preventive defenses*

There are five kinds of preventive defenses: *legislation, security controls, training, organizational policies*, and *organizational procedures*.

Legislation

This approach focuses on deterring, disincentivizing, or discouraging social engineering attacks, by increasing the personal risk to the attacker. For example, the HP pretexting scandal (Baer, 2008) led to the US Telephone Records and Privacy Act (2006), which criminalizes the employment of fraudulent tactics to persuade telephone companies to release phone records, with a punishment of up to 10 years of prison. Prior to this Act, only pretexting for financial records was illegal under the Financial Services Modernization Act (1999). Similar to pretexting, impersonation (i.e., false identity) is illegal under several laws in the United States, when it is used to cause harm or gain benefits. One of these laws is 18 US Code § 912, which criminalizes the impersonation of an officer or servant of the US Government. Charges under this law can carry a maximum sentence of 3 years and/or a fine. It is difficult to quantify the impact of legislation on criminal activities, in part because legislation assists with the allocation of resources for prevention as well as increasing awareness of an issue which in turn reduces its incidence (Akirav, 2018).

Access controls

One approach to defending against social engineering attacks like tailgating is to employ credential-based access in controlled areas (Abeywardana, Pfluegel, & Tunnicliffe, 2016; Redmon, 2005; Tipton, 2009).

This mechanism often requires individuals to scan their badge and/or enter their personal identification number (PIN) to access a facility in question. Although this approach can be effective in preventing random individuals from tailgating, it does not stop all tailgating because it can be bypassed when an attacker is accompanied by an authorized individual (Cheh, Thakore, Chen, Temple, & Sanders, 2019). In order to prevent this attack, authorized individuals need guidance in dealing with situations where their risk perception level might be low.

Training

Training (e.g., security education, awareness, resistance training) is a widely employed defense against social engineering attacks in the physical world (Abeywardana et al., 2016; Anderson, 2020; Gragg, 2003; Mitnick & Simon, 2003). One study shows that training can reduce social engineering victimization from 62.5% to 37% (Bullée, Montoya, Pieters, Junger, & Hartel, 2015). Training can also help individuals recognize patterns, which can be leveraged to identify social engineering attacks, teach strategies against ongoing attacks, and improve threat appraisal and risk perceptions. Training on policies can improve policy compliance (Pahnila, Siponen, & Mahmood, 2007; Soomro, Shah, & Ahmed, 2016). Training on strategies against tailgating, shoulder surfing, baiting, and reverse social engineering can reduce such attacks (Jansen & van Schaik, 2017; Wang, Li, & Rao, 2017).

Organizational policies

Organizational policies are widely employed defense against social engineering attacks in the physical world (Mitnick & Simon, 2003; Tipton, 2009). Organizational policies define expected behaviors and identify information that needs protection (Gragg, 2003). These policies also help reduce uncertainty by defining acceptable practices (Greenlees, 2009) and serve as deterrents against specific behaviors (Redmon, 2005). For some social engineering attacks, policies may be the only alternative to mitigate them. For example, since dumpster diving is legal (as long as there is no trespassing) (Wingo, 1997), establishing corporate policies to define the proper destruction of corporate materials might be the best strategy against dumpster diving. However, establishing policies often face a range of challenges. For policies to be enforceable, they must be implemented and monitored. The effectiveness can be affected by multiple factors: one is the culture of an organization subculture (Da Veiga & Martins, 2017; Flores & Ekstedt, 2012); another is the attitude in an organization toward compliance and social influence (Carmichael, Morisset, & Groß, 2018). This is because policy enforcement requires the collaboration of the members of an organization. For example, enforcing a policy targeting tailgating requires individuals to challenge other individuals suspicious of tailgating (Redmon, 2005) (i.e., exert social influence) and complying with the requirement to display their identification (i.e., display compliance attitude). Simulating Influencing Human Behaviour in Security (SHRUBS) (Carmichael et al., 2018) is a tool that examines how psychological aspects and interactions (e.g., beliefs, social norms, the influence of authority figures) can affect the global compliance of a policy. Such an analysis can help identify areas of intervention to improve compliance. Policies can also help minimize the loss incurred by social engineering attacks. For example, a two-factor authentication policy can mitigate the risk when a PIN is compromised via shoulder surfing.

Organizational procedures

An organizational procedure provides predefined, step-by-step instructions on addressing a situation, such as strategies for coping with a threat in real time. Procedures should be in line with policies and should be part of a training program. In some contexts, procedures are referred to as Social Engineering Land Mines (SELM) (Gragg, 2003). An SELM is an action that deters an ongoing social engineering attack by surprising the attacker. A "Justified Know-it-all" SELM is a person who knows the associated security risks and can handle suspicious events. Other SELM are "Call-Back" (Flores & Ekstedt, 2012) and "Please-Hold" (Ghafir, Prenosil, Alhejailan, & Hammoudeh, 2016) procedures. An extensive list of procedures can be found in Mitnick and Simon (2003) under the "Verification and authorization procedure" section. Procedures can thwart the "assisting" step in physical reverse social engineering by providing legitimate resources and assistance to targets when they encounter a problem.

3.3.2 *Proactive defense*

An effective proactive defense is *audit and compliance*. Audit provides an opportunity to measure the effectiveness of a policy and make adjustments to improve the security posture. Audit can help uncover unidentified security weaknesses, which may be caused by poor implementations of policies or the lack of specificity in policies. One study on auditing (Cheh et al., 2019) shows that tailgating often occurs when employees escorted visitors into the service rooms.

4 Characterizing social engineering attack model, techniques, and defenses in cyberspace

Different from the physical world, social engineering attacks in cyberspace are characterized by computer-mediated interactions between a victim and an attacker, which have an effect on the outcome of the interactions (e.g., trust building; Riegelsberger et al., 2003).

4.1 Social engineering attack model in cyberspace

As highlighted in Fig. 1, we consider the following social engineering model of attack in cyberspace.

4.1.1 Attacker

The goal and approach of a social engineering attacker in cyberspace are similar to their counterpart in the physical world. However, there are significant differences. First, cyberspace provides an attacker with more resources to personalize attacks, while providing better identity protection mechanisms. Second, the possible anonymity of digital channels allows the attacker to protect their identity while potentially operating in different jurisdictions, which can help with evading legal woes of their actions. Digital channels also allow an attacker to approach multiple victims at once, lowering the cost of waging social engineering attacks and increasing the odds of finding a victim. As a consequence, social engineering attacks with a low response rate are still profitable (Herley, 2012). Third, it is easier to project credibility (for creating trust) in cyberspace than in the physical world because most individuals rely on basic heuristics to judge credibility. Most individuals associate credibility in cyberspace with superficial attributes, such as the professional appearance of a website (Dhamija, Tygar, & Hearst, 2006), or the presence of high quality and rich content (Kim, Kishore, & Sanders, 2005). Note that credibility reduces suspicion, improves message persuasion (Wathen & Burkell, 2002), and increases victims' susceptibility to social engineering attacks (Dhamija et al., 2006; Hirsh, Kang, & Bodenhausen, 2012; Jakobsson, 2007). An attacker's credibility is characterized by the following attributes: *commonality*, *reputation*, and *trustworthiness*, which are elaborated below.

Commonality can be easily established online because an attacker can use the information in social media and websites to build common ground with a victim. Information like bias, beliefs, norms, and dialects of a community is useful for an attacker. Using this kind of information, an attacker can impersonate a community member or an acquaintance in online forums or social media groups.

Reputation is often based on one's network of associates. One method for improving others' perceived reputation of an attacker in cyberspace is to increase the attacker's social media connections with reputable individuals. An attacker can build on a perceived commonality to entice reputable individuals to accept an invitation to connect. Another enticement for attracting reputable individuals in social media is the size of the social network in question. An attacker can project an extensive social network through the use of bots and fake personas.

Trustworthiness is the perception that the other party is acting in good faith. To project trustworthiness, an attacker can incorporate artifacts (e.g., links, images, graphics) in messages. For example, security indicators like Secure Sockets Layer (SSL) padlocks (Jakobsson, 2007) or images of organizations which voucher for an individual's trustworthiness, such as Better Business Bureau (BBB) or Federal Deposit Insurance Corporation (FDIC) logos. An attacker can also include URLs that appear to originate from a known trusted site. An attacker can use a URL that resembles legitimate, well-known URLs (e.g., www.paypa1.com, where the letter "l" is substituted by number "1" (Dhamija et al., 2006)). Another approach is to generate a benign-looking malicious URL, such as a long URL which, when shortened by the browser, appears to be benign. For example, an attacker may use https://myaccount.google.com-securitysettingpage.tk (Heartfield & Loukas, 2018 b). Finally, an attacker may exploit a victim's trust in a third party, which may be a service or entity trusted by the victim and may provide a communication channel between the victim and the attacker (e.g., dating or employment sites).

4.1.2 *Message*

As previously mentioned, the purpose of the attacker is to encourage peripheral processing of messages. Like in the physical world, an attacker can exploit the same psychological principles to tailor messages, namely *persuasion* (Ferreira & Lenzini, 2015; Van Der Heijden & Allodi, 2019; Wright, Jensen, Thatcher, Dinger, & Marett, 2014), *scamming* (Ferreira, Coventry, & Lenzini, 2015), *incentives and motivators* (Cain, Edwards, & Still, 2018; Halevi et al., 2013, 2015; Herley, 2012; Workman, 2007), and *visceral triggers* (Wang, Herath, Chen, Vishwanath, & Rao, 2012).

4.1.3 *Victim*

In cyberspace, environment affects a victim's trust and risk perception (Riegelsberger et al., 2003). Risk perception is an individual's assessment of the risk involving an action, and risk perception affects how much risk an individual is willing to accept (Workman, 2007). A victim in cyberspace is also characterized by the following attributes: *domain expertise, domain knowledge,* and *vigilance.*

First, expertise in domains other than cybersecurity does not reduce one's susceptibility to social engineering attacks in cyberspace (Montañez, Golob, & Xu, 2020). This is because they often rely on visual elements and emotions when making decisions involving risks (Kumaraguru et al., 2006). In addition, online risk perceptions of nonexperts are shaped by the perceived benefit of an activity, and online activities that are considered beneficial are perceived as less risky and performed more often (Byrne et al., 2016).

Second, applying domain knowledge to recognize social engineering attacks in cyberspace is cognitively demanding because developing domain knowledge often involves learning patterns that indicate malicious intents. Common pattern identification techniques (e.g., URL parsing) require one to deal with technical complexities, which is cognitively demanding, error prone, and may encourage undue trust (Anderson, 2020). This is reasonable because when the risk is high but the situation is difficult to evaluate, trust is an alternative to reduce the complexity of decision making (Riegelsberger et al., 2003).

Third, vigilance requires attention, which is affected by two components, namely attention *switching* and *maintenance* (Wogalter, 2018). Attention switching is the process of redirecting attention from one task to another, whereas attention maintenance is the process of dedicating cognitive resources to processing information. Attention switching is a precursor of attention maintenance and leads to central processing (Wang et al., 2012). Salient stimuli trigger attention switching. In ordinary circumstances, an individual's attention is directed to perform the primary task, which supports one's goals (e.g., managing emails, visiting websites, or searching for information online). To detect deception cues in a social engineering message, one must redirect their attention to notice inconsistencies in the message, where the inconsistencies must be salient enough to be detected and trigger the switch. However, a digital environment in cyberspace provides little stimuli to detect deception cues. Moreover, the lack of audio and visual cues makes online detection harder (Lewis & George, 2008). If the deceptive message cues are not salient and suspicion is not triggered, the message is processed via the peripheral route and the attacker victimizes a target (Vishwanath et al., 2011). As a consequence, vigilance is often not triggered, while noting that peripheral route is also affected by individual characteristics (e.g., computer habits; Vishwanath et al., 2018).

4.2 Social engineering attack techniques in cyberspace

As highlighted in Fig. 1, there are more social engineering attack techniques in cyberspace than the physical world. We divide social engineering attack techniques in cyberspace into four categories: *contextualization*, *masquerading*, *physical-based access*, and *digital reverse engineering*, which are elaborated below (Fig. 3).

4.2.1 Contextualization

Contextualization is a message framing technique in which an attacker presents itself as a member of an "in-group" or a community of interest by including details in the message relevant to the group. Examples of details can be current events (Al-Hamar et al., 2010; Pritom, Schweitzer, Bateman, Xu, & Xu, 2020) or topics specific to a community (Luo, Zhang, Burd, & Seazzu, 2013). In addition to enhancing the attacker's credibility,

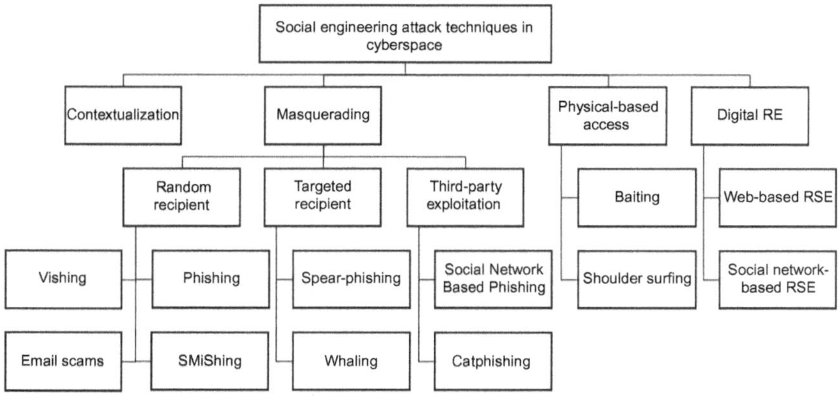

FIG. 3 Taxonomy of social engineering attack techniques in cyberspace, which are richer than their counterpart in the physical world.

contextualization also provides a reason for reducing risk perceptions. Indeed, studies show that contextualization increases social engineering response and compliance rates (Goel, Williams, & Dincelli, 2017; Luo et al., 2013). This technique shares some characteristics with the pretexting attack technique in the physical world because it, like pretexting, also provides a false pretense to establish contact.

4.2.2 *Masquerading*

Masquerading is a family of techniques by which an attacker assumes an online persona to project credibility and trustworthiness. The difference among these techniques lies in the communication channel and the victim selection. Masquerading is similar to impersonation in the physical world and can be further divided into the following three subcategories: *random recipient*, *targeted recipient*, and *third-party exploitation*, where each subcategory has multiple specific techniques.

Random recipient

Attack techniques in this subcategory attempt to contact a large number of potential victims at random. In a crafted message, the attacker usually offers an incentive to motivate the recipients to perform an action. Since the potential victims are selected at random, incentives are based on the general characteristics of the target population. The difference among the techniques in this subcategory is the transmission medium that is used to carry out the attack, including email, voice over phone, and text over phone. This subcategory has four specific attack techniques: *phishing*, *SMiShing*, *vishing*, and *email scams*.

First, phishing is the most commonly used attack technique over email. A typical phishing email contains a link that directs a victim to a malicious website for downloading a malicious software (e.g., ransomware) or stealing a victim's credentials. The goal of the message is to motivate the recipient to click the link, for which incentives and urgency cues are widely present, while noting that incentives can vary by groups (e.g., offering free academic products is an effective incentive among college students; Abraham & Chengalur-Smith, 2010) and urgency requests are used to urge targets to click a link to avoid losing services (Al-Hamar et al., 2010; Vishwanath et al., 2011; Yeboah-Boateng & Amanor, 2014).

Second, SMiShing uses text messages or simple messaging service (i.e., SMS) to send social engineering messages. Like phishing, a message includes a link which can cause a target to download malicious content to a mobile device. Unlike phishing, SMS messages have a character count limit, which may force an attacker to leverage psychological elements to trigger an immediate response. For this reason, SMiShing messages include incentive (e.g., explicit content), urgency cues (e.g., an account expiration message), and visceral triggers (e.g., promise of a sexual encounter) to encourage victim compliance (Yeboah-Boateng & Amanor, 2014). The prevalence of SMiShing is due in part to the lack of user awareness of risks and the limited security features available to mobile platform.

Third, vishing (or IVR) uses voice over phone as a medium. It is common in recent years as advancement in Voice Over IP (VoIP) allows an attacker to originate a call from any country or area code, regardless of its physical geographical location (i.e., caller ID spoofing) (Yeboah-Boateng & Amanor, 2014). Caller ID spoofing also allows an attacker to align the message story with observable cues to increase their projected credibility. In addition, the anonymity in the Internet makes tracking the source of a call difficult (Griffin & Rackley, 2008). An attacker typically assumes the persona of an authority figure and uses urgency cues to get the victim to release personal information. Since legitimate advertising calls are common, it is difficult for most people to distinguish between vishing and legitimate company calls.

Fourth, email scams attempt to establish relationships with victims on the basis of a common benefit, such as a future financial gain. A well-known attack using this technique is the Nigerian "419" email scam (Rege, 2009; Stajano & Wilson, 2009). The crafted message targets the most gullible and susceptible (Herley, 2012).

Targeted recipient

In this subcategory, an attacker tailors or personalizes messages to preselected targets. This subcategory has two specific techniques: *spear-phishing* and *whaling*. First, spear phishing personalizes a message based on a distinct characteristic of a victim, such as interest or hobby (Al-Hamar et al., 2010), personality traits (Halevi et al., 2015; Heartfield & Loukas, 2018 b), or organizational context (Pienta, Thatcher, & Johnston, 2020). It is typically used when an attacker believes that the victim has access to a desirable asset, such as a company's financial information or transaction systems. Second, whaling targets high profile individuals (e.g., C-suite executives, celebrities, wealthy individuals) (Koyun & Al Janabi, 2017). It requires an attacker to demonstrate intimate knowledge of the victim (e.g., private and public life) as proof of authenticity (Pienta et al., 2020). Such information may be acquired by exploiting a victim's network of associates, friends, and family members. Whaling is often used on CEO frauds and ghost invoice attacks (Junger, Wang, & Schlömer, 2020).

Third-party trust exploitation

This subcategory of attack techniques targets a specific community hosted by a third-party infrastructure as an attacker can exploit the trust the individuals have of the third party (Allodi, Chotza, Panina, & Zannone, 2019). This subcategory includes the following specific attack techniques: *social network-based phishing*, *anglerphising*, and *catphishing*. First, social network-based phishing exploits social media websites to contact potential victims, by creating fake profiles to establish communications with the victims. The resulting attacks are often difficult to detect because there are fewer details in the message about the origin of the message and because users often interpret artifacts such as social media connections, friend networks, and profile photos as evidence of credibility (Vishwanath, 2017). Second, anglerphishing selects victims based on their social media activities with respect to a brand (O'Hagan, 2018). For example, an attacker can send direct messages to victims based on their likes, comments, or tags about a brand. The messages usually contain a link to malicious sites that mirror the brand site. Third, catphishing selects victims from online dating sites, by seeking to establish an affection bond and build trust by maintaining frequent communications with a victim (Rege, 2009). Once trust is formed and risk perceptions are lowered, the attacker can request financial assistance. If the victim fails to comply, the attacker uses emotional blackmail or extortion (Rege, 2009).

4.2.3 Physical-based access

This category of attack techniques requires a physical channel to execute, but the objective is to gain access to a network. This category has two specific attack techniques: *baiting* and *shoulder surfing*. First, baiting plants a USB drive containing malicious code at a location that can be easily found (e.g., parking lot or coffee shop; Krombholz, Hobel, Huber, & Weippl, 2015; Salahdine & Kaabouch, 2019; Wagenaar, Pavlov, &

Yannick, 2011). When the USB drive is connected to a computer, the malicious code is executed and gives the attacker access to a system of interest. The effectiveness of baiting varies: the success rate for college students varies between 45% and 98% (Tischer et al., 2016); the success rate in business setting is about 20% (Mearian, 2011). Second, shoulder surfing attempts to gain access to a network by stealing network credentials. An attacker steals a user's credentials (e.g., password or passcode) by watching the user typing them (Mitnick & Simon, 2003; Redmon, 2005). The credentials are then used to access the network. Since this attack technique requires an attacker to have physical access to the physical space of a victim (Ghafir et al., 2016), this technique may be executed together with physical impersonations.

4.2.4 *Digital reverse social engineering (RSE)*

Similar to reverse social engineering in the physical world, this category relies on creating conditions to trigger a victim to reach out to the attacker. The difference is that in digital reverse social engineering, an attacker does not have to sabotage a system but simply create the appearance that the system is having problems, while noting that reaching out to the attacker also results in a higher level of trust toward the attacker (Irani, Balduzzi, Balzarotti, Kirda, & Pu, 2011). This category has two specific attack techniques: *web-based RSE* and *social network-based RSE*. First, web-based RSE is commonly used to propagate malicious code when a victim encounters a social engineering message and navigates according to the malicious ads (Nelms, Perdisci, Antonakakis, & Ahamad, 2016; Vadrevu & Perdisci, 2019), search result (search engine poisoning) (Abraham & Chengalur-Smith, 2010), or a link in webpost (community of interest) (Vadrevu & Perdisci, 2019). For example, malicious ads often contain a message which informs the victim of a missing update, application plug-in, or extension to display the content they want to view, which all can include a malicious payload. Legitimate but compromised websites are usually used for malicious ads posting (Abraham & Chengalur-Smith, 2010). As another example, search engine poisoning injects malicious websites into the search results. Second, social network-based RSE exploits social media features to recommend a connection between a victim and an attacker, such as friend recommendations, demographic search, or a victim's page visitor log (Huber, Kowalski, Nohlberg, & Tjoa, 2009; Irani et al., 2011). For example, demographic search can be exploited to connect a victim to the attacker based on the victim's demographic data. Social media RSE is less known but effective because individuals are likely to communicate with someone recommended by social media platforms (Bilge, Strufe, Balzarotti, & Kirda, 2009).

4.3 Social engineering defenses in cyberspace

As described in the methodology, social engineering defenses can be divided into preventive, proactive, active, reactive, and adaptive defenses. Most social engineering defense technologies focus on implementing the latter three defenses, and as a result, they affect human decision making in similar ways. With an average detection rate of 99%, these technologies almost eliminate user contact with cyber threats. However, this effectiveness has unintended consequences detrimental to user security performance. When the event rate is less than 1%, users are more likely to fail to identify social engineering attacks (Sawyer & Hancock, 2018). Therefore, the social engineering messages that avoid detection by the technologies are more likely to be successful. This finding on the effect of low event rate in social engineering detection is similar to the effect of signal detection of low-frequency events and decrements on vigilance (Davies & Parasuraman, 1982; Proctor & Vu, 2010). These social engineering defenses also share approaches to detection. Detection in technology implementation follows one of two approaches: static analysis or dynamic analysis. Static analysis relies on the presence of specific elements in the message (e.g., keywords, URLs, syntax, sender address, etc.) associated with known social engineering attacks. Dynamic analysis relies on statistical analysis and patterns in a message to detect suspicious content. This analysis often involves machine learning and natural language processing to detect suspicious activity. Dynamic analysis social engineering defenses might incorporate static analysis (Fette, Sadeh, & Tomasic, 2007), or use existing network logs for cross-validation (Ho, Sharma, Javed, Paxson, & Wagner, 2017) to improve detection. Finally, the social engineering defense technologies do not require user involvement.

In what follows, we characterize the social engineering defenses that have been presented in the literature against social engineering attacks in cyberspace, including *preventive defenses*, *proactive defenses*, and *reactive defenses*, while noting that *reactive defenses* do not have their counterpart in the physical world.

4.3.1 *Preventive defense*

It would be ideal that social engineering attacks in cyberspace, like other kinds of attack, can be completely prevented because attacks would never succeed. Even though this is not achievable, we should always aim to prevent attacks from succeeding to the extent possible. Preventive defenses against social engineering attacks in cyberspace include: *legislation, access control, training, organizational policy, system and user interface design*, and *sandboxing*.

Legislation

In the United States, laws governing cyberspace are an extension to the laws discussed in Section 3.3.1. The effectiveness of these laws have not been understood because the attribution of social engineering attacks in cyberspace remains to be a hard problem to solve, making it difficult to enforce these laws. Moreover, it is not clear whether the existing laws are adequate in covering all possible kinds of social engineering attacks in cyberspace.

Access control

Many cybersystems use authentication for access control, and password is the most common method of authentication. However, password management policies often fail to account for human cognitive limitations. For example, a policy requiring frequent password changes can increase insecure practices, such as generating less secure passwords or reusing passwords (Adams & Sasse, 1999); a policy requiring excessively long and complex passwords can increase memory and cognitive loads. An alternative to password is two-factor authentication (2FA), which can be based on "something you know and something you have" (e.g., a token and passcode or PIN). In addition to enhancing authentication, 2FA can help mitigate phishing social engineering attacks (Brian, 2018).

Training

Training aims to increase individuals' skills for coping with social engineering attacks (Arachchilage & Love, 2014; Downs et al., 2006; Halevi et al., 2016; Pattinson et al., 2012; Van Schaik et al., 2017; Wright & Marett, 2010). Training that focuses on providing information on social engineering attacks might not be sufficient because warning users of potential social engineering attacks is not enough. For example, a study (Junger et al., 2017) on the effect of priming and warning on personal information sharing shows that no significant difference between the control group and other groups exposed to priming and warning. Another study (Jansen & van Schaik, 2017) shows that exposing individuals to strong fear-appeal messages on phishing increases their protection motivation, which has a direct effect on behavioral intent. Alternatives to in-classroom security education and awareness training are resistance training and security role-play games. Resistance training is the direct exposure of users to a planned social engineering attack, is efficient at quantifying secure behavior instead of secure intent, and can reinforce classroom training and increase security alertness and demonstrate individuals' susceptibility to social engineering attacks (Flores & Ekstedt, 2013; Gragg, 2003). Finally, games are another alternative to effective training (Aladawy, Beckers, & Pape, 2018; Irvine, Thompson, & Allen, 2005; Newbould & Furnell, 2009; Olanrewaju &

Zakaria, 2015). The advantage of games is that they can expose an individual to a wide array of social engineering attacks (e.g., phishing, pop-up, baiting), levels of sophistication, and various scenarios. Unfortunately, effectiveness of training decreases over time, perhaps because of the memorability of training contents and participants' cognitive engagement (Jampen, Gür, Sutter, & Tellenbach, 2020; Reinheimer et al., 2020).

Organizational policies

In cyberspace, policies can help shape the landscape of an enterprise network and reduce its attack surface. As discussed in Section 3.3.1, policies can help define acceptable practices and deter unwanted user behaviors. Policies are ineffective when they are counter to human nature (Sasse, Brostoff, & Weirich, 2001) or organizational culture and norms (Da Veiga & Martins, 2017; Flores & Antonsen, 2013). In an organization where information security attitudes are more relaxed, employees might have difficulties adhering to strict security control policies. Some indicators of information security attitudes are information protection, data handling, training, management, security behavior, and compliance. Other factors that affect security policy compliance from a user perspective are incorrect understanding of the applicability of the policy, difficulties in applying policies, or unrealistic and unattainable policies (Kirlappos, Beautement, & Sasse, 2013). In some organizations, policies that are in line with the dominant individual attitudes can be beneficial. Individual attitudes toward policies are affected by their levels of technical competence—users with higher technical competence are less likely to favor security policy enhancements (Pfleeger & Caputo, 2012). In large organizations with a high level of technical competence, delegating responsibility for system updates to users can lead to better attitudes and higher compliance than centralized system management.

System and user interface design

In recent years, interest in human-centered design and usable security has increased. These approaches can help promote users' secure behaviors by integrating security early in the system design process and communicating security information to enhance secure behaviors. A common approach to communicate information to the user in applications is to leverage the user interface (UI), which can be a good mechanism to deliver stimuli to trigger suspicion in users. Although this approach might seem straightforward, developing intuitive UIs that promote security is a challenging task because designers must consider how nontechnical users interact with the system and interpret system information (Condori-Fernandez, Suni-Lopez, Munante, & Daneva, 2020). Another challenge is to account for temporal psychological attributes that can increase users' susceptibility to social engineering attacks, such as inattentional blindness (Pfleeger & Caputo, 2012) and limited cognitive capacity from workload and stress (Caputo, Pfleeger, Freeman, & Johnson, 2014).

One of the earlier attempts to leverage UI to communicate security information were browser security toolbars. A browser security toolbar is a browser extension that uses a passive security warning system. It provides a central location to view all security information for a website. A passive warning alerts the user of the security issues but does not interrupt the user's operations. Multiple studies have demonstrated the inefficiency of this approach (Alsharnouby, Alaca, & Chiasson, 2015; Cranor, Egelman, Hong, & Zhang, 2007; Egelman, Cranor, & Hong, 2008; Wu, Miller, & Garfinkel, 2006). Security toolbar inefficiencies owe in part to the lack of understanding of how users make online security decisions. Users determine a website's legitimacy by the appearance of the content in the main content area rather than the information provided in a small area where the security toolbar is displayed (Wu et al., 2006). In addition to message location, security messages must be salient to grab a user's attention. Toolbar's intuitiveness and level of difficulty for attackers to manipulate also affect its effectiveness (Alsharnouby et al., 2015).

In recent years, the approach to communicate online security information to users has shifted to *active warnings*, which differ from *passive warnings* in that they display security alert messages in the main content area of the browser, namely the place where users direct most of their attention (Alsharnouby et al., 2015). In addition, interrupting users from performing their primary task forces them to switch their attention to the warning. Active warnings have shown to be more effective in promoting secure behavior than passive warnings. One study (Egelman et al., 2008) shows that 79% of the participants receiving an active warning message follow the advice, which is in sharp contrast to the 13% when using a passive warning message.

Sandboxing

A sandbox is an isolated environment inside another computing environment. It creates a layer of isolation between the main computing environment and the application running in the sandbox, effectively preventing a malicious application running in the sandbox from gaining access to a system's resources outside the sandbox. Sandbox can help prevent social engineering attacks. For example, in web browsers, sandboxes can prevent the manipulation of visual components of a website. In social engineering attacks that involve the download of malicious code, sandboxes can also help mitigate them (Heartfield & Loukas, 2018 a, 2018 b).

4.3.2 *Proactive defenses*

We classify these defenses into two categories: *audit and compliance* and *security (threat) intelligence.*

Audit and compliance

Auditing is a mechanism to test the effectiveness of policies and can offer insights into users' secure behaviors that cannot be captured otherwise. One study on testing the effectiveness of security policies (Orgill, Romney, Bailey, & Orgill, 2004) shows that participants willingly provide their usernames (81%) and passwords (59%) to the auditor, who is able to acquire an all-day access card to the entire facility and access sensitive information (e.g., credit cards, copy of a master key). Penetration tests can supplement audits. When an audit involves a penetration test, it is important to consider the psychological toll that can take on victimized individuals. Several options are available to minimize the possible psychological toll on the victims (Dimkov et al., 2010; Jakobsson & Finn, 2007).

Security (threat) intelligence

Security intelligence is the sharing of information about threats and actors to prevent or mitigate an attack. Security intelligence can be generated by different sources. Currently, most of its use focus on cyber investigations. However, it can also be used for proactive defense. One approach to generate security intelligence is through crowd sourcing (Vincent, 2019). In this approach, the users share information about malicious social engineering attack encounters to warn others. Sharing information about a recent or ongoing social engineering attack with the users has the potential of increasing user's vigilance and protection attitudes (Jansen & van Schaik, 2017). Security intelligence is most beneficial when it also includes a coping strategy to protect against the threat.

4.3.3 *Reactive defenses*

We classify these defenses into two categories: *network-based detection, behavior-based detection,* and *real-time detection.*

Network-based detection

Most social engineering defense technologies identify patterns associated with known social engineering attacks. In most networks, social engineering defense technologies are layered (Flores & Ekstedt, 2013), and each technology follows a different detection approach. One specific approach is to combine static and dynamic analysis to detect social engineering attacks (e.g., emails). Artificial intelligence/machine learning (AI/ML) has been widely used to detect such attacks (see, e.g., Gutierrez et al., 2018; Hamid & Abawajy, 2011; Pritom, Schweitzer, Bateman, Xu, & Xu, 2020; Pritom et al., 2020; Xu, Zhan, Xu, & Ye, 2013, 2014). Cognitive features, like cognitive vulnerability triggers, can be leveraged to detect and prioritize suspicious emails based on their likelihood of success (Van Der Heijden & Allodi, 2019). The study shows that in a financial institution, successful phishing emails contain more cognitive vulnerability triggers with a message topic relevant to the members of the organization.

Behavior-based detection

This defense leverages behavioral patterns of entities (e.g., application, system, or user) over an extended period to detect attacks (Stringhini & Thonnard, 2015). For example, user email behaviors can be derived from the previous emails and user email habits, including writing style (e.g., word patterns, use of unique words), composition style (e.g., email activity patterns, recurrent URLs in the emails), and social patterns (e.g., common email interactions, personal address list). This defense can be effective against CEO frauds (Junger et al., 2020), which use compromised email accounts of executives to request the execution of an unauthorized financial transaction. This is because an attacker's email behavior would be different from an executive's. This approach can also be used to detect fake social media entities, for example, by using honeypots to identify social media spammers (Lee, Caverlee, & Webb, 2010). Fake media entities can also be detected using the mathematical properties of their social graph, while leveraging social psychological factors (e.g., persuasion mechanism, enticement) in modeling the social graph (Alvisi, Clement, Epasto, Lattanzi, & Panconesi, 2013).

Real-time detection

In web-based social engineering attacks, dynamic rendering of webpage content is often leveraged to avoid detection via various kinds of evasion techniques (e.g., code obfuscation or behavior modification based on the platform), while leveraging AI/ML techniques (Xu et al., 2013, 2014). Dynamic rendering content changes on each session based on user properties like geographical region, language, or browsing history. Digital reverse social engineering (RSE) attacks use dynamic rendering content. Examples of RSE vectors that use dynamic rendering are search results, web advertisements, and web posts. In addition to facilitating evasion, dynamic content in RSE can tailor attacks based on victims' preferences and leverage attention-grabbing elements (Nelms et al., 2016).

5 Contrast analysis

Our contrast analysis is driven by the following question: (i) What are common to, and different between, social engineering attacks in the physical world and in cyberspace? (ii) Which defenses designed for the physical world can be adapted to defenses for cyberspace, and vice versa? (iii) What defenses must be specifically designed to cope with social engineering attacks in cyberspace because they are unique to this domain?

5.1 Social engineering attacks in the physical world vs. cyberspace

Social engineering attacks in the physical world and their counterpart in cyberspace have much in common. First, they both use messages containing persuasive content, visceral triggers, and deception elements. Second, they exploit the same psychological attributes to conduct attacker-victim interactions. For example, the perceived credibility of an attacker is crucial to the success of the attack, whereas triggering individuals' suspicion is key to achieve effective defense. Third, humans are often blamed for system failures. However, the lack of human perspective in the design results in solutions that are incoherent with human cognitive processes, and that introduces weakness in the system after its release (Leveson, 1995).

Insight 1. *Future system designs should strive to achieve resilience against social engineering attacks or* social engineering resistance by design.

There are significant differences between them. First, unlike social engineering attacks in the physical world, persuasive content used in cyberspace can be intercepted and analyzed for linguistic elements before it reaches the user. This mediation offers opportunities for designing new defenses in cyberspace based on message persuasiveness. Second, there is a difference in how compliance is affected by the environment: achieving compliance in the physical world can be aided through visual elements and interactions, but compliance in cyberspace is achieved through establishing relationships with a victim. Third, effective defenses in the physical world and cyberspace would be quite different because domain knowledge significantly reduces scam victimization in the physical world but this does not hold true in cyberspace. Fourth, social engineering attacks in cyberspace are not restricted to geographical boundaries, which make it difficult to design preventive defenses such as legislation.

Insight 2. *While social engineering attacks in the physical world and their counterpart in cyberspace have much in common, there are significant differences. This means that we need to design effective defenses in both worlds, respectively.*

5.2 Adapting defenses from one world to the other

Discussions in Sections 3.3 and 4.3 suggest that the following defenses in the physical world may be adapted to cyberspace. First, the physical world provides mechanisms to verify the credibility of an individual or organization, partly owing to legislative requirements. Although these mechanisms have limitations, they have contributed to reducing the prevalence of social engineering attacks in the physical world. This has implications for designing defenses in cyberspace, where a credibility verification process must safeguard against attacker manipulations. Moreover, defenses in cyberspace must also be compatible with human cognitive models of credibility in the physical world. Second, known social engineering attacks in the physical world could be leveraged to derive threat detection patterns in cyberspace. For example, the ability to reach a wider audience and the low cost of operations facilitate the migration of criminal enterprises from the physical world to cyberspace (Langenderfer & Shimp, 2001). Evidence of this migration is the variance of the recruiting scams (Allodi et al., 2019; Garcia, Bessani, Gashi, Neves, & Obelheiro, 2014; Stajano & Wilson, 2009). This situation presents a unique opportunity to build resilient cybersystems by leveraging the knowledge of social engineering attacks in the physical world.

Insight 3. *The strategies and processes that are designed to prevent social engineering attacks in the physical world can be leveraged to design effective defenses in cyberspace that are compatible with human cognition models.*

Discussions in Sections 3.3 and 4.3 also suggest that the following defenses in cyberspace may be adapted to the physical world. First, the human-centered design approach in cyberspace can bring new insights into our understanding of human information processing from a security and protection perspective. Some of these insights might be transferable to the physical world to gain a deeper understanding of crime prevention. Second, existing detection mechanisms in cyberspace might help proactively warn of potential scams and frauds in the physical world. As criminal enterprises operate across the physical world and cyberspace, detection in cyberspace could serve as an early-warning mechanism for similar attacks in the physical world. For example, knowing that social engineering attackers are leveraging significant events occurring in the physical world (Pritom et al., 2020), defenders can monitor for keywords associated with these events to actively detect new social engineering attacks.

Insight 4. *Reactive defense in cyberspace might help identify attack patterns and provide early warning for social engineering attack in the physical world.*

5.3 Designing defenses in cyberspace

Since our focus is on defending against social engineering attacks in cyberspace, our study suggesting the following directions for the future research. First, we need to understand the root causes to human susceptibility to social engineering attacks in cyberspace. This is critical to design effective defenses. Second, there is a lack of stimuli in cyberspace to detect deception cues and trigger vigilance. Therefore, it is important to investigate how to help users detect deception cues and trigger their vigilance. For this purpose, it is critical to understand how online information is processed by humans and how to trigger an individual's protective response online. Third, how can we develop effective domain knowledge to effectively train individuals to resist social engineering attacks in cyberspace? Self-efficacy (i.e., one's ability to address a situation) plays an important role in protection motivation, but a user's confidence in handling a threat does not equate to social engineering attack outcomes. The causes of these deviations are not well understood. An answer could provide an insight into how to promote online protection behavior.

Insight 5. *There are fundamental problems in understanding and addressing social engineering attacks in cyberspace that remain largely open.*

6 Conclusions

We have presented a unified terminology and methodology for characterizing and understanding social engineering attacks in the physical world and cyberspace. Grounded in individual and social cognition, the methodology included a novel model for describing social engineering attacker-victim interactions, including the psychological factors that are relevant to social engineering attacks. The methodology also guided us to systematize social engineering attacks and defenses in the physical world and cyberspace. In particular, our contrast analysis of the social engineering attacks and defenses in the physical world and cyberspace led to a number of insights, which shed light on the future research. As shown, our understanding of social engineering attacks, especially in cyberspace, is superficial as fundamental problems are yet to be answered.

Acknowledgments

The first author is also affiliated with the MITRE Corporation, which is provided for identification purposes only and is not intended to convey or imply MITRE's concurrence with, or support for, the positions, opinions, or viewpoints expressed by the author. This work was supported in part by ARO Grant No. W911NF-17-1-0566, NSF Grant nos. 2122631 (1814825) and 2115134, and Colorado State Bill 18-086. Approved for public release; distribution unlimited. Public release case number 21-1666.

References

Abbasi, A., Zahedi, F. M., & Chen, Y. (2016). Phishing susceptibility: The good, the bad, and the ugly. In *2016 IEEE conference on intelligence and security informatics (ISI)* (pp. 169–174). IEEE.

Abeywardana, K. Y., Pfluegel, E., & Tunnicliffe, M. J. (2016). A layered defense mechanism for a social engineering aware perimeter. In *2016 SAI computing conference (SAI)* (pp. 1054–1062).

Abraham, S., & Chengalur-Smith, I. (2010). An overview of social engineering malware: Trends, tactics, and implications. *Technology in Society, 32*(3), 183–196.

Adams, A., & Sasse, M. A. (1999). Users are not the enemy. *Communications of the ACM, 42*(12), 40–46. https://doi.org/10.1145/322796.322806.

Akirav, O. (2018). A model for determining legislative significance and effectiveness. *The Theory and Practice of Legislation, 6*(3), 343–361.

Aladawy, D., Beckers, K., & Pape, S. (2018). PERSUADED: Fighting social engineering attacks with a serious game. In *International conference on trust and privacy in digital business* (pp. 103–118).

Alexander, M. (2016). *Methods for understanding and reducing social engineering attacks*. April.

Alexander, M., & Wanner, R. (2016). Methods for understanding and reducing social engineering attacks. *SANS Institute, 1*, 1–32.

Al-Hamar, M., Dawson, R., & Guan, L. (2010). A culture of trust threatens security and privacy in Qatar. In *2010 10th IEEE international conference on computer and information technology* (pp. 991–995).

Allodi, L., Chotza, T., Panina, E., & Zannone, N. (2019). The need for new antiphishing measures against spear-phishing attacks. *IEEE Security & Privacy, 18*(2), 23–34.

Alsharnouby, M., Alaca, F., & Chiasson, S. (2015). Why phishing still works: User strategies for combating phishing attacks. *International Journal of Human-Computer Studies, 82*, 69–82.

Alvisi, L., Clement, A., Epasto, A., Lattanzi, S., & Panconesi, A. (2013). SoK: The evolution of sybil defense via social networks. In *2013 IEEE symposium on security and privacy* (pp. 382–396).

Anderson, R. (2020). *Security engineering: A guide to building dependable distributed systems*. John Wiley & Sons.

Arachchilage, N. A. G., & Love, S. (2014). Security awareness of computer users: A phishing threat avoidance perspective. *Computers in Human Behavior, 38*, 304–312.

Ariu, D., Frumento, E., & Fumera, G. (2017). Social engineering 2.0: A foundational work. In *Proceedings of the computing frontiers conference* (pp. 319–325).

Baer, M. H. (2008). Corporate policing and corporate governance: What can we learn from Hewlett-Packard's pretexting scandal? In *University of Cincinnati Law Review, Corporate Law Symposium*.

Bilge, L., Strufe, T., Balzarotti, D., & Kirda, E. (2009). All your contacts are belong to us: Automated identity theft attacks on social networks. In *Proceedings of the 18th international conference on World Wide Web* (pp. 551–560). New York, NY: Association for Computing Machinery. https://doi.org/10.1145/1526709.1526784.

Bohm, M. (2011). *Why Russians don't smile.* April.

Brian, K. (2018). *Google: Security keys neutralized employee phishing—Krebs on security.* Accessed 10 February 2021.

Bullee, J.-W., Montoya, L., Junger, M., & Hartel, P. (2017). Spear phishing in organisations explained. *Information & Computer Security, 25*(5), 593–613.

Bullée, J.-W. H., Montoya, L., Pieters, W., Junger, M., & Hartel, P. H. (2015). The persuasion and security awareness experiment: Reducing the success of social engineering attacks. *Journal of Experimental Criminology, 11*(1), 97–115.

Buller, D. B., & Burgoon, J. K. (1996). Interpersonal deception theory. *Communication Theory, 6*(3), 203–242.

Byrne, Z. S., Dvorak, K. J., Peters, J. M., Ray, I., Howe, A., & Sanchez, D. (2016). From the user's perspective: Perceptions of risk relative to benefit associated with using the Internet. *Computers in Human Behavior, 59*, 456–468.

Cacioppo, J. T., & Petty, R. E. (1984). *ACR North American Advances, 11*, 673–675.

Cain, A. A., Edwards, M. E., & Still, J. D. (2018). An exploratory study of cyber hygiene behaviors and knowledge. *Journal of Information Security and Applications, 42*, 36–45.

Caputo, D. D., Pfleeger, S. L., Freeman, J. D., & Johnson, M. E. (2014). Going spear phishing: Exploring embedded training and awareness. *IEEE Security Privacy, 12*(1), 28–38. https://doi.org/10.1109/MSP.2013.106.

Carmichael, P., Morisset, C., & Groß, T. (2018). Shrubs: Simulating influencing human behaviour in security. In *Proceedings of the 8th workshop on socio-technical aspects in security and trust* (pp. 1–11).

Cheh, C., Thakore, U., Chen, B., Temple, W. G., & Sanders, W. H. (2019). Leveraging physical access logs to identify tailgating: Limitations and solutions. In *2019 15th European dependable computing conference (EDCC)* (pp. 127–132).

Cho, J.-H., Cam, H., & Oltramari, A. (2016). Effect of personality traits on trust and risk to phishing vulnerability: Modeling and analysis. In *2016 IEEE international multi-disciplinary conference on cognitive methods in situation awareness and decision support (CogSIMA)* (pp. 7–13).

Cialdini, R. B., & Cialdini, R. B. (2007). *Vol. 55. Influence: The psychology of persuasion.* Collins.

Condori-Fernandez, N., Suni-Lopez, F., Munante, D., & Daneva, M. (2020). How can personality influence perception on security of context-aware applications? In *International workshop on socio-technical aspects in security and trust* (pp. 3–22).

Cranor, L. F., Egelman, S., Hong, J. I., & Zhang, Y. (2007). Phinding phish: An evaluation of anti-phishing toolbars. In *NDSS* (pp. 1–19).

Darwish, A., El Zarka, A., & Aloul, F. (2012). Towards understanding phishing victims' profile. In *2012 international conference on computer systems and industrial informatics* (pp. 1–5).

Da Veiga, A., & Martins, N. (2017). Defining and identifying dominant information security cultures and subcultures. *Computers & Security, 70*, 72–94.

Davies, D. R., & Parasuraman, R. (1982). *The psychology of vigilance.* Academic Press.

Dhamija, R., Tygar, J. D., & Hearst, M. (2006). Why phishing works. In *Proceedings of the SIGCHI conference on human factors in computing systems* (pp. 581–590).

Dimkov, T., Van Cleeff, A., Pieters, W., & Hartel, P. (2010). Two methodologies for physical penetration testing using social engineering. In *Proceedings of the 26th annual computer security applications conference* (pp. 399–408).

Dove, M. (2020). *The psychology of fraud, persuasion and scam techniques: Understanding what makes us vulnerable.* Routledge.

Downs, J. S., Holbrook, M. B., & Cranor, L. F. (2006). Decision strategies and susceptibility to phishing. In *Proceedings of the second symposium on usable privacy and security* (pp. 79–90).

Duffield, G., & Grabosky, P. (2001). The psychology of fraud. *Trends and Issues in Crime and Criminal Justice, 1*(199), 1–6.

Egelman, S., Cranor, L. F., & Hong, J. (2008). You've been warned: An empirical study of the effectiveness of web browser phishing warnings. In *Proceedings of the SIGCHI conference on human factors in computing systems* (pp. 1065–1074).
Ferreira, A., Coventry, L., & Lenzini, G. (2015). Principles of persuasion in social engineering and their use in phishing. In *International conference on human aspects of information security, privacy, and trust* (pp. 36–47).
Ferreira, A., & Lenzini, G. (2015). An analysis of social engineering principles in effective phishing. In *2015 workshop on socio-technical aspects in security and trust* (pp. 9–16).
Fette, I., Sadeh, N., & Tomasic, A. (2007). Learning to detect phishing emails. In *Proceedings of the 16th international conference on world wide web* (pp. 649–656).
Flores, W. R., & Antonsen, E. (2013). The development of an instrument for assessing information security in organizations: Examining the content validity using quantitative methods. In *CONF-IRM* (p. p. 44).
Flores, W. R., & Ekstedt, M. (2012). A model for investigating organizational impact on information security behavior. In *WISP 2012 proceedings*.
Flores, W. R., & Ekstedt, M. (2013). Countermeasures for social engineering-based malware installation attacks. In *CONF-IRM* (p. p. 23).
Garcia, M., Bessani, A., Gashi, I., Neves, N., & Obelheiro, R. (2014). Analysis of operating system diversity for intrusion tolerance. *Software: Practice and Experience, 44*(6), 735–770.
Gavett, B. E., Zhao, R., John, S. E., Bussell, C. A., Roberts, J. R., & Yue, C. (2017). Phishing suspiciousness in older and younger adults: The role of executive functioning. *PLoS One, 12*(2), e0171620.
Ghafir, I., Prenosil, V., Alhejailan, A., & Hammoudeh, M. (2016). Social engineering attack strategies and defence approaches. In *2016 IEEE 4th international conference on future internet of things and cloud (FiCloud)* (pp. 145–149).
Goel, S., Williams, K., & Dincelli, E. (2017). Got phished? Internet security and human vulnerability. *Journal of the Association for Information Systems, 18*(1), 2.
Gragg, D. (2003). A multi-level defense against social engineering. *SANS Reading Room, 13*, 1–21.
Greenlees, C. (2009). *Engineering & Technology, 4*(13), 55–57.
Griffin, S. E., & Rackley, C. C. (2008). Vishing. In *Proceedings of the 5th annual conference on information security curriculum development* (pp. 33–35).
Grimes, G. A., Hough, M. G., & Signorella, M. L. (2007). Email end users and spam: Relations of gender and age group to attitudes and actions. *Computers in Human Behavior, 23*(1), 318–332.
Gutierrez, C. N., Kim, T., Della Corte, R., Avery, J., Goldwasser, D., Cinque, M., & Bagchi, S. (2018). Learning from the ones that got away: Detecting new forms of phishing attacks. *IEEE Transactions on Dependable and Secure Computing, 15*(6), 988–1001.
Guyon, H., Falissard, B., & Kop, J.-L. (2017). Modeling psychological attributes in psychology–an epistemological discussion: Network analysis vs. latent variables. *Frontiers in Psychology, 8*, 798.
Halevi, T., Lewis, J., & Memon, N. (2013). A pilot study of cyber security and privacy related behavior and personality traits. In *Proceedings of the 22nd international conference on World Wide Web* (pp. 737–744).
Halevi, T., Memon, N., Lewis, J., Kumaraguru, P., Arora, S., Dagar, N., & Chen, J. (2016). Cultural and psychological factors in cyber-security. In *Proceedings of the 18th international conference on information integration and web-based applications and services* (pp. 318–324).
Halevi, T., Memon, N., & Nov, O. (2015). Spear-phishing in the wild: A real-world study of personality, phishing self-efficacy and vulnerability to spear-phishing attacks. https://doi.org/10.2139/ssrn.2544742. Available at SSRN: https://ssrn.com/abstract=2544742 (January 2, 2015).

Hamid, I. R. A., & Abawajy, J. (2011). Phishing email feature selection approach. In *2011 IEEE 10th international conference on trust, security and privacy in computing and communications* (pp. 916–921).

Han, Y., Lu, W., & Xu, S. (2021). Preventive and reactive cyber defense dynamics with ergodic time-dependent parameters is globally attractive. *IEEE TNSE, 8*(3), 2517–2532.

Harrison, B., Svetieva, E., & Vishwanath, A. (2016). Individual processing of phishing emails: How attention and elaboration protect against phishing. *Online Information Review, 40*(2), 265–281.

Heartfield, R., & Loukas, G. (2018a). Detecting semantic social engineering attacks with the weakest link: Implementation and empirical evaluation of a human-as-a-security-sensor framework. *Computers & Security, 76*, 101–127.

Heartfield, R., & Loukas, G. (2018b). Protection against semantic social engineering attacks. In *Versatile cybersecurity* (pp. 99–140). Springer.

Herley, C. (2012). Why do Nigerian scammers say they are from Nigeria? In *Proceedings of 11th annual workshop on the economics of information security (WEIS'2012)*. Germany: Berlin. 25-26 June, 2012, Available at: http://weis2012.econinfosec.org/papers/Herley_WEIS2012.pdf.

Hirsh, J. B., Kang, S. K., & Bodenhausen, G. V. (2012). Personalized persuasion: Tailoring persuasive appeals to recipients' personality traits. *Psychological Science, 23*(6), 578–581.

Ho, G., Sharma, A., Javed, M., Paxson, V., & Wagner, D. (2017). Detecting credential spearphishing in enterprise settings. In *26th {USENIX} security symposium ({USENIX} security 17)* (pp. 469–485).

Howe, A. E., Ray, I., Roberts, M., Urbanska, M., & Byrne, Z. (2012). The psychology of security for the home computer user. In *2012 IEEE symposium on security and privacy* (pp. 209–223).

Huber, M., Kowalski, S., Nohlberg, M., & Tjoa, S. (2009). Towards automating social engineering using social networking sites. In *2009 international conference on computational science and engineering: Vol. 3* (pp. 117–124).

Hutchins, E., Cloppert, M., & Amin, R. (2011). Intelligence-driven computer network defense informed by analysis of adversary campaigns and intrusion kill chains. In *International conference on information warfare and security*.

Indrajit, R. E. (2017). Social engineering framework: Understanding the deception approach to human element of security. *International Journal of Computer Science Issues (IJCSI), 14*(2), 8.

Irani, D., Balduzzi, M., Balzarotti, D., Kirda, E., & Pu, C. (2011). Reverse social engineering attacks in online social networks. In *International conference on detection of intrusions and malware, and vulnerability assessment* (pp. 55–74).

Irvine, C. E., Thompson, M. F., & Allen, K. (2005). Cyberciege: Gaming for information assurance. *IEEE Security & Privacy, 3*(3), 61–64.

Jakobsson, M. (2007). The human factor in phishing. In *Privacy & security of consumer information* (pp. 1–19).

Jakobsson, M., & Finn, P. (2007). *IEEE Technology and Society Magazine, Special Issue on Usability and Security, 26*(1), 46–58.

Jalali, M. S., Bruckes, M., Westmattelmann, D., & Schewe, G. (2020). Why employees (still) click on phishing links: Investigation in hospitals. *Journal of Medical Internet Research, 22*(1), e16775.

Jampen, D., Gür, G., Sutter, T., & Tellenbach, B. (2020). Don't click: Towards an effective anti-phishing training. A comparative literature review. *Human-Centric Computing and Information Sciences, 10*(1), 1–41.

Jansen, J., & van Schaik, P. (2017). Persuading end users to act cautiously online: Initial findings of a fear appeals study on phishing. In *HAISA* (pp. 1–11).

Junger, M., Montoya, L., & Overink, F. J. (2017). Priming and warnings are not effective to prevent social engineering attacks. *Computers in Human Behavior, 66*, 75–87.

Junger, M., Wang, V., & Schlömer, M. (2020). Fraud against businesses both online and offline: Crime scripts, business characteristics, efforts, and benefits. *Crime Science, 9*(1), 1–15.

Kim, Y. J., Kishore, R., & Sanders, G. L. (2005). From DQ to EQ: Understanding data quality in the context of e-business systems. *Communications of the ACM, 48*(10), 75–81.

Kirlappos, I., Beautement, A., & Sasse, M. A. (2013). "Comply or Die" is dead: Long live security-aware principal agents. In *International conference on financial cryptography and data security* (pp. 70–82).

Klein, G. A., & Calderwood, R. (1991). Decision models: Some lessons from the field. *IEEE Transactions on Systems, Man, and Cybernetics, 21*(5), 1018–1026.

Koyun, A., & Al Janabi, E. (2017). Social engineering attacks. *Journal of Multidisciplinary Engineering Science and Technology (JMEST), 4*(6), 7533–7538.

Krombholz, K., Hobel, H., Huber, M., & Weippl, E. (2015). Advanced social engineering attacks. *Journal of Information Security and applications, 22*, 113–122.

Kumaraguru, P., Acquisti, A., & Cranor, L. F. (2006). Trust modelling for online transactions: A phishing scenario. In *Proceedings of the 2006 international conference on privacy, security and trust: Bridge the gap between PST technologies and business services* (p. p. 11). ACM.

Langenderfer, J., & Shimp, T. A. (2001). Consumer vulnerability to scams, swindles, and fraud: A new theory of visceral influences on persuasion. *Psychology & Marketing, 18*(7), 763–783.

Lawson, P. A., Crowson, A. D., & Mayhorn, C. B. (2018). Baiting the hook: Exploring the interaction of personality and persuasion tactics in email phishing attacks. In *Congress of the international ergonomics association* (pp. 401–406). Springer.

Lea, S. E. G., Fischer, P., & Evans, K. M. (2009). *The psychology of scams: Provoking and committing errors of judgement.*

Lee, K., Caverlee, J., & Webb, S. (2010). The social honeypot project: Protecting online communities from spammers. In *Proceedings of the 19th international conference on World Wide Web* (pp. 1139–1140). New York, NY: Association for Computing Machinery. https://doi.org/10.1145/1772690.1772843.

Leveson, N. G. (1995). *Safeware: System safety and computers.* ACM.

Lewis, C. C., & George, J. F. (2008). Cross-cultural deception in social networking sites and face-to-face communication. *Computers in Human Behavior, 24*(6), 2945–2964.

Lin, T., Capecci, D. E., Ellis, D. M., Rocha, H. A., Dommaraju, S., Oliveira, D. S., & Ebner, N. C. (2019). Susceptibility to spear-phishing emails: Effects of internet user demographics and email content. *ACM Transactions on Computer-Human Interaction (TOCHI), 26*(5), 32.

Lin, Z., Lu, W., & Xu, S. (2019). Unified preventive and reactive cyber defense dynamics is still globally convergent. *IEEE/ACM ToN, 27*(3), 1098–1111.

Luo, X. R., Zhang, W., Burd, S., & Seazzu, A. (2013). Investigating phishing victimization with the Heuristic–Systematic model: A theoretical framework and an exploration. *Computers & Security, 38*, 28–38.

Mandiant. (2013). *Apt1 report.* Accessed 8 July 2016.

McAllister, D. J. (1995). Affect-and cognition-based trust as foundations for interpersonal cooperation in organizations. *Academy of Management Journal, 38*(1), 24–59.

Mearian, L. (2011). *Government tests show security's people problem.* https://www.computerworld.com/article/2510014/government-tests-show-security-s-people-problem.html. Accessed 27 July 2011.

Mitnick, K. D., & Simon, W. L. (2003). *The art of deception: Controlling the human element of security.* John Wiley & Sons.

Montañez, R., Golob, E., & Xu, S. (2020). Human cognition through the lens of social engineering cyberattacks. *Frontiers in Psychology, 11*, 1755. Available at: https://www.frontiersin.org/article/10.3389/fpsyg.2020.01755.

Nelms, T., Perdisci, R., Antonakakis, M., & Ahamad, M. (2016). Towards measuring and mitigating social engineering software download attacks. In *25th {USENIX} security symposium ({USENIX} security 16)* (pp. 773–789).

Nelson, R. (2001). *Vol. 10. Methods of hacking: Social engineering*. The Institute for Systems Research, University of Maryland. http://www.academia.edu/4903480/Methods_of_Hacking-socialEngineering,_diakses.

Newbould, M., & Furnell, S. (2009). Playing safe: A prototype game for raising awareness of social engineering. In *Australian information security management conference* (p. p. 4).

O'Hagan, L. (2018). Angler phishing: Criminality in social media. In *5th European conference on social media ECSM 2018* (p. p. 190).

Olanrewaju, A.-S. T., & Zakaria, N. H. (2015). Social engineering awareness game (SEAG): An empirical evaluation of using game towards improving information security awareness. In *5th International conference on computing and informatics (ICOCI)* (pp. 11–13).

Orgill, G. L., Romney, G. W., Bailey, M. G., & Orgill, P. M. (2004). The urgency for effective user privacy-education to counter social engineering attacks on secure computer systems. In *Proceedings of the 5th conference on information technology education* (pp. 177–181).

Pahnila, S., Siponen, M., & Mahmood, A. (2007). Employees' behavior towards is security policy compliance. In *2007 40th Annual Hawaii international conference on system sciences (HICSS'07)* (p. p. 156b). IEEE.

Pattinson, M., Jerram, C., Parsons, K., McCormac, A., & Butavicius, M. (2012). Why do some people manage phishing e-mails better than others? *Information Management & Computer Security*, *20*(1), 18–28.

Pendleton, M., Garcia-Lebron, R., Cho, J., & Xu, S. (2016). A survey on systems security metrics. *ACM Computing Surveys*, *49*(4), 62:1–62:35.

Pfleeger, S. L., & Caputo, D. D. (2012). Leveraging behavioral science to mitigate cyber security risk. *Computers & Security*, *31*(4), 597–611.

Pienta, D., Thatcher, J. B., & Johnston, A. (2020). Protecting a whale in a sea of phish. *Journal of Information Technology*, *35*(3), 214–231.

Pritom, M., Schweitzer, K., Bateman, R., Xu, M., & Xu, S. (2020). Data-driven characterization and detection of COVID-19 themed malicious websites. In *IEEE ISI'2020*.

Pritom, M. M. A., Schweitzer, K. M., Bateman, R. M., Xu, M., & Xu, S. (2020). Characterizing the landscape of COVID-19 themed cyberattacks and defenses. In *2020 IEEE international conference on intelligence and security informatics (ISI)* (pp. 1–6). IEEE.

Proctor, R. W., & Vu, K.-P. L. (2010). Cumulative knowledge and progress in human factors. *Annual Review of Psychology*, *61*, 623–651.

Purkait, S., Kumar De, S., & Suar, D. (2014). An empirical investigation of the factors that influence internet user's ability to correctly identify a phishing website. *Information Management & Computer Security*, *22*(3), 194–234.

Rajivan, P., & Gonzalez, C. (2018). Creative persuasion: A study on adversarial behaviors and strategies in phishing attacks. *Frontiers in Psychology*, *9*, 135.

Redmiles, E. M., Chachra, N., & Waismeyer, B. (2018). Examining the demand for spam: Who clicks? In *Proceedings of the 2018 CHI conference on human factors in computing systems* (p. p. 212). ACM.

Redmon, K. C. (2005). *Mitigation of social engineering attacks in corporate America*. Greenville: East Carolina University.

Rege, A. (2009). What's love got to do with it? Exploring online dating scams and identity fraud. *International Journal of Cyber Criminology*, *3*(2), 494–512. Available at: https://www.cybercrimejournal.com/AunshulIJCCJuly2009.pdf.

Reinheimer, B., Aldag, L., Mayer, P., Mossano, M., Duezguen, R., Lofthouse, B., & Volkamer, M. (2020). An investigation of phishing awareness and education over time: When and how to best remind users (pp. 259–284).

Riegelsberger, J., Sasse, M. A., & McCarthy, J. D. (2003). The researcher's dilemma: Evaluating trust in computer-mediated communication. *International Journal of Human-Computer Studies*, *58*(6), 759–781.

Rocha Flores, W., Holm, H., Svensson, G., & Ericsson, G. (2014). Using phishing experiments and scenario-based surveys to understand security behaviours in practice. *Information Management & Computer Security*, *22*(4), 393–406.

Salahdine, F., & Kaabouch, N. (2019). Social engineering attacks: A survey. *Future Internet, 11*(4), 89.

Sasse, M. A., Brostoff, S., & Weirich, D. (2001). Transforming the 'weakest link'—A human/computer interaction approach to usable and effective security. *BT Technology Journal, 19*(3), 122–131.

Sawyer, B. D., & Hancock, P. A. (2018). Hacking the human: The prevalence paradox in cybersecurity. *Human Factors, 60*(5), 597–609.

Sharevski, F., Treebridge, P., Jachim, P., Li, A., Babin, A., & Westbrook, J. (2019). Social engineering in a post-phishing Era: Ambient tactical deception attacks. *arXiv preprint arXiv:1908.11752*. https://doi.org/10.1145/3368860.3368863.

Sheng, S., Holbrook, M., Kumaraguru, P., Cranor, L. F., & Downs, J. (2010). Who falls for phish? A demographic analysis of phishing susceptibility and effectiveness of interventions. In *Proceedings of the SIGCHI conference on human factors in computing systems* (pp. 373–382).

Simon, H. A. (1976). *Administrative behavior: A study of decision-making processes in administrative organization*. New York: Free Press.

Soomro, Z. A., Shah, M. H., & Ahmed, J. (2016). Information security management needs more holistic approach: A literature review. *International Journal of Information Management, 36*(2), 215–225.

Stajano, F., & Wilson, P. (2009). *Understanding scam victims: Seven principles for systems security*. Accessed 6 June 2021.

Stringhini, G., & Thonnard, O. (2015). That ain't you: Blocking spearphishing through behavioral modelling. In *International conference on detection of intrusions and malware, and vulnerability assessment* (pp. 78–97). Springer.

Strom, B. (2018). *ATT&CK 101*. Accessed 25 May 2021.

Tembe, R., Zielinska, O., Liu, Y., Hong, K. W., Murphy-Hill, E., Mayhorn, C., & Ge, X. (2014). Phishing in international waters: Exploring cross-national differences in phishing conceptualizations between Chinese, Indian and American samples. In *Proceedings of the 2014 symposium and bootcamp on the science of security* (p. p. 8). ACM.

Thompson, S. T. (2006). Helping the hacker? Library information, security, and social engineering. *Information Technology and Libraries, 25*(4), 222–225.

Tipton, H. F. (2009). *Official (ISC) 2 guide to the CISSP CBK*. CRC Press.

Tischer, M., Durumeric, Z., Foster, S., Duan, S., Mori, A., Bursztein, E., & Bailey, M. (2016). Users really do plug in USB drives they find. In *2016 IEEE symposium on security and privacy (SP)* (pp. 306–319). IEEE.

Vadrevu, P., & Perdisci, R. (2019). What you see is not what you get: Discovering and tracking social engineering attack campaigns. In *Proceedings of the internet measurement conference* (pp. 308–321).

Van Der Heijden, A., & Allodi, L. (2019). Cognitive triaging of phishing attacks. In *28th {USENIX} security symposium ({USENIX} security 19)* (pp. 1309–1326).

Van Schaik, P., Jeske, D., Onibokun, J., Coventry, L., Jansen, J., & Kusev, P. (2017). Risk perceptions of cyber-security and precautionary behaviour. *Computers in Human Behavior, 75*, 547–559.

Vincent, A. (2019). Don't feed the phish: How to avoid phishing attacks. *Network Security, 2019*(2), 11–14.

Vishwanath, A. (2017). Getting phished on social media. *Decision Support Systems, 103*, 70–81.

Vishwanath, A., Harrison, B., & Ng, Y. J. (2018). Suspicion, cognition, and automaticity model of phishing susceptibility. *Communication Research, 45*(8), 1146–1166.

Vishwanath, A., Herath, T., Chen, R., Wang, J., & Rao, H. R. (2011). Why do people get phished? Testing individual differences in phishing vulnerability within an integrated, information processing model. *Decision Support Systems, 51*(3), 576–586.

Wagenaar, D., Pavlov, D., & Yannick, S. (2011). *USB baiting*. Universite van Amsterdam.

Wang, J., Herath, T., Chen, R., Vishwanath, A., & Rao, H. R. (2012). Research article phishing susceptibility: An investigation into the processing of a targeted spear phishing email. *IEEE transactions on professional communication, 55*(4), 345–362.

Wang, J., Li, Y., & Rao, H. R. (2017). Coping responses in phishing detection: An investigation of antecedents and consequences. *Information Systems Research, 28*(2), 378–396.

Warm, J. S., Matthews, G., & Finomore, V. S. (2018). Vigilance, workload, and stress. In *Performance under stress* (pp. 131–158). CRC Press.

Wathen, C. N., & Burkell, J. (2002). Believe it or not: Factors influencing credibility on the web. *Journal of the American Society for Information Science and Technology, 53*(2), 134–144.

Wingo, H. (1997). Dumpster diving and the ethical blindspot of trade secret law. *Yale Law & Policy Review, 16*(1), 195–219.

Wogalter, M. S. (2018). Communication-human information processing (c-hip) model. In *Forensic human factors and ergonomics* (pp. 33–49). CRC Press.

Workman, M. (2008). Wisecrackers: A theory-grounded investigation of phishing and pretext social engineering threats to information security. *Journal of the American Society for Information Science and Technology, 59*(4), 662–674.

Workman, M. (2007). Gaining access with social engineering: An empirical study of the threat. *Information Systems Security, 16*(6), 315–331.

Wright, R. T., Jensen, M. L., Thatcher, J. B., Dinger, M., & Marett, K. (2014). Research note—Influence techniques in phishing attacks: An examination of vulnerability and resistance. *Information Systems Research, 25*(2), 385–400.

Wright, R. T., & Marett, K. (2010). The influence of experiential and dispositional factors in phishing: An empirical investigation of the deceived. *Journal of Management Information Systems, 27*(1), 273–303.

Wu, M., Miller, R. C., & Garfinkel, S. L. (2006). Do security toolbars actually prevent phishing attacks? In *Proceedings of the SIGCHI conference on human factors in computing systems* (pp. 601–610).

Xu, L., Zhan, Z., Xu, S., & Ye, K. (2013). Cross-layer detection of malicious websites. In *Third ACM conference on data and application security and privacy (CODASPY'13)* (pp. 141–152).

Xu, L., Zhan, Z., Xu, S., & Ye, K. (2014). An evasion and counter-evasion study in malicious websites detection. In *IEEE CNS* (pp. 265–273).

Xu, S. (2014). Cybersecurity dynamics. In *Proc. HotSoS'14* (pp. 14:1–14:2).

Xu, S. (2019). Cybersecurity dynamics: A foundation for the science of cybersecurity. In *Proactive and dynamic network defense* (pp. 1–31).

Xu, S. (2020). The cybersecurity dynamics way of thinking and landscape (invited paper). In *ACM workshop on moving target defense.*

Xu, S., Lu, W., & Xu, L. (2012). Push- and pull-based epidemic spreading in networks: Thresholds and deeper insights. *ACM Transactions on Autonomous and Adaptive Systems, 7*(3), 32:1–32:26.

Xu, S., Lu, W., Xu, L., & Zhan, Z. (2014). Adaptive epidemic dynamics in networks: Thresholds and control. *ACM TAAS, 8*(4).

Xu, S., Yung, M., & Wang, J. (2021). Seeking foundations for the science of cyber security. *Information Systems Frontiers.* https://doi.org/10.1007/s10796-021-10134-8.

Yeboah-Boateng, E. O., & Amanor, P. M. (2014). Phishing, smishing & vishing: An assessment of threats against mobile devices. *Journal of Emerging Trends in Computing and Information Sciences, 5*(4), 297–307.

Zheng, R., Lu, W., & Xu, S. (2018). Preventive and reactive cyber defense dynamics is globally stable. *IEEE TNSE, 5*(2), 156–170.

CHAPTER

2

A dual integrated dynamic intrusion detection system (DID-IDS) for protection against network and social engineering attacks

Abubakar Bello[a], *Hany Alashwal*[b], *and Ahmed A. Moustafa*[c,d]

[a]Cyber Security and Behaviour, School of Social Sciences, Western Sydney University, Sydney, NSW, Australia, [b]College of Information Technology, United Arab Emirates University, Abu Dhabi, United Arab Emirates, [c]School of Psychology, Faculty of Society and Design, Bond University, Gold Coast, Queensland, Australia, [d]Department of Human Anatomy and Physiology, the Faculty of Health Sciences, University of Johannesburg, South Africa

1 Introduction

As machines are becoming more connected (e.g., due to advances in automation and the Internet of Things), cyberattacks are expected to be increasing. Cyberattacks can impact smart power grids (Hong et al., 2017; Ten et al., 2011), secure water treatment plants (Adepu et al., 2019), medical devices (Beavers & Pournouri, 2019), computer networks, and smart homes (Arabo, 2015), among many others. For example, more than 10 billion dollars were lost due to malware in 2007, and it is increasing over time (Teoh et al., 2018). According to Kim et al. (2018), social engineering attacks are the least detected today, although they account for over a quarter of cyberattacks (Bowen, Devarajan, & Stolfo, 2014).

https://doi.org/10.1016/B978-0-323-90570-1.00013-9

Accordingly, the field of cyber security should develop more sophisticated intrusion detection methods to help secure the overly connected world of machines.

For simplicity, in this article, we divide cyberattacks into two types: (a) standard attacks on computer systems, which, for example, involve using brute-force methods to find a system password that may work, IP address spoofing, or traffic interception and (b) social engineering methods, such as phishing, which involves attacking a system user's mind to give away important information, such as passwords (Lohani, 2019). The difference between the two kinds is that the first can be done without directly interacting with computer system users, while the second involves interaction with system users. In social engineering attacks, for example, Mbaziira and Jones (2016) found that cybercriminals often rely on linguistic skills to deceive computer system users. While there have been research studies on the prevalence of social engineering hacking methods, there are only a few studies on detecting social engineering attacks (Hoeschele, 2006), which we discuss below. These two types of cyberattacks rely on using different techniques. Standard computer system attacks often involve brute-force search methods, while social engineering often involves the use of psychological techniques and linguistic skills to convince computer system users to give away secure information.

2 Detection of information system and social engineering intrusion attacks

There is an intrusion detection method for each kind of cyber security attack. Each intrusion detection system often works on a different dataset. Specifically, computer system detection methods often work on system users' logs. User access logs contain numerical and textual information such as passwords used, failed attempted to log in, changes to files, and so on (Hong et al., 2017). Intrusion detection systems of standard computer system attacks attempt to detect password attacks, IP address spoofing, finding unusual behavior in a network, such as increased traffic, or traffic interception methods. In contrast, social engineering detection methods work on analyzing linguistic data, such as emails, social media, and phone messages (Hoeschele, 2006). Because of the nature of the datasets used in each intrusion detection method, computer system detection methods often use neural networks and other machine learning methods to find anomalous behavior in user logs, while social engineering detection methods often use natural language processing methods to detect anomalous emails or messages (see Fig. 1 for explanation). Below, we discuss each kind of detection method in some detail as well as their limitations.

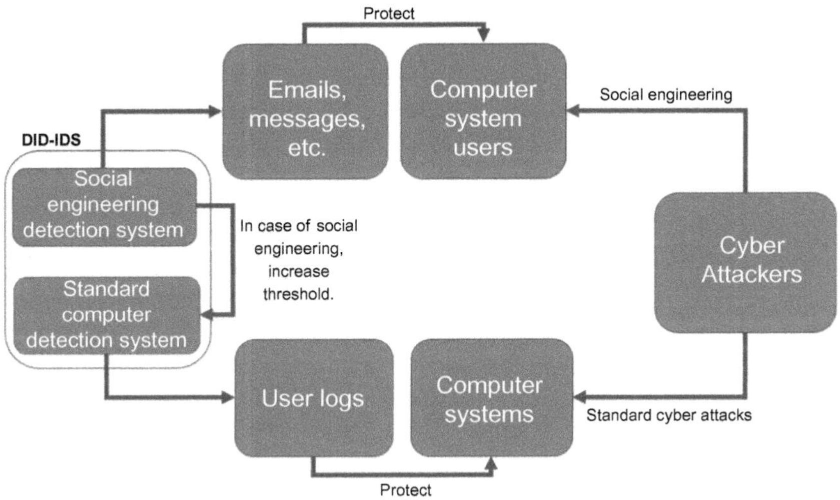

FIG. 1 Explanation of different kinds of cyberattacks and corresponding intrusion detection systems. Standard cyberattacks refer to password attacks, IP address spoofing, or traffic interception methods. DID-IDS, dual integrated dynamic intrusion detection system.

3 Prior computer system intrusion detection systems

As mentioned above, most computer system intrusion detection systems work on user logs. There are several computer system anomaly detection methods, including statistical or neural network models. Many statistical techniques employ change point detection methods, that is, detect if there is any change in the sequence of input data in the user logs or changes in traffic in a network (Tartakovsky et al., 2013). In addition, there have been several neural network models of computer system intrusion detection. This system often works by raining the network to learn (encode) some normal patterns, such as normal network behavior. In the second phase, the network is presented with both old and novel patterns. The network is then tested if it can respond differently to both types of patterns. Some other neural network models were applied to continuous data, such as time-series data, and thus have the capacity to detect changes in traffic data.

4 Prior social engineering intrusion detection methods

Some prior work relies on training system analysts and computer uses to detect social engineering attacks (Dolan, 2004; Heartfield & Loukas, 2018; Nicholson et al., 2017). These methods are, however, not automated and can be very costly and time-consuming. However, there have several natural language processing algorithms to detect social engineering attacks (Lansey et al., 2019; Peng et al., 2018; Sawa et al., 2016; Tsinganos et al., 2018) in emails as well as in online chat environments. Machine learning methods can be used for deception detection and breach of security (Mbaziira & Jones, 2016). This can be done by training an algorithm on the differences between standard and phishing emails. Mbaziira and Jones (2016) used the Enron email dataset which contains 500,000 emails and found that deceptive messages often include fewer ratios of unique words, higher frequencies of adjectives and adverbs, fewer self-references, less complex sentences, in comparison to non-deceptive messages. Along these lines, Richardson et al. (2018) used the University of New Brunswick Information Security Centre of Excellence (ISCX) Intrusion Detection Evaluation Dataset and transformed them into a sequence of languages. By using machine learning methods, they then applied detection methods and reported less common words, which are arguably anomalous.

5 A new dual integrated dynamic intrusion detection system (DID-IDS)

As discussed above, most computer science research focuses on developing intrusion detection systems to find anomalous attacks in user logs or network traffic, yet few studies focused on the detection of social engineering attacks. However, as discussed above, social engineering attacks (e.g., phishing, malware) account for more than 25% of cyber security attacks (Bowen et al., 2014; Teoh et al., 2018).

Accordingly, here, we argue that firms and organizations should develop and use dual intrusion dynamic intrusion detection systems (DID-IDS) that works at both levels: computer systems (i.e., detect anomalous use logs and traffic) and computer system users (i.e., to detect social engineering attacks). This integrated system should use machine learning methods to analyze both user logs and emails. These two systems should also work in tandem (see Fig. 1). For example, if an intrusion detection algorithm detects an increase in social engineering attacks, then it should alert the intrusion detection system at a computer systems level to increase the threshold so that it can detect potential cyberattacks and an increase in network traffic (see Fig. 1). It is predicted that if social engineering attack is successful, a cybercriminal may then overuse the computer systems, and thus lead to an increase in traffic. Thus, increasing the threshold of a computer system detection system may then monitor an increase in traffic. Such a dynamic threshold can be more efficient than a static or fixed threshold, as reported in prior studies (Faizal et al.,

2009; Kalmuk et al., 2016). Below, we discuss the benefits of using the DID-IDS over each social engineering or computer system intrusion detection alone.

One benefit of the DID-IDS is to decrease the amount of time needed by system analysts, as the social engineering intrusion detection system will help detect potential attacks on a computer system. While this may increase processing time, this then helps increase security and decrease the number of working hours of system analysts. This is important as there are several limitations of excessive reliance on system analysts. For example, system analysts may show limited situational awareness due to the large amount of data they have to deal with, which has led to several security breaches (Aggarwal et al., 2018; Barford, 2010; Dutt et al., 2013; Knott et al., 2013; Mancuso et al., 2014; Tyworth et al., 2013; Veksler et al., 2018). Importantly, the DID-IDS can potentially assist system analysts in detecting potential cyberattacks.

As another benefit of the DID-IDS, there have been some studies on deceiving cyber attackers, for example, by making them work more in order to be able to attack a network or IT systems (Almeshekah & Spafford, 2014; Frederick, 2012; Michael et al., 2004; Ormrod, 2014; Rowe, 2007; Stech & Heckman, 2011). According to some studies (Ferguson-Walter et al., 2017; Ferguson-Walter & LaFon, 2015), it is important to increase attacker workload while decreasing that of the computer system users. Here, we argue that the DID-IDS will make cyber attackers work harder as they will need to attack two integrated systems instead of one. Accordingly, the DID-IDS will help decrease the number of attacks on a computer system.

Furthermore, there is another benefit of using the DID-IDS. One of the main problems in cybersecurity is how to minimize false positives and false negatives (Agnaou et al., 2016). While false positives can increase the workload of security analysts (as they often have to sort through these potential attacks to determine if they are safe or not), false negatives can increase the number of attacks on the security system (Murtaza et al., 2013). The DID-IDS should be able to decrease false positives and false negatives, as it provides additional security and thus decreases false negatives. Importantly, given that increasing threshold of computer system detection is done in an intelligent way (i.e., when there are some potential social engineering attacks), it will then help minimize false positives. Future DID-IDS should be developed and tested to confirm or disconfirm their potential benefits mentioned here.

In sum, the DID-IDS decreases the amount of time needed by system analysts, makes cyber attackers work more, and minimizes false positives and false negatives. Future work should aim to combine existing computer and social engineering intrusion detection systems to develop an efficient DID-IDS.

References

Adepu, S., Palleti, V. R., Mishra, G., & Mathur, A. (2019). *Investigation of cyber attacks on a water distribution system*. ICS Security.

Aggarwal, P., Frédéric, M., Gonzalez, M. C., & Dutt, V. (2018). Understanding cyber situational awareness in a cyber security game involving recommendation. *International Journal on Cyber Situational Awareness, 3*, 29.

Agnaou, A., Abou EL Kalam, A., Ouahman, A., & De Montfort, M. (2016). Automated technique to reduce positive and negative false from attacks collected through the deployment of distributed honeypot network. *(IJCSIS) International Journal of Computer Science and Information Security, 14*(9).

Almeshekah, M., & Spafford, E. (2014). Planning and integrating deception into computer security devices. In *New security paradigms workshop*.

Arabo, A. (2015). Cyber security challenges within the connected home ecosystem futures. *Procedia Computer Science, 61*, 227–232.

Barford, P. (2010). Cyber SA: Situational awareness for cyber defense. In P. Liu, S. Jajodia, V. Swarup, & C. Wang (Eds.), *Advances in information security: Vol. 46. Cyber situational awareness* (pp. 3–13). Boston, MA: Springer.

Beavers, J., & Pournouri, S. (2019). Recent cyber attacks and vulnerabilities in medical devices and healthcare institutions. In *Blockchain and clinical trial* (pp. 249–267).

Bowen, B. M., Devarajan, R., & Stolfo, S. (2014). Measuring the human factor of cyber security. *IEEE 2011 Conference on Technology for Homeland Security*. IEEE.

Dolan, A. (2004). *Social engineering*. SANS Reading Room.

Dutt, V., Ahn, Y., & Gonzalez, C. (2013). Cyber situation awareness: Modeling detection of cyber attacks with instance-based learning theory. *Human Factors: The Journal of the Human Factors and Ergonomics Society, 55*(3), 605–618.

Faizal, M. A., Zaki, M., Shahrin, S., Robiah, Y., Siti, R. S., & Nazrulazhar, B. (2009). Threshold verification technique for network intrusion detection system. *(IJCSIS) International Journal of Computer Science and Information Security, 2*(2).

Ferguson-Walter, K., & LaFon, D. (2015). Deception for cyber defense: A case study. *Journal of Sensitive Cyber Research and Engineering, 3*(1), 45–58.

Ferguson-Walter, K. J., LaFon, D. S., & Shade, T. B. (2017). Friend or faux: Deception for cyber defense. *Journal of Information Warfare, 16*(2), 28–42.

Frederick, E. (2012). Testing deception tactics in response to cyberattacks. In *Paper presented at the proceedings of the National symposium on moving target research*.

Heartfield, R., & Loukas, G. (2018). Detecting semantic social engineering attacks with the weakest link: Implementation and empirical evaluation of a human-as-a-security-sensor framework. *Computers & Security, 76*, 101–127.

Hoeschele, M. (2006). *Detecting social engineering*. A Thesis Submitted to the Faculty of Purdue University, CERIAS Tech Report 2006-15.

Hong, J., Das, S., Kim, T., & Dabbish, L. (2017). *Social cybersecurity: Applying social psychology to cybersecurity*.

Kalmuk, A., Granichin, O., Granichina, O., & Ding, M. (2016). A dynamic threshold based algorithm for change detection in autonomous systems. *IFAC-PapersOnLine, 49*(13), 141–145.

Kim, M., Song, C., Kim, H., & Park, D. (2018). *Catch me, yes we can!—Pwning social engineers using natural language processing techniques in real-time*. https://www.ics.uci.edu/~harris/harris_blackhat18.pdf.

Knott, B. A., Mancuso, V. F., Bennett, K., Finomore, V., McNeese, M., & McKneely, J. A. (2013). Human factors in cyber warfare: Alternative perspectives. In *Paper presented at the in proceedings of the human factors and ergonomics society annual meeting, Los Angeles, CA*.

Lansey, M., Polatidis, N., & Kapetanakis, S. (2019). SEADer: A social engineering attack detection method based on natural language processing and artificial neural networks. In *Paper presented at the international conference on computational collective intelligence ICCCI 2019: Computational collective intelligence*.

Lohani, S. (2019). Social engineering: Hacking into humans. *International Journal of Advanced Studies of Scientific Research, 4*(1).

Mancuso, M., Christensen, J. C., Cowley, J., & Finomore, V. (2014). Human factors in cyber warfare II: Emerging perspectives. *Proceedings of the Human Factors and Ergonomics Society Annual Meeting, 58*(4), 415–418.

Mbaziira, A., & Jones, J. (2016). *A text based deception detection model for cybercrime*. International Conference on Technology and Management.

Michael, J., Rowe, N., Auguston, M., Drusinsky, D., Rothstein, H. S., & Wingfield, T. C. (2004). *Phase II report on intelligent software decoys: Intelligent software decoy tools for cyber counterintelligence and security countermeasures.*

Murtaza, S. S., Khreich, W., Hamou-Lhadj, A., & Couture, M. (2013). A host-based anomaly detection approach by representing system calls as states of kernel modules. In *Paper presented at the IEEE 24th international symposium on software reliability engineering (ISSRE).*

Nicholson, J., Coventry, L., & Briggs, P. (2017). Can we fight social engineering attacks by social means? Assessing social salience as a means to improve phish detection. In *Paper presented at the proceedings of the thirteenth symposium on usable privacy and security (SOUPS). Santa Clara, CA, USA.*

Ormrod, D. (2014). The coordination of cyber and kinetic deception for operational effect: Attacking the C4ISR interface. In *Paper presented at the proceedings of the 2014 IEEE military communications conference, 6–8 October.*

Peng, T., Harris, I., & Sawa, Y. (2018). Detecting phishing attacks using natural language processing and machine learning. In *Paper presented at the IEEE 12th international conference on semantic computing (ICSC).*

Richardson, B. D., Radford, B. J., Davis, S. E., Hines, K., & Pekarek, D. (2018). Anomaly detection in cyber network data using a cyber language approach. In *Proceedings of the American statistical association symposium on data science and statistics.*

Rowe, N. C. (2007). Experiments with a testbed for automated defensive deception planning for cyber-attacks. In *Paper presented at the proceedings of the 2nd international conference in I-Warfare and security.*

Sawa, Y., Bhakta, R., Harris, I. G., & Hadnagy, C. (2016). Detection of social engineering attacks through natural language processing of conversations. In *Paper presented at the IEEE Tenth international conference on semantic computing (ICSC).*

Stech, F., & Heckman, K. (2011). Scientometrics of deception, counter-deception, and deception detection in cyber-space. *PsychNology Journal, 9*(2), 79–122.

Tartakovsky, A. G., Polunchenko, A. S., & Sokolov, G. (2013). Efficient computer network anomaly detection by changepoint detection methods. *IEEE Journal of Selected Topics in Signal Processing, 7*(1), 4–11.

Ten, C., Hong, J., & Liu, C. (2011). Anomaly detection for cybersecurity of the substations. *IEEE Transactions on Smart Grid, 2*(4), 865–873.

Teoh, T. T., Chiew, G., Franco, E. J., Ng, P. C., Benjamin, M. P., & Goh, Y. J. (2018). Anomaly detection in cyber security attacks on networks using MLP deep learning. In *International conference on smart computing and electronic enterprise (ICSCEE), Shah Alam* (pp. 1–5).

Tsinganos, N., Sakellariou, G., Fouliras, P., & Mavridis, I. (2018). Towards an automated recognition system for chat-based social engineering attacks in enterprise environments. In *Paper presented at the proceedings of the 13th international conference on availability, reliability and security.*

Tyworth, M., Giacobe, N. A., Mancuso, V. F., McNeese, M. D., & Hall, D. L. (2013). A human-in-the-loop approach to understanding situation awareness in cyber defence analysis. *EAI Endorsed Transactions on Security and Safety, 13*(1–6).

Veksler, V. D., Buchler, N., Hoffman, B. E., Cassenti, D. N., Sample, C., & Sugrim, S. (2018). Simulations in cyber-security: A review of cognitive modeling of network attackers, defenders, and users. *Frontiers in Psychology, 9*, 691. https://doi.org/10.3389/fpsyg.2018.00691.

CHAPTER

3

Working from home users at risk of COVID-19 ransomware attacks

Anthony An Duong[a], *Abubakar Bello*[b], *and Alana Maurushat*[a]

[a]School of Social Sciences, Western Sydney University, Sydney, NSW, Australia, [b]Cyber Security and Behaviour, School of Social Sciences, Western Sydney University, Sydney, NSW, Australia

1 Introduction

The year 2020 has seen an increase in online usage as many would find themselves working and studying from home. This was due to the COVID-19 pandemic where it has forced many people to stay at home. It became a situation where accessing the internet has become a necessary requirement as working from home user must be able to continue their organizational duties as well as to be up to date with their prescribed news about the COVID-19 pandemic.

First accounts on the coronavirus were reported on December 2019 and has gradually spread to multiple countries around the world (Nebehay, 2020). In extreme circumstances, the coronavirus is very dangerous as it can cause respiratory infection which may lead to death (World Health Organisation, 2020). As of March 2020, the COVID-19 cases have risen to 7,50,000 and death toll was around 36,000 (The Guardian, 2020). Due to the alarming cases and death tolls, it has led many organizations to move their employees to work from home to reduce the spread of the virus and to adhere to quarantine rules as well. However, sadly, on October 15, 2020, the global death toll has reached more than one million (Worldometer, 2020). The year 2020 was indeed a difficult year for many people. Many working from home users would have to face the new reality to work online.

https://doi.org/10.1016/B978-0-323-90570-1.00001-2

Many working from home users are required to use an electronic device of their own choice so that they can continue their studies or organizational duties. Furthermore, maintaining communication with others during the COVID-19 pandemic has become a necessary mission as well. For working from home user to be able to continue their studies, work, and communication with others, it does require them to be able to connect to the internet which has led to cyber-security concerns. As such, the concerns are mainly around the fact that many working from home user may not have a strong understanding that they may have weak security control measures (KPMG, 2020). Having a weak security home environment could lead to cyber-criminals exploiting those vulnerabilities. The COVID-19 pandemic has pushed many people to work from home. However, it is believed that the COVID-19 pandemic would open more doors for cyber-criminals (Coker, 2020) as there are many people working online who may not be fully aware that they can be vulnerable to potential cyber-attacks.

One of the cyber-attacks that cyber-criminals would most likely launch during the COVID-19 pandemic is ransomware. Ransomware is one of many dangerous malicious software that can infect a working from home user's electronic device. As such, it has the capabilities to prevent the user from having access to their respective data. Access to their data would be provided if the user is able to pay the demanded ransom to the cyber-criminals (Richardson & North, 2017). The year 2020 has seen an increase in ransomware usage by cyber-criminals (Skybox Security, 2020). The increase in ransomware usage is an indicator that cyber-criminal is using this opportunity to exploit working from home users during the COVID-19 pandemic. As such, many working from home users could be susceptible to the ransomware attack which can produce many challenges ahead.

Many of the challenges against ransomware may also include that victims and organization may not be able to identify genuine sources, mobile phone mitigation challenges against ransomware, ransomware has strong encryption algorithm, increase in digital extortion, targeting more organization than individuals, and most concerning that many victims are willing to pay the ransom to the cyber-criminals. Ransomware is the most dangerous malware that can be used by cyber-criminals during the COVID-19 pandemic. However, cyber-criminals must use an attack vector to deliver the ransomware to working from home user.

The attack vectors that cyber-criminals can use to deliver ransomware are social engineering, phishing, trojan, remote desktop protocol, drive-by download, and malvertising. Social engineering requires the cyber-criminals to interact with the target to build trust for the purpose of gaining information (Krombholz et al., 2015). Social engineers may also send phishing emails in the hope that victims would interact with the malicious contents where the possibilities of installing a trojan can happen. Trojan is normally considered a legitimate software (Grace, 2020) although it could potentially be ransomware itself. Cyber-criminals may wish to use a remote desktop protocol to remotely access (Arntz, 2018) a working from home user's electronic device to install ransomware. Drive-by download and malvertising are other ways for cyber-criminals to distribute ransomware. During the COVID-19 pandemic, working from home users may face the possibility of ransomware attacks via the six attack vectors. Furthermore, this had led to the need to understand how it is important to implement strong security measures against any ransomware attacks during the COVID-19 pandemic.

Normally mitigation against ransomware attacks would normally work if the individual and organization have strong comprehension that their data is extremely important, and action must be taken to protect their assets. There are existing defense mechanisms that working from home user can implement in their own home environment. Web content filtering, black and white listing, seeking professional consultation, intrusion detection system, and firewall are examples of existing defense mechanism against ransomware attacks although they have control gaps within them. Thus, a mitigation model against ransomware is developed to assist working from home user to understand the security measures that are required to mitigate against potential ransomware. This paper starts with an overview of ransomware followed by issues and challenges with ransomware, ransomware attack vectors, existing defense mechanisms and control gaps, and mitigation model against ransomware.

2 Overview of ransomware

2.1 What is ransomware

Ransomware is a dangerous malware that has the capabilities to prevent users from accessing their data (refer to Fig. 1). Normally, if a user wishes to have access to their data again, they are required to pay the ransom so that the cyber-criminal can provide them with a private key to decrypt their data (Richardson & North, 2017). Normally, cyber-criminals

FIG. 1 Stages in ransomware.

would prefer the victim to pay the ransom in a form of Bitcoin as these are usually encrypted and it would make it harder to trace the transaction (Newby & Razmazma, 2016). Ransomware has a dangerous encryption algorithm that can prevent the user from accessing their data, although it is also capable to disrupt other electronic devices.

Normally, ransomware is used to target personal computer devices as they do generally attract cyber-criminals. However, ransomware can have the abilities to also attack mobile phone devices. As such, it works by altering the PIN on the personal mobile devices forcing the victim to pay the ransom so that they can receive the new PIN to unlock it (Zetter, 2015). Due to the malware abilities to target personal computers and mobile phone devices, ransomware may be a preferable choice for cyber-criminals to use to launch an attack onto their victims who are working from home. Not only the malware can encrypt victim's data, but it does also tend to force their victims to pay the ransom as well. As such, symantec have estimated that ransomware annually extorted millions of their victims and has noted that forwarding the ransom to cyber-criminals would not guarantee that their data will be released back to them (Zetter, 2015). Over the years, cyber-criminals have used ransomware to cause a lot of disruption to their targets. Ransomware has become more sophisticated as well as more modernized for cyber-criminals. To understand how ransomware grew to be the most dangerous and atrocious malware, it is important to discuss it origins as well as how it grew over the years. The first ransomware was first utilized in 1989 which was called the AIDS Trojan ransomware.

2.2 Ransomware evolution

In 1989, AIDS Trojan ransomware was the first-known ransomware that was organized to be deployed through 20,000 infected floppy disks (Johansen, 2019). The intention was to maliciously target the staff members at the AIDS research project and World Health Organization (WHO) (Trautman & Ormerod, 2018). From the infect computer, the AIDS ransomware would encrypt data files as well as hide the directory folders that was in the C drive. AIDS silently stay dormant in the computer systems and after the system undergoes 90 restarts, the ransomware would get activated. Due to the technology at the time, the AIDS ransomware could not propagate on a large scale as there were less computer connectivity (Yaqoob et al., 2017). The introduction of ransomware was indeed terrifying at the time. AIDS ransomware has set an example where a malicious software is capable to lock people out from accessing their files. As technology continues to advance, so would ransomware.

The AIDS ransomware was the beginning of the making of many more potential ransomwares. Among many of the ransomware programs that were later introduced, cyber-criminals had attempted to find ways to improve it from its predecessor. But at the same time would implement different ways to deliver the malware to it victims. In 2005, Trojan.Gpcoder was introduced and spread via phishing emails (Sjouwerman, 2015). Luckily at the time, Trojan.Gpcoder has very weak encryption and it was very easy to overcome its challenges (Richardson & North, 2017). Trojan. Gpcoder was not the most sophisticate ransomware but nevertheless has open more door for new ransomware to be introduced.

In the following year, Trojan.Cryzip was introduced in early 2006 which had the capabilities to copy files from the victims and deleting the original version from the user electronic device (Richardson & North, 2017) which may be the first kind of ransomware to do so. Locker ransomware had begun to surface in 2007 where the attack had begun to spread towards the United States and Europe (Zetter, 2015). Ransomware started to become a major issue around 2011 as 30,000 new strain was introduced in the first quarter, another 30,000 introduced in the second quarter, and third quarter saw another 60,000 new strains (Sjouwerman, 2015). Citadel is a toolkit that was introduced in 2012 which can dispense and produce ransomware (Segura, 2016). Up until 2012, ransomware was growing at a rapid pace and has become the most used malware by cyber-criminals and has grown to be more atrocious over the following years.

In 2013, CryptoLocker was introduced where it used a private and public cryptography to encrypt victim's data (Richardson & North, 2017). Initially, CryptoLocker was delivered via Gameover ZeuS banking Trojan botnet but later cyber-criminals used emails to distribute the malware where they claimed to be FedEx or UPS (Zetter, 2015). Victim would normally have up to 72h to pay the ransom (Sjouwerman, 2015). Between September 2013 and May 2014, approximately 5,00,000 victims had the CryptoLocker ransomware infection (Cannell, 2016). The end of 2015 was devastating as the FBI reported that all victims of ransomware had paid approximately 27 million dollars to their extortionist (Cannell, 2016) and this would no doubt continue to rise over the next coming years as ransomware becomes more advanced in terms of encryption as well as extortion. It was clear at this point that ransomware was indeed becoming more weaponized by cyber-criminals and has continued to extort more victims. This would have had a negative financial and emotional impact on the victims. The COVID-19 pandemic presents several opportunities for cyber-criminals as many people are working from home and are connected to the cyberspace.

The COVID-19 pandemic has forced many organizations and their respective employees to work from home. Due to the unexpected adaptation to work from home, it is possible that working from home user are not prepared for a new transition but at the same time may lack understanding of the cyber-risk during the COVID-19 pandemic. As such, cyber-criminals would see this as a great opportunity to exploit working from home users during the COVID-19 pandemic. The first half of 2020 saw an alarming 72% increase in ransomware attack growth (Skybox Security, 2020). The possibility of such an enormous growth may have been due to the weak security controls as well as the possibility of users being lured into phishing emails during the COVID-19 pandemic (KPMG, 2020).

The introduction of ransomware has indeed disrupted many businesses, organizations, and victims. The malware itself has grown to become the most popular tool to be used by cyber-criminals. However, cyber-criminals would normally decide to use either a crypto or locker ransomware as these are the two common types used by extortionists.

2.3 Ransomware types

2.3.1 *Crypto-ransomware*

If the user were infected with crypto-ransomware and chooses to remove it from their media storage device, this would still leave their data encrypted (Richardson & North, 2017). However, crypto-ransomware would leave the critical system files untouched which still allows the users to have some access to their device as the cyber-criminals would need the victim to pay the ransom via their infect electronic device (Savage, Coogan, & Lau, 2015). By using crypto ransomware, the victim would have to pay the ransom in bitcoins (Richardson & North, 2017) as this type of payment is untraceable (Rosenberg, 2015). This type of ransomware would be the first choice for cyber-criminal as the aim is to prevent the user from accessing their data as well as become unable to trace the cyber-criminal's whereabouts.

2.3.2 *Locker-ransomware*

The second type is locker-ransomware where it prevents the user from accessing their electronic devices (Savage et al., 2015). The data itself that is on the user's electronic device would remain untouched. In this case, the user may choose to easily remove their media storage from their electronic device, making the likelihood that their data is not damage or encrypted. Usually the payment would be present in payment vouchers where the victim is required to send the payment codes (Savage et al., 2015). Therefore, this type of ransomware is usually not the best choice for cyber-criminals as it is not the most effective way to extort their victims to pay the ransom.

Crypto-ransomware and locker-ransomware are the two types that can be used be cyber-criminals to extort their victims. It has been discussed that has grown to become a popular tool to be used by cyber-criminals which would produce many challenges ahead. These challenges against ransomware attacks would most likely be related to weak security controls measures within a person's home environment as well as not realizing how easy it is for cyber-criminals to exploit those vulnerabilities.

3 Challenges and issues with ransomware

3.1 Internet of thing challenges against ransomware

In the digital age, individuals have started to embrace the benefits of using the Internet of Things (IoT). IoT has become an important aspect in people's lives as it provides many ways to assist with their daily tasks or needs. As such, IoT has given individuals the ability to connect to their smart cars (Al-Fuqaha et al., 2015), homes (Ahmed et al., 2016), parking (Perera et al., 2013), utility meters (Sanduleac et al., 2016), healthy monitors (Ghosh et al., 2016), and other potential electronic applications. However, this may provide a several opportunities for cyber-criminals to exploit as IoT devices are programmed to store sensitive and personal information. Basically, IoT devices are data collectors and stores information like age, name, location, health details, and much more where the possibility of cyber-criminals stealing one's identity is quite high (NortonLifeLock Employee, 2019). The concerns are mainly on not having strong defensive control for IoT or for users' respective electronics devices. Developing security as well as designing preventative measures against ransomware attacks was perhaps considered challenging as well as quite alarming (Jing et al., 2014). The COVID-19 pandemic has forced most people to work from home but at the same time provided users with little protection against ransomware attacks.

Working from home users would utilize electronic or IoT devices to continue to perform their personal or organization duties. IoT does contain sensitive information which is why it has provided a high level of uncertainly as cyber-criminals may choose to launch a ransomware attack onto any desirable targets. Ransomware is much different from other traditional malware attacks as it has the capabilities to affect large working environments leading financial losses as well suffering from an information breach (Bertino & Islam, 2017). Once the breach is successful, the ransomware would hold the victim's data hostage, may only allow restrict access to their electronic devices, demanding the ransom, and once the ransom is paid cyber-criminals may choose to decrypt the victim's data (Yaqoob et al., 2017). If the user has failed to pay the ransom, the cyber-criminals may wish to extend the payment deadline or permanently remove all data from the victim's electronic device (Nassi et al., 2017).

3.2 Victims or organization may not know how to identify genuine sources

Considering that working from home has pushed people to move onto the internet, it could be quite difficult for them to identify genuine sources as cyber-criminals do have ways to lure working from user into their malicious deceptions. These malicious deceptions may lead the victim to install and execute a ransomware. As such, the challenges that arises with ransomware is that they can surreptitiously infiltrate computers using various unauthentic, malicious, and fraudulent applications (Bugeja et al., 2017). There are cases where the software would falsely generate an alarm that would make the user believe that their systems has been compromised and require them to pay the money for data recovery (Yaqoob et al., 2017).

3.3 Mobile phone mitigation challenges

Across the globe, ransomware has been identified as a threat to network and computer security (Gazet, 2010). Due to its covert nature, it has spread to hand-held mobile phone devices. Mobile detection approaches have not been effective to detect ransomware (Andronio et al., 2015) leaving billions of phone users to be vulnerable to ransomware attacks (Popoola, Ojewande, Sweetwilliams, John, & Atayero, 2017).

3.4 Ransomware has strong encryption algorithm

For cyber-criminals to extort their targets, they would usually launch a ransomware attack that can encrypt the files on the electronic device of the victim. Encrypting victims' files has indeed become one of the many threats of ransomware as they are not usually easy to decrypt. By encrypting the victim's files, the cyber-criminals would demand a high ransom so that they can decrypt the user's files. The ransomware would encrypt the full hard drive or Computer's Master File Table (MFT). Cyber-criminals would normally use trojans to deliver ransomware to their victims as it helps disguise the files as authentic. Within any computer security environment, advanced encryption can in fact be helpful to effectively protect sensitive information. But has been integrated into ransomware which has become a malicious tool for cyber-criminals to use and became a serious cyber-threat as ransomware are equipped with enhance algorithms to lock victim's data (Popoola et al., 2017)

3.5 An increase in digital extortion

The number of online users and services would indeed continue to grow over many years to come. Due to the number of online users and services on the internet growing exponentially, it would indeed increase digital extortion (Bhardwaj, 2017). Over the years, ransomware has been rated as the highest cyber-threat because it has caused lots of disruption to many businesses (Brewer, 2016). Approximately 80% of ransomware attacks can discover vulnerabilities in flash application that are found in many businesses which can destructively propagate itself across the network holding many users or businesses hostages (Popoola et al., 2017). There are many working from home users and online services during the COVID-19 pandemic. Many would not realize that the cyber-criminals would also occupy the cyber-space who can exploit this opportunity to digitally further extort and to have a powerful financial gain from online businesses and working from home users.

3.6 Ransomware attacks are targeting more organizations than individuals

Nowadays, many cyber-criminals would choose to focus to target large organization than individuals. The shift from individual to large organization is that organization tends to hold enormous form of information and may have a large attack surface. For instance, in 2016, the Hollywood Presbyterian Medical Centre was attacked, and had to shut down due to being bit by a Crypto-ransomware. It had encrypted all information and files in their computer systems and had denied access to medical staff from retrieving patients' sensitive and personal health archives (Everett, 2016). The Methodist Hospital is another example where Kentucky was lucky enough to retrieve and recover patient archives. The cyber-criminals had used stolen administrative credentials and had infect the Methodist Hospital network servers with SamSam ransomware (Popoola et al., 2017).

3.7 Many are willing to pay the ransom

If an organization was attacked by ransomware and has no form of backup, they may have little choice but to pay the ransom to the cyber-criminals to avoid the data being lost. As such, it was found that approximately 46% of organization had encountered a form of ransomware attack, and about 57% of those were medium-size organization and 53% were large organization. Many of these organizations were more than willing to pay the ransom (Popoola et al., 2017). Without a data recovery in place, organization would have little choice but to obey the cyber-criminal's order and pay the ransom in exchange for their data.

In 2016, it was reported by Federal Bureau of Investigation (FBI) that one billion US dollar was loss from ransomware attacks (Popoola et al., 2017). The losses have demonstrated that many victims would be more than happy to pay the ransom so that they can have access to their data again. Approximately 40% of victims would indeed pay the ransom. In addition, it was found that 3 out of 4 ransomware attacks, cyber-criminals were more than happy to negotiate the ransom with the victim but usually would give on average a 28% reduction from the original demanded ransom (Popoola et al., 2017). Due to the alarming number of victims and organizations paying the ransom to cyber-criminals, it may demonstrate there are very little security controls measures against ransomware attacks. It is important for working from home user to be prepare and anticipate any ransomware attack, especially during the COVID-19 pandemic. For ransomware to be delivered and executed correctly, it does require the victim to either interact with the cyber-criminals directly, indirectly or may accidentally download and execute an unknown program or software. To spread and distribute ransomware, cyber-criminals may either use one or a combination of attack vectors to lure their victims.

4 Ransomware attack vectors

Ransomware is normally spread and distribute via social engineering, phishing, trojans, remote desktop protocol, drive-by downloads, and malvertising. Social engineering is to develop a relationship with their victim to extract personal and sensitive information from them. Other attacks vectors such as phishing, Trojan, Remote Desktop Protocol, Drive-by Downloads, and Malvertising can cause the same level of disruption but is reliant to be delivered via social engineering tactics. Due to the nature of social engineering, the victim may not be able to know whether they are communicating to a trustworthy source. Due to the COVID-19 pandemic, it is possible for cyber-criminals to distribute ransomware to working from home user by using the either of the six attack vectors.

4.1 Social engineering

The purpose of social engineering is to build trust but at the same time access and obtain information from potential victims and organization. Normally, social engineers would manipulate, influence, and deceive their targets to give away sensitive information (Krombholz et al., 2015). Employees may fall into social engineer's trap if they believe that they were communicating with a trustworthy source or the user has failed to detect the deception from the cyber-criminals (Qin & Burgoon, 2007).

Usually, email is a common medium that social engineers would use to communicate to their targets (Krombholz et al., 2015). It is known to be the most convenience way to deceive their targets to visit a suspicious source where the victim may unintentionally install ransomware. In the digital era, it has given social engineers the ability to deceive and trick their victims. However, all cyber-criminals would have different intentions and would choose to adopt many different types of approaches to trick their targets. These approaches are typically either physical, social, reverse, technical, and socio technical.

4.1.1 Types of social engineering

1. Physical approaches

The physical approaches would require the social engineer to physically obtain information by hand normally in a form of a piece of paper. There are some circumstances where employee would be required to dispose piles of paper into a recycling bin. In this case, it would no doubt makes it more convenience for social engineers to find personal and sensitive information. This is commonly known as dumpster diving as it requires the social engineer to physically search through the recycling bin for information (Granger, 2001). Other physical approaches would involve theft and extortion (Krombholz et al., 2015). However, due to the COVID-19 pandemic, social engineers may be reluctant to use this approach. But nevertheless, there is a possibility that working from home user may decide to throw out documents. The electronic device specifications could be found within a recycling bin. After the social engineers have found the exact specification from the target's electronic device, they may wish to practice how to install ransomware onto the model device before initiating the real attack.

2. Social approaches

This approach would require heavy reliance on socio-psychological methods like Cialdini's principles of persuasion so that social engineers can trick their victims (Krombholz et al., 2015). To increase the success of the attack, social engineer would develop a relationship with the victim. An electronic device such as a phone is commonly used with this type of social engineering (Granger, 2001). As such, it is possible for working from home user to encounter this type of approach from social engineers.

3. Reverse social engineering

Reverse social engineering has three distinctive steps. The first step is to disrupt, destroy, and damage the target's computer system which may involve disconnecting an employee from the organization's network. The second step requires the social engineers to advertise themselves making the victim believe that they are capable to resolve the problem. The final step is to convince the victim to give their password or install a software so that they can fix the issue (Krombholz et al., 2015). The software that could be install may very well be ransomware and working from home users may face this possibility.

4. Technical approaches

This method would normally be initiated over the internet. The goal is to gather information that they can find on the internet. Normally, search engines and social network platform are valuable information and tools for social engineers to use. If it was found that a user who uses the same password across multiple platforms or accounts, a social engineer may unlawfully access those accounts (Granger, 2001). Considering that many working from home user are online during the COVID-19 pandemic and are exchanging information with each other, it would create a pool of opportunities for cyber-criminals to harvest that information.

5. Socio-technical approaches

Social-technical approaches involve combining several approaches of social engineering. An attacker may wish to create a malware-infected media storage and place it in a location where the victim would most likely use it (Krombholz et al., 2015). A device like a USB stick may contain a malware like trojan (Stasiukonis, 2006) leading to the possibility of ransomware to be installed. Social engineers may wish to add labels such as "confidential" to lure victim to use the media storage (Krombholz et al., 2015).

4.2 Phishing

Phishing is an attack vector that can deceive the victim to install a ransomware. This attack vector requires the use of impersonation to gather usernames, passwords, and other credentials from their victims (Aburrous et al., 2010). Since there are many people working from home, cyber-criminals may choice to use phishing attacks to trick the victim to install ransomware. Depending on the circumstance and intention, the cyber-criminal may wish to use either email, spear, whaling, and smishing and vishing.

4.2.1 Types of phishing

1. Email phishing

Normally in a form of a spam message that claims to be from a legitimate organization. The spam email would most high likely have an embedded link, which requires user to interact with the link, to redirect them to an unauthentic webpage (Almomani et al., 2013). Approximately, 80% of ransomware attacks are normally distributed via email phishing attacks (Solomon, 2019). Maze ransomware is an example of a ransomware that was deliver via phishing emails (Gendre, 2020).

2. Spear-phishing

Spear-phishing is very similar to email-phishing. However, this method is all about targeting a specific individual or organization (Amro, 2018). Ryuk ransomware is an example that was delivered via spear-phishing where it had targeted a single organization and had demanded a high ransom for data retrieval (Checkpoint, 2020).

3. Whaling-phishing

Whaling-phishing is design to target high profile individuals. This method requires a large amount of time to gather information about target and may need to develop trust with them (Amro, 2018). Zeppelin ransomware is an example that was delivered via whaling-phishing (BlackBerry, 2019).

4. Smishing and vishing

Smishing and vishing does not use any form of emails phishing. But is heavy reliant on the user mobile phones devices. As such, smishing requires the cyber-criminals to forward a short messaging service (SMS) or text messages via phone to their victim. While vishing uses a phone call or voice mail to lure the victim into giving the attackers their personal information. Filecoder ransomware is an example where users interact with the link which install a malicious app that would be presented to the victim as an adult game simulator (Osborne, 2019).

4.3 Trojan

Trojan is a type of malware that can disguised itself as a legitimate software (Grace, 2020). The malware itself does require a medium so that it can infect the user's computer. It can be accomplished when a user receives a phishing email believing that it is legitimate without realizing that they have accidentally downloaded the attachment. If the user decides to execute the file, the trojan malware can spread itself, affecting files and damaging one's electronic device (Alison, 2020). As previous discussed in ransomware overview, AIDS trojan ransomware is an example as well as Trojan.Gpcoder and Trojan.Cryzip.

4.4 Remote desktop protocol

Remote Desktop Protocol is being able to remotely access an electronic device for the purpose of accessing the desktop, files, and other applications (Arntz, 2018). For working purposes, working from home user may require to remotely access another colleague's electronic device to resume and commerce organizational duties. Due to the COVID-19 pandemic, cyber-criminals may disguise themselves as authentic IT support to have remote access to an individual electronic device. It is possible for cyber-criminals to use RDP to deliver Ryuk ransomware to the target electronic device (Checkpoint, 2020).

4.5 Drive-by download

Drive-by download is where the user's unintentionally visits a perfectly disguise website that downloads a file or software on their electronic device without their consent (Cassetto, 2019). If a user were to interact with a phishing email link, it may redirect them to a malicious drive-by download webpage where they may unintentionally download ransomware. Bad rabbit ransomware was introduced in 2017 and spread via drive-by download where users had visited a genuine website without knowing that it was already compromised (Mamedov et al., 2017).

4.6 Malvertising

Malvertising disguises themselves as unauthentic online advertisement. It distributes malware and other forms of threat to user by installing a malicious code on the electronic device, then sends information to cyber-criminal command and control (C&C) servers (Lastline, 2018). Cyber-criminals may wish to install ransomware install onto the user's electronic device. Sodinokibi Ransomware is an example that was deliver via malvertising (Trend Micro, 2019).

5 Existing defense mechanism and control gaps

Normally, defense mechanism is normally designed to mitigate against any cyber-risk. A control gap may occur only when there are very little to zero control measures that are in place. Without having proper controls in place or outlined, it may not be effective to prevent a risk from occurring (Knowledge Leader, 2019). There are mitigation techniques that can be effective against ransomware attacks. But mitigation against ransomware can only be effective if only the individuals and organization realize how precious their assets are and how protecting it is the first necessary step for the organization. As such, there are times where an individual or organization information is compromised and this can be due to many reasons. But it is mainly due to not frequently practising cyber-security defense measures, understanding that the cyber-world is always changing, and to act accordingly to those changes.

Most of our information is stored digitally and it is quite easy to access them. For cyber-criminals to gain access to an organization information and infrastructure, they would usually interact or trick employees to give them information. It is important to realize that human do make errors as they are the weakest link within any organization (Bhayani, 2019). Especially, if they do not understand the consequences of cyber-attacks, how it can negatively impact their lives, and affect the reputation of the organization. Web content filtering, black and white listing, third-party consultation, intrusion detection system, and firewall are some examples of existing defense mechanism that can help detect and evade ransomware but may have controls gaps that can be challenging for working from home users.

5.1 Web content filtering (WCF)

WCF can be used to mitigate against ransomware attacks as it is designed to be embedded into a web-browsing application that is capable to screen active contents from webpages that user visited. The purpose is to see if the contents have codes that could potentially install and run a malicious payload onto the user's electronic device. As such, suspicious webpages that runs and execute Java codes can possibly harm users' electronic devices (Australian Cyber Security Centre, 2017) are filtered and blocked. However, WCF may block a website that is non-malicious which may irritate working from home users. They may need to apply the necessary changes to the filtering system so that they can access the exact website to resume working from home duties. In addition, there are approximately three to five million brand new websites that are introduced onto the internet each week which can be difficult to filter out the malicious and non-malicious contents. WCF can help working from home users to be safe from malicious websites but may prohibit the use of some websites which may be important for some to resume working from home duties.

5.2 Black and white listing

There are many ways for ransomware to be installed and executed onto the victim electronic device and organization may wish to obtain a black and white list. Blacklisting is a list of known suspicious but potentially malicious Uniform Resource Locator (URL) (Sheng et al., 2009). Visiting a suspicious website may result ransomware being installed onto the victim or organization electronic devices. That is why it is helpful for organization to obtain a blacklist as they can identify which URL that can harm, damage, or access their information without their permission. However, this approach may not be effective to identify and detect vishing, and smishing attacks (Shahriar et al., 2015). In addition, the AV-TEST Institute have reported that over 3,50,000 malicious contents where introduce every day and may be difficult to track and black-list those contents (Bisawa, 2020).

Whitelisting is the reverse of blacklisting as it is about providing users websites that are trustworthy for them to access (Shahriar et al., 2015). It can be quite effective to detect potential smishing attacks (Mahmoud & Mahfouz, 2012). Due to new number of websites being introduced each week, keeping the white list up to date can be challenging, and may affect the operation for the organization (Bisawa, 2020). Black and white listing does present similar control gaps to WCF.

5.3 Third-party mitigation

Seeking professional consultation can be essential as the victim or organization may not know any mitigation or recovery technique that has occur from ransomware attacks (Shinde et al., 2016). This type of assistant can be important especially if the victim or organization does not know what type of ransomware that they had encountered. In addition, third-party consultation does provide a deep analysis of the current situation that the victim and organization are currently facing, such as the damage that was done by the ransomware. They may also provide to their victim and/or organization appropriate mitigation procedures that can control and prevent the situation from escalating or occurring again. However, as previously mentioned in ransomware attack vectors, social engineers may attempt to use reverse social engineering. As such, there is a possibility that the third-party entity may could be a social engineer themselves, leading them to have access to the organization network as well exploit other vulnerabilities (Infosec Resources, 2018).

5.4 Intrusion detection system

An intrusion detection system (IDS) is another existing defense mechanism that can detect potential ransomware attacks. It is a hardware appliance or software application that has the capabilities to screen the traffic that is moving through the individuals or organizations network.

Furthermore, IDS can also alert users when it has detected a suspicious action and other potentials threats within the network that could harm the individual (Pratt, 2018). However, IDS would usually require experienced personnel to understand how to resolve the issue as the software application or hardware appliance is not designed to prevent the incident from happening on its own. Furthermore, it cannot process any encrypted packets which is a concern as cyber-criminals may surreptitiously pass the encrypted packets through the individuals or organizations network (Rapid7, 2017).

5.5 Firewall

Working from home user may wish to use firewall as a layer to prevent any potentials ransomware attacks. It is a firmware or software that is design to prevent any illegal access to an individuals or organizations network, inspect outgoing or incoming traffic, and identify and block potential threats (Rouse, 2020). However, firewall cannot protect users from opening or interacting with potential phishing emails (Cockerham, 2018) which may lead them to intentionally install ransomware.

The five existing defense mechanisms can be used in a working from home environment. However, it does have potential control gaps that can be challenging for working from home users as they may lack the understanding that each existing defense mechanism may have its own limitation. Therefore, it is extremely practical to have a cyber-awareness training model for employees to practice and to be aware that frequent practices would strengthen the weakest link that are presented in within the organization.

6 Mitigation model against ransomware

To avoid the possibilities of cyber-criminals installing ransomware onto the victim's electronic devices, this section would discuss how to mitigate against the six attack vectors. To have a successful outcome, it is important for the IT department to explain and demonstrate to working from home user the concept of each attack vectors, how to appropriately respond to each of them, and how to implement the mitigation model against each of the attack vectors. Refer to Fig. 1 for the general flowchart for how to implement the model to teach Working from Home (WFH) users to mitigate against the six attacks.

6.1 Recommended mitigation against social engineering

To avoid being a victim of social engineering attacks, one must realize that their information is the most valuable assets. Protecting the physical and digital security are the steps that employees must take to avoid being a victim of social engineering. It is important for all members within an organization to know that the likelihood of being a victim of a social engineer attack is possible. As such, implementing a strong physical security will build a strong physical structure that would prevent social engineer from obtaining personal and sensitive information from the victim.

1. Do not touch anything suspicious or unknown

As such, working from home users must understand that they should not touch or use anything that does not belong to them as it may contain malware. Saleem and Hammoudeh (2018) explains that one must not plug-in any unknown or suspicious USB or other electronic devices. They must forward the unknown electronic device to their respective IT department so that they can analyze it.

2. Shredding documents

If there was a decision where the employee was instructed to throw away the physical copies into a recycling bin, the chance of a dumpster diver obtaining that information could happen. If it is necessary to throw away the physical copies, employee must use a paper shredder. The purpose of shredding one's documents is to protect against identify theft, produce high confidence from customer, and prevent any possible privacy losses (Data Safe, 2019) such as user's device specification.

3. Other required physical protection

Other ways to strengthen the physical security is to report any suspicious activities as well as to install a CCTV surveillance for the purpose of monitoring one's working environment (Saleem & Hammoudeh, 2018). In the case for physical approach of social engineering, the aspects of having physical security will reduce the likelihood of being a potential victim of social engineering. However, it is evenly as important for organization to strengthen the digital security aspect as well.

4. Digital protection

Most of the information that are stored in an organization would be stored digitally. This can provide employees the capability and convenience to work remotely (Rivera, 2019). However, despite the convenience of having information ready to be access remotely, working from home user must implement a strong digital security to protect one's information. Saleem and Hammoudeh (2018) suggest that having antimalware, spam protection guard, and firewall are examples of digital security which can be effective against social engineering which can be implemented within a home environment. Also, employees are encouraged to use passphrase instead of passwords so that it is complex for cyber-criminals to hack but at the same time easy for the user to remember (Australian Cyber Security Centre, 2020b).

6.1.1 How to implement "Recommended mitigation against social engineering"

Referencing the stages from Fig. 2, Table 1 provides a snapshot of how the IT department can implement the "Recommended mitigation against social engineering" as well as how to help and educate WFH user how to mitigate against social engineering attacks.

By implementing the above recommendation from Table 1, it would greatly benefit working from home users and organizations. As such, this implementation is there to help them to understand the dangers as well as prepared them to mitigate against social engineering attacks.

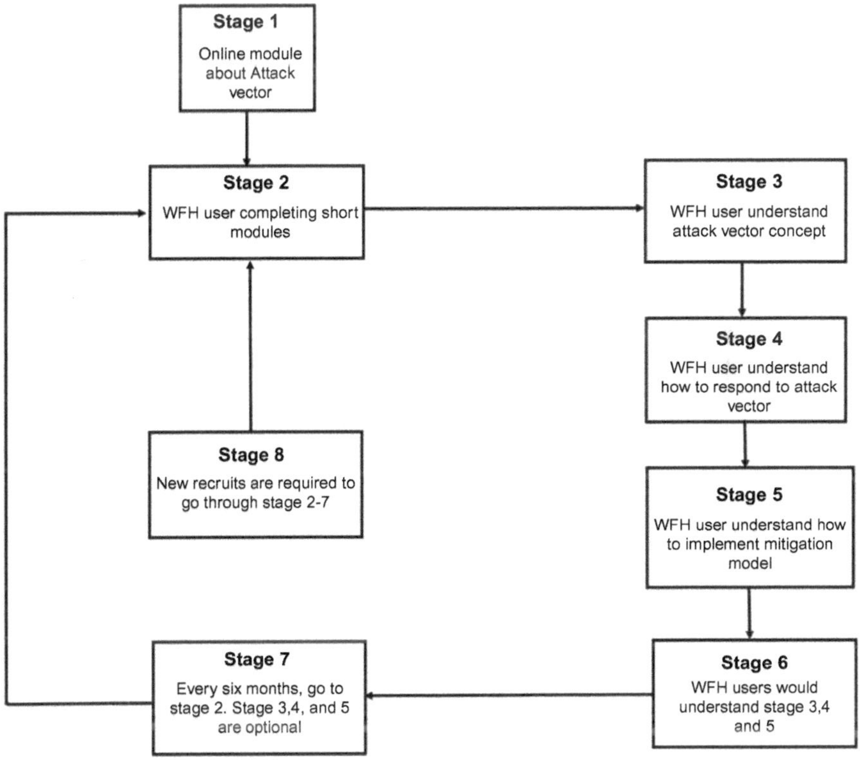

FIG. 2 Implementation flow chart to teach working from home (WFH) users.

TABLE 1 Implementing "Recommended mitigation against social engineering" to WFH users.

Stages	Description
Stage 1	The IT department are required to organize a short online module with quizzes and answers to teach working from home users about social engineering attacks and its potential to spread ransomware
Stage 2	All working from home users are required to complete the short modules with quizzes and answers before moving to stage 3
Stage 3	Using any online face-to-face medium, the IT department are required to personally teach working from home users about the concept of social engineering. Repeat if working from home users have not yet understand the concept of social engineering
Stage 4	Using any online face-to-face medium, the IT department must demonstrate how to respond appropriately to potential social engineering attacks. For instance, the IT may say that if you are asked to give your username and password to suspicious sources, one must be able to deny it request. Repeat if working from home users have not understood how to respond appropriately to social engineering attacks
Stage 5	Using any online face-to-face medium, the IT department are required to go through all the "Best practices against social engineering" with working from home users. For instance, IT may show to working from home user how to correctly install a prescribe antimalware and how to update it regularly. Repeat if working from home user have not understood how to appropriately apply the security control measures from "Best practices against social engineering"
Stage 6	At this stage, working from home user would now have a strong comprehension about the concept of social engineering, how to appropriately respond to it, and how to appropriately apply the security controls measures from "Best practices against social engineering"
Stage 7	Every 6 months, the IT department must go through stage 2 with working from home users to ensure that they are up to date with their skills. Stage 3, 4 and 5 are other optional stages working from home user may wish to go through again to help rejuvenate their training
Stage 8	In the future, an organization may recruit new working from home users. The IT department are required to go through stage 2–7 with the new working from home users

6.2 Phishing drill simulation

Social engineers may wish to use other tactics to gain personal/sensitive information from their victims. The most common method that social engineer would use is to send out many emails to potential victims to obtain their private information. In this case, it is known as phishing where its purpose is to obtain the victims usernames, passwords, and other credentials (Aburrous et al., 2010). As such, it is important that individuals who are working from home understand the consequences of phishing emails and how it can potentially affect their personal lives and the reputation of the organization as well. To prevent such event from happening, it is advisable that the organization IT department organize a phishing drill simulation to mitigate against phishing attacks.

The phishing drill simulation should be created and run by the IT departments of the organization. The purpose of this phishing mitigation drill is to educate everyone in the organization about the dangers of phishing attacks and to help prepare them against ransomware attacks via phishing. This would require working from home user to be exposed to examples of phishing emails, that are not intended to be maliciously dangerous but rather asking them to identify the real and phishing emails (Scarfone et al., 2008). Hence, the phishing mitigation drills should be implemented into the awareness and training program to help enhance working from home user to react and recognize malicious email addresses, links, and attachments (Sittig & Singh, 2016).

6.2.1 How to implement "Phishing Drill Simulation"

Referencing the stages from Fig. 2, Table 2 provides a snapshot of how the IT department can implement the "Phishing Drill Simulation" as well as help and educate WFH user how to mitigate against phishing attacks.

The above-recommended implementation from Table 2 would greatly benefit working from home users and organizations. It is to help them to understand the dangers as well as prepared them to mitigate against phishing attacks.

6.3 Secure remote desktop protocol

It is important to configure an individual's remote desktop protocol to prevent cyber-criminals having the opportunity to access one's electronic device. In doing so, would also avoid the possibilities of ransomware being installed (Whitney, 2019). In order to prevent one's remote desktop protocol being exposed to cyber-criminals, users must follow the recommend outline plan.

1. Change RDP Port

Change port 3389 to another port is a necessary step (Cobb, 2020). This would provide the reassurance that users would not appear on a list of possible attacks (Arntz, 2018).

2. Restrict and limit access

To prevent any of the consequences from remote desktop attacks, it is important to only allow remote desktop port to be access by trusted parties

TABLE 2 Implementing "Phishing Drill Simulation" to WFH users.

Stages	Description
Stage 1	The IT department are required to organize a short online module with quizzes and answers to teach working from home users about phishing attacks and its potential to spread ransomware
Stage 2	All working from home users are required to complete the short modules with quizzes and answers before moving to stage 3
Stage 3	Using any online face-to-face medium, the IT department are required to personally teach working from home users about the concept of phishing. Repeat if working from home users have not yet understand the concept of phishing
Stage 4	Using any online face-to-face medium, the IT department must now demonstrate how to respond appropriately to potential phishing emails. For instance, the IT may say if there is anyone who were to encounter a phishing email, one must not interact with the contents within the email. Repeat if working from home users have not understood how to respond appropriately to phishing emails
Stage 5	Using any online face-to-face medium, the IT department are required to go through the "phishing drill simulation" with working from home users. For instance, IT must show at least 10 real emails and 10 fake phishing emails to working from home users. All working from home users are required to identify and explain with reason which email is real and phishing. Repeat if working from home user were not able to correctly identify a real and phishing email from the "phishing drills simulation"
Stage 6	At this stage, working from home user would now have a strong comprehension about the concept of phishing attacks, how to appropriately respond to it, and how to identify real and phishing emails from the "phishing drill simulation"
Stage 7	Every 6 months, the IT department must go through stage 2 with working from home users to ensure that they are up to date with their skills. Stage 3, 4 and 5 are other optional stages working from home user may wish to go through again to help retrain themselves
Stage 8	In the future, an organization may recruit new working from home users. The IT department are required to go through stage 2–7 with the new working from home users

only and by doing so would prevent any unnecessary malicious attempts (Cobb, 2020). For windows 10 platform, to create a user group for the purpose of remote access, the user may wish to proceed to do the following:

1. Navigate to the "setting" application
2. Go to "system" and select "Remote Desktop"
3. Select "Select users that can remotely access this PC"
4. Then the user will see a "Remote Desktop Users" dialog window box
5. Click the "add" button, then click onto "Advanced" button

6. Click "Find now" and select the user you wish to have access to the "Remote Desktop Users" group. After that click "Ok"
7. Then click "OK" (Admin, 2019)

3. **Use VPN and principle of least privilege**

Due to the number of employees working from home, using Remote Desktop may be necessary. An organization must highly consider using a VPN to not only enable remote access but would not expose the organization electronic device to the internet. In addition, it is very important to only ensure those employees can have access to Remote Desktop but at the same time only grant certain permission that are required for them to complete their work (Cobb, 2020).

4. **Up to date software and patches**

Cyber-criminals tend to exploit vulnerabilities from outdated software. Working from home users must have the latest updates so that they can have an up-to-date version for both server and client software (Cobb, 2020).

5. **Strong passphrase**

It is recommended to implement two-factor authentication and strong password when using RDP (Cobb, 2020). In addition, passphrase is normally a lot better than passwords as they are unique, complex, and longer phrases, and quite easy to remember (Australian Cyber Security Centre, 2020b). Diceware was created by Arnold Reinhold, IT guru, which was developed to generate a passphrase for users from a list of 7776 words where each corresponding phrase would have a 5-digit number next to it (Lemonnier & Latto, 2020). The user would have to:

1. Install Diceware: can be open in a text editor application.
2. Users must roll a die at least 5 times or roll 5 dice all at once and placed the number in order that it was generated.
3. Search for the 5-digit number in the list that match the number you have rolled.
4. Repeat the process again at least 9 times to generate 9 unique word paraphrase.
5. Memorize the new paraphrase (Reinhold, 2020).

Users should be advised that they must also include capitals, symbols, and numbers into the paraphrase so that it would increase security. But it is not recommended to use the same paraphrase across different service as if one account get compromised than other account that the user is using would also get compromised (Lemonnier & Latto, 2020).

6.3.1 *How to implement "Secure Remote Desktop Protocol"*

Referencing the stages from Fig. 2, Table 3 provides a snapshot of how the IT department can implement the "Secure Remote Desktop Protocol" as well as help and educate WFH user how to mitigate against remote desktop protocol attacks.

The above-recommended implementation from Table 3 would greatly benefit working from home users and organizations as it helps them to understand the dangers as well as how to mitigate against remote desktop protocol attacks.

TABLE 3 Implementing "Secure Remote Desktop Protocol" to WFH users.

Stages	Description
Stage 1	The IT department is required to organize a short online module with quizzes and answers to teach working from home users about remote desktop protocol attacks and its potential to spread ransomware
Stage 2	All working from home users are required to complete the short modules with quizzes and answers before moving to stage 3
Stage 3	Using any online face-to-face medium, the IT department are required to personally teach working from home users about the concept of remote desktop protocol. Repeat if working from home users have not yet understand the concept of Remote Desktop Protocol
Stage 4	Using any online face-to-face medium, the IT department must now demonstrate how to respond appropriately to potential remote desktop protocol attacks. For instance, the IT may say that if a suspicious source asks the working from home user to change their port number to 3389 where one should not comply with the demand. Repeat if working from home users have not understood how to respond appropriately to potential remote desktop protocol attacks
Stage 5	Using any online face-to-face medium, the IT department are required to go through all the "Secure Remote Desktop Protocol" with working from home users. For instance, IT may show to working from home user how to correctly use Diceware to generate their new passphrase. Repeat if working from home user have not understood how to appropriately apply the security measures from "Secure Remote Desktop Protocol"
Stage 6	At this stage, working from home user would now have a strong comprehension about the concept of remote desktop protocol, how to appropriately respond to it, and how to appropriately apply the security measures from "Secure Remote Desktop Protocol"
Stage 7	Every 6 months, the IT department must go through stage 2 with working from home users to ensure that they are up to date with their skills. Stage 3, 4 and 5 are other optional stages working from home user may wish to go through again to help retrain themselves
Stage 8	In the future, an organization may recruit new working from home users. The IT department is required to go through stage 2–7 with the new working from home users

6.4 Closed gate approach to Trojan

Trojan is a type of malware that can disguised themselves as legitimate software (Grace, 2020). Once they are in the user's electronic device, they may not know its existents. This type of malware has the capabilities to record things on the victim electronic device and send back information to the cyber-criminals (Anderson, 2019). In this case, trojan can act as an installer for ransomware. To avoid this, users are encouraged to follow the recommendation below.

1. Up to date software and patches

To avoid cyber-criminals exploiting vulnerabilities that are found from outdate patches, it is recommended that users have the latest update for their operating systems, software, and browser (Anderson, 2019).

2. Firewall

It is recommended to have configurate one's firewall to control incoming malicious internet traffic and prevent the possibility of Trojans being installed onto the user's electronic device (Anderson, 2019). It was discussed that firewall would have its own unique control gaps. However, the IT department would be responsible to teach working from home users to understand function of a firewall and how to manually detect a possible trojan.

3. Antimalware

Antimalware is to ensure that one has not installed Trojan onto one's electronic device and to scan other programs and files to see if they are safe to execute. It is important to have the latest patches for antimalware (Anderson, 2019).

4. Never open suspicious attachments, or install and execute them

It is recommended to never interact with suspicious emails and their attachments as the user may unintentionally download Trojan (Anderson, 2019). This also applies for installing and running programs that comes from vague and suspicious sources. Always double check the sources before interacting with them.

6.4.1 How to implement "Closed Gate approach to Trojan"

Referencing the stages from Fig. 2, Table 4 provides a snapshot of how the IT department can implement the "Closed Gate Approach to Trojan" as well as help and educate WFH user how to mitigate against trojan attacks.

TABLE 4 Implementing "Closed Gate approach to Trojan" to WFH users.

Stages	Description
Stage 1	The IT department is required to organize a short online module with quizzes and answers to teach working from home users about trojan attacks and its potential to spread ransomware
Stage 2	All working from home users are required to complete the short modules with quizzes and answers before moving to stage 3
Stage 3	Using any online face-to-face medium, the IT department are required to personal teach working from home users about the concept of trojan. Repeat if working from home users have not yet understand the concept of trojan
Stage 4	Using any online face-to-face medium, the IT department must now demonstrate how to respond appropriately to potential trojan attacks. For instance, IT may say that if a suspicious source asks the working from home user to open and install the program from the email, one must not comply with the demand. Repeat if working from home users have not understood how to respond appropriately to potential trojan attacks
Stage 5	Using any online face-to-face medium, the IT department are required to go through all the "Never Open the Gate for Trojan" with working from home users. For instance, IT may show to working from home user how to correctly configurate one's firewall to prevent the possibility of trojans being installed onto the electronic device. Repeat if working from home user have not understood how to appropriately apply the security measures from "Never Open the Gate for Trojan"
Stage 6	At this stage, working from home user would now have a strong comprehension about the concept of trojan, how to appropriately respond to it, and how to use and appropriately apply the security measures from "Never Open the Gate for Trojan"
Stage 7	Every 6 months, the IT department must go through stage 2 with working from home users to ensure that they are up to date with their skills. Stage 3, 4 and 5 are other optional stages working from home user may wish to go through again to help retrain themselves
Stage 8	In the future, an organization may recruit new working from home users. The IT department are required to go through stage 2–7 with the new working from home users

The above-recommended implementation from Table 4 would greatly benefit working from home users and organizations as it helps them to understand the dangers as well as how to mitigate against trojan attacks.

6.5 Drive-by downloads controls

It is important for users must be aware that ransomware that can be spread via drive-by downloads. There are times where it can be quite difficult to pick up as its purpose is to surreptitiously download itself onto the user's electronic device. To avoid any unnecessary frustration and consequences of being victimize by ransomware, it is recommended to do the following:

1. Up to date software and patches

Is important to have an up-to-date software to avoid cyber-criminals exploiting vulnerabilities from users (Levinson, 2012) as well as keeping other application regularly updated.

2. Web-filtering

Web filtering software can prevent users from going into compromised websites as they have built-in mechanism that can detect and determine if a user has come across an unsafe website (Levinson, 2012). Web filtering does have its own unique limitation. It is important for the IT department to teach working from home users to understand function as well as how to manually detect a potential drive-by download page.

3. Disable java

Within the user's PDF reader preference list, it is important that users disable JavaScript and for IT departments to remove any Java that are in their systems (Levinson, 2012).

4. BLADE—Block all drive-by download exploits

BLADE, developed by SRI International and Georgia Tech, is a windows immunization system that can block drive-by download from occurring and infecting the user's systems (Levinson, 2012).

6.5.1 How to implement "Drive-by Downloads Controls"

As seen in the stages in Fig. 2, Table 5 provides a snapshot of how the IT department can implement the "Drive-by Downloads Controls" as well as help and educate WFH user how to mitigate against drive-by download attacks.

The above recommended implementation from Table 5 would greatly benefit working from home users and organizations as it helps them to understand the dangers as well as how to mitigate against drive-by download attacks.

6.6 Malvertising interaction controls

Malvertising disguises themselves as legitimate advertisement on a website. This attack vector is like drive-by download but requires the user to interact with the malicious advertisement. Once a user interacts with a

TABLE 5 Implementing "Drive-by downloads controls" to WFH users.

Stages	Description
Stage 1	The IT department is required to organize a short online module with quizzes and answers to teach working from home users about drive-by download attacks and its potential to spread ransomware
Stage 2	All working from home users are required to complete the short modules with quizzes and answers before moving to stage 3
Stage 3	Using any online face-to-face medium, the IT department are required to personal teach working from home users about the concept of drive-by download. Repeat if working from home users have not yet understand the concept of drive-by download
Stage 4	Using any online face-to-face medium, the IT department must now demonstrate how to respond appropriately to potential drive-by download attacks. For instance, IT may say that a phishing email may have an embedded suspicious link which could lead them to a drive-by download page and one must not press on. Repeat if working from home users have not understood how to respond appropriately to potential drive-by download attacks
Stage 5	Using any online face-to-face medium, the IT department is required to go through all the "Do not Drive-by download" with working from home users. For instance, IT may show to working from home user how to correctly use and install Web filtering software. Repeat if working from home user have not understood how to appropriately apply the security measures from "Do not Drive-by download"
Stage 6	At this stage, working from home user would now have a strong comprehension about the concept of drive-by download, how to appropriately respond to it, and how to use and appropriately apply the security measures from "Do not Drive-by download"
Stage 7	Every 6 months, the IT department must go through stage 2 with working from home users to ensure that they are up to date with their skills. Stage 3, 4 and 5 are other optional stages working from home user may wish to go through again to help retrain themselves
Stage 8	In the future, an organization may recruit new working from home users. The IT department are required to go through stage 2–7 with the new working from home users

malicious advertisement, it can surreptitiously install a ransomware into victim's electronic device. To prevent this from happening, users must follow the below recommendation:

1. Up to date software and patches

It is important that users realize that they should remove vulnerabilities that are in their electronic devices. This can be as simple as updating one's application, operating systems, and web browser. In addition, users must also remove any flash or java software that they do not usually to prevent being exploited by malvertising (Lake, 2020).

2. Safe practice—Think and look before clicking

In general, users must be cautions when interacting and clicking onto anything on the internet. As such, it is recommended that users should never interact with pop-up advertisement, vague notices, or any suspicious advertisement. In doing so would lower one's chance of being victimize by ransomware (Panda Security, 2019).

3. Ad blockers

Ad blockers are design to filter any malicious advertisement that the user may encounter. Its main function is to protect individuals from clicking and viewing malicious advertisement. Furthermore, ad blocker can reduce cookies number being saved onto the electronic device, protect individual's privacy, and load webpages faster (Panda Security, 2019).

4. Antimalware

It is still recommended for users to have antimalware install onto their electronic device so that the security software can scan one's systems for potential malware or to perform regular check-ups (Panda Security, 2019).

6.6.1 How to implement "Malvertising Interaction Controls"

Based on stages shown in Fig. 2, Table 6 provides a snapshot of how the IT department can implement the "Malvertising Interaction Controls" as well as help and educate WFH user how to mitigate against malvertising attacks.

The above-recommended implementation from Table 6 would greatly benefit working from home users and organizations as it helps them to understand the dangers as well as how to mitigate against malicious advertisement attacks.

The mitigation model against ransomware attacks would provide working from home user and organization the necessary training and understanding to avoid a devastating ransomware attack. However, it is important to always frequently backup their data. Not only it can help assist with the restoration process but can help avoid business disruption and for the organization to continue working uninterrupted (Australian Cyber Security Centre, 2020a). In addition, this would also prevent the possibility that the target paying the demanded ransom as well. There are different types of data backup methods that working from home users and organizations can choose to implement. Full, differential, and incremental are examples of data backup strategies that can be used within any working environment.

TABLE 6 Implementing "Malvertising Interaction Controls" to WFH users.

Stages	Description
Stage 1	The IT department is required to organize a short online module with quizzes and answers to teach working from home users about malvertising attacks and its potential to spread ransomware
Stage 2	All working from home users are required to complete the short modules with quizzes and answers before moving to stage 3
Stage 3	Using any online face-to-face medium, the IT department are required to personal teach working from home users about the concept of malvertising. Repeat if working from home users have not yet understand the concept of drive-by download
Stage 4	Using any online face-to-face medium, the IT department must now demonstrate how to respond appropriately to potential drive-by download attacks. For instance, IT may say that one should never interact with pop-up advertisement, vague notices, or any suspicious advertisement. Repeat if working from home users have not understood how to respond appropriately to potential malvertising attacks
Stage 5	Using any online face-to-face medium, the IT department are required to go through all the "Never interact with any Malvertising" with working from home users. For instance, IT may show to working from home user how to correctly use and install Ad blockers. Repeat if working from home user have not understood how to appropriately apply the security measures from "Never interact with any Malvertising"
Stage 6	At this stage, working from home user would now have a strong comprehension about the concept of malvertising, how to appropriately respond to it, and how to use and appropriate apply the security measures from "Never interact with any Malvertising"
Stage 7	Every 6 months, the IT department must go through stage 2 with working from home users to ensure that they are up to date with their skills. Stage 3, 4 and 5 are other optional stages working from home user may wish to go through again to help retrain themselves
Stage 8	In the future, an organization may recruit new working from home users. The IT department are required to go through stage 2–7 with the new working from home users

7 Regular and consistent data backup

7.1 Full back up

The first type that working from home user or organization may wish to implement is a full back up plan. This type would usually resolve around backing up all data that is on the system and this includes all hard copies data. Implementing full backup does allow the user to easily find files but can consume a lot of time to perform this type of backup (HP Invent, 2017).

An organization may have terabytes worth of data to backup which can devour a lot of drive to backup as well as being very time consuming. On the other hand, full back up can be used by working from home users as they do not usually have huge amount of data to backup.

7.2 Differential back up

The second type is a differential back up where it only stores data that have undergone changes since its previous full back up. Compare to performing a full back up, it does not consume a lot of time. It can be performed daily which would produce the most up to date data sets that can be restore for future use. Differential back up can take a lot of capacity to be stored but as long as it can be managed onto one media device, it should not impose any problems for the users (HP Invent, 2017).

7.3 Incremental back up

The incremental backup would require storing all files which has undergone certain changes since its previous backup. This would take up less storage space in the media device as it stores files that have change or been created since its prior backup. However, the restoration process is complex as it requires multiple drives to restore the data (HP Invent, 2017) which may cause the process to be slow.

It is important for working from home users and organizations to develop a habit to secure and back up their data. The three back up methods can be use in any circumstances. However, it is encouraged for working from home user and organization to use a full back up method as it ensures that all files have at least a one back up for the restoration process. The process to restore the full back up can be time consuming. However, in the event of a ransomware attack, a full back up ensures working from home users and organizations data is safe and can resume studying and working duties.

8 Conclusion

During the COVID-19 pandemic, the cyberspace has become a necessary dimension for users to communicate and connect with others. However, it is a period where many would find themselves to be in a position of hardship as many would be working and interacting with others from the comfort of their own home via their unique electronic devices. The electronic devices that working from home user are using would have sensitive and personal information on it. During the COVID-19 pandemic, connecting to the internet is a requirement but due to the influx of online users, there is will many cyber-criminals who would choose to exploit other during the COVID-19 pandemic. As such, many working from home user would enter the cyberspace without realizing that there may be vulnerabilities within their own home environment which can be a cybersecurity concern. Due to the sudden change to work from home, many would not have realized that strengthen their security measures is the key to prevent cyber-criminals from exploiting their vulnerabilities.

By strengthen one's home environment, it would have prevented many cyber-criminals from launching malicious attack during the COVID-19 pandemic. Ransomware has become a lucrative choice for cyber-criminal to use as it is malicious software that can infect and prevent a working from home user's from accessing their data. Furthermore, there are working from home users that are susceptible to the ransomware attack because their electronic devices do have sensitive information can be used against them. Using powerful encryption, cyber-criminals would hold onto working from home user's data until they pay them the demand ransom. Ransomware was a popular choice for cyber-criminals to use as there are many online users that can be extorted during the COVID-19 pandemic. This in turn would produce many challenges for working from home users.

Working from home would produce many challenges against ransomware. Some of these challenges may include that many were not able to identify genuine sources which may have led to the installation of a ransomware. Mobile phone users may be target for ransomware attacks as there are very little mitigation measures for hand-held devices. Ransomware has evolved to a point where it has strong encryption to prevent the user from assessing their data. Malwares have evolved over the years which has led to more digital extortion where organization has come a desirable choice to be targeted by cyber-criminals. Many individuals working from home users may have to face the difficult decision to pay the ransom to regain access to their data. For ransomware to be installed correctly, cyber-criminals must use an attack vectors to deliver it to the target.

The dangerous malware is reliant upon the delivery of the six attack vectors which are social engineering, phishing, trojan, remote desktop protocol, drive-by download, and malvertising. Without having a strong security measures against ransomware, many working from home user would face the possibility of digital extortion from cyber-criminals. There are existing defense mechanisms that can mitigate ransomware attacks but has its own control gaps. However, organization's IT department would be responsible to cyber-train their employees to protect their assets and reputation. The mitigation model against ransomware is developed to assist working from home user to mitigate against potential ransomware attacks. In addition, individuals and organization must continue to back up their data to ease disruption. However, future work must focus on education around danger of ransomware and how to mitigate it.

Since ransomware attacks have increase over the years it has led to more victims being extorted by cyber-criminals. As such, it is recommended that future work must start to focus more on teaching and educating working from home users about the dangers of ransomware attacks and how to mitigate such an event from occurring. Future work may also implement a ransomware mitigation program into secondary school and tertiary education which would also further cyber-enhance future generation to be prepared for potential ransomware attacks. In doing so, it would encourage more people to actively protect their data as well as produce an outcome where many cyber-users understand the concept of ransomware and would hopefully strengthen future generation from any potential cyber-attacks, especially during the COVID-19 pandemic.

The COVID-19 pandemic has opened more doors for online user to enter the cyberspace. But will contain a haven of cyber-criminals who would see this as an opportunity to hurt others. During the COVID-19 pandemic or other similar circumstance, one must be able to strengthen their security control measures as well as to prepare themselves that they could be a possible victim of any ransomware attacks.

References

Aburrous, M., Hossain, M. A., Dahal, K., & Thabtah, F. (2010). Predicting phishing websites using classification mining techniques with experimental case studies. In *2010 seventh international conference on information technology: New generations* (pp. 176–181). IEEE.

Admin. (2019). *Add user to remote desktop users group in windows 10*. Top Password Software. Retrieved October 9, 2020, from https://www.top-password.com/blog/add-user-to-remote-desktop-users-group-in-windows-10/.

Ahmed, E., Yaqoob, I., Gani, A., Imran, M., & Guizani, M. (2016). Internet-of-things-based smart environments: state of the art, taxonomy, and open research challenges. *IEEE Wireless Communications, 23*(5), 10–16.

Al-Fuqaha, A., Guizani, M., Mohammadi, M., Aledhari, M., & Ayyash, M. (2015). Internet of things: A survey on enabling technologies, protocols, and applications. *IEEE Communication Surveys and Tutorials, 17*(4), 2347–2376.

Alison, G. (2020). *What is a Trojan? Is it a virus or is it malware?*. Norton. Retrieved October 9, 2020 https://us.norton.com/internetsecurity-malware-what-is-a-trojan.html.

Almomani, A., Gupta, B. B., Atawneh, S., Meulenberg, A., & Almomani, E. (2013). A survey of phishing email filtering techniques. *IEEE Communication Surveys and Tutorials, 15*(4), 2070–2090.

Amro, B. (2018). *Phishing techniques in mobile devices*. arXiv. preprint arXiv:1802.04501.

Anderson, S. (2019). *How to defend your PC and devices against a Trojan Horse virus*. Safety Detectives. Retrieved October 20, 2020 from https://www.safetydetectives.com/blog/what-is-a-trojan-horse-and-how-to-protect-against-it/.

Andronio, N., Zanero, S., & Maggi, F. (2015). Heldroid: Dissecting and detecting mobile ransomware. In *International symposium on recent advances in intrusion detection* (pp. 382–404). Cham: Springer.

Arntz, P. (2018). *How to protect your RDP access from ransomware attacks*. Malwarebytes. Retrieved October 9, 2020, from https://blog.malwarebytes.com/security-world/business-security-world/2018/08/protect-rdp-access-ransomware-attacks/.

Australian Cyber Security Centre. (2017). *Strategies to mitigate cyber security incidents—Mitigation details*. Retrieved October 9, 2020, from https://www.cyber.gov.au/publications/strategies-to-mitigate-cyber-security-incidents-mitigation-details.

Australian Cyber Security Centre. (2020a). *Data backup and restoration*. Retrieved October 9, 2020, from https://www.cyber.gov.au/acsc/view-all-content/guidance/data-backup-and-restoration.

Australian Cyber Security Centre. (2020b). *Using a phrase or sentence, not one word, as your password*. Retrieved October 9, 2020, from https://www.cyber.gov.au/acsc/view-all-content/guidance/comparison-password-vs-passphrase.

Bertino, E., & Islam, N. (2017). Botnets and internet of things security. *Computer, 50*(2), 76–79.

Bhardwaj, A. (2017). Ransomware: A rising threat of new age digital extortion. In *Online banking security measures and data protection* (pp. 189–221). IGI Global.

Bhayani, S. (2019). *Human beings are the weakest link in the data protection*. Entrepreneur. Retrieved October 9, 2020, from Process https://www.entrepreneur.com/article/328264.

Bisawa, V. (2020). *Application whitelisting vs. application blacklisting: Pros and cons*. Colortokens. Retrieved October 22, 2020 from https://colortokens.com/blog/application-whitelisting-application-blacklisting-pros-cons/.

BlackBerry. (2019). *Zeppelin: Russian ransomware targets high profile users in the U.S. and Europe*. Retrieved October 9, 2020, from https://blogs.blackberry.com/en/2019/12/zeppelin-russian-ransomware-targets-high-profile-users-in-the-us-and-europe.

Brewer, R. (2016). Ransomware attacks: Detection, prevention and cure. *Network Security, 2016*(9), 5–9.

Bugeja, J., Jacobsson, A., & Davidsson, P. (2017). An analysis of malicious threat agents for the smart connected home. In *2017 IEEE international conference on pervasive computing and communications workshops (PerCom Workshops)* (pp. 557–562). IEEE.

Cannell, J. (2016). *Cryptolocker Ransomware: What you need to know*. Malwarebytes. Retrieved October 9, 2020, from https://blog.malwarebytes.com/101/2013/10/cryptolocker-ransomware-what-you-need-to-know/.

Cassetto, O. (2019). *Drive by downloads: What they are and how to avoid them*. Exabeam. Retrieved October 20, 2020 from https://www.exabeam.com/information-security/drive-by-download/.

Checkpoint. (2020). *Ryuk ransomware attack*. Retrieved October 9, 2020, from https://www.checkpoint.com/cyber-hub/threat-prevention/ransomware/ryuk-ransomware/.

Cobb, M. (2020). 10 RDP security best practices to prevent cyberattacks. In *Search security*. Retrieved October 9, 2020, from https://searchsecurity.techtarget.com/tip/10-RDP-security-best-practices-to-prevent-cyberattacks.

Cockerham, R. (2018). *What are the advantages and disadvantages of using a firewall?*. Techwalla. Retrieved October 22, 2020 from https://www.techwalla.com/articles/what-are-the-advantages-and-disadvantages-of-using-a-firewall.

Coker, J. (2020). *Cyber-criminals change tactics to exploit #COVID19*. Info Security. Retrieved October 9, 2020, from https://www.infosecurity-magazine.com/news/cybercriminals-change-tactics/.

Data Safe. (2019). *9 reasons to shred your documents*. Retrieved October 20, 2020 from https://www.destruction.com/2019/09/04/9-reasons-to-shred-your-documents/.

Everett, C. (2016). Ransomware: To pay or not to pay? *Computer Fraud & Security, 2016*(4), 8–12.

Gazet, A. (2010). Comparative analysis of various ransomware virii. *Journal in Computer Virology, 6*(1), 77–90.

Gendre, A. (2020). *Ransomware examples: Recent attacks and distribution methods*. Vade Secure. Retrieved October 9, 2020, from https://www.vadesecure.com/en/ransomware-examples-recent-attacks-and-distribution-methods/.

Ghosh, A. M., Halder, D., & Hossain, S. A. (2016). Remote health monitoring system through IoT. In *2016 5th international conference on informatics, electronics and vision (ICIEV)* (pp. 921–926). IEEE.

Grace, A. (2020). *What is a Trojan? Is it a virus or is it malware?*. Norton. Retrieved October 20, 2020 from https://us.norton.com/internetsecurity-malware-what-is-a-trojan.html.

Granger, S. (2001). *Social engineering fundamentals, part I: Hacker tactics*. SecurityFocus.

HP Invent. (2017). *The importance of protecting your data*. Retrieved October 2020, 20 from https://myconsole.com.au/wp-content/uploads/2017/04/hp_why_backup.pdf.

Infosec Resources. (2018). *The advantages & disadvantages of outsourcing incident response*. Retrieved on October 22, 2020, from https://resources.infosecinstitute.com/advantages-disadvantages-outsourcing-incident-response/#gref.

Jing, Q., Vasilakos, A. V., Wan, J., Lu, J., & Qiu, D. (2014). Security of the Internet of Things: Perspectives and challenges. *Wireless Networks, 20*(8), 2481–2501.

Johansen, A. G. (2019). *What is ransomware and how to help prevent ransomware attacks*. Norton. Retrieved October 9, 2020, from https://us.norton.com/internetsecurity-malware-ransomware-5-dos-and-donts.html.

Knowledge Leader. (2019). *Control gap remediation methodology training guide*. Retrieved October 9, 2020, from https://www.knowledgeleader.com/knowledgeleader/content.nsf/web+content/gucontrolgapremediationmethodologytrainingpresentation.

KPMG. (2020). *The rise of ransomware during COVID-19*. Retrieved October 9, 2020, from https://home.kpmg/au/en/home/insights/2020/05/rise-of-ransomware-during-covid-19.html.

Krombholz, K., Hobel, H., Huber, M., & Weippl, E. (2015). Advanced social engineering attacks. *Journal of Information Security and Applications, 22*, 113–122.

Lake, J. (2020). *What is malvertising and how can you defend against it?*. Comparitech. Retrieved October 20, 2020 from https://www.comparitech.com/blog/information-security/malvertising/.

Lastline. (2018). *Malvertising: What you need to know*. Retrieved October 20, 2020 from https://www.lastline.com/blog/how-malvertising-works-what-you-need-to-know/.

Lemonnier, J., & Latto, N. (2020). *How to create a strong password—That you won't forget*. AVG. Retrieved October 9, 2020, from https://www.avg.com/en/signal/how-to-create-a-strong-password-that-you-wont-forget.

Levinson, M. (2012). *6 ways to defend against drive-by downloads*. CIO. Retrieved October 9, 2020, from https://www.cio.com/article/2448967/6-ways-to-defend-against-drive-by-downloads.html.

Mahmoud, T. M., & Mahfouz, A. M. (2012). SMS spam filtering technique based on artificial immune system. *International Journal of Computer Science Issues (IJCSI), 9*(2), 589.

Mamedov, O., Sinitsyn, F., & Ivanov, A. (2017). *Bad Rabbit ransomware*. SecureList. Retrieved October 20, 2020 from https://securelist.com/bad-rabbit-ransomware/82851/.

Nassi, B., Shamir, A., & Elovici, Y. (2017). *Oops!.. I think I scanned a malware*. arXiv preprint arXiv:1703.07751.

Nebehay, S. (2020). *WHO says it no longer uses 'pandemic' category, but virus still emergency*. Reuters. Retrieved October 9, 2020, from https://www.reuters.com/article/uk-china-health-who-idUKKCN20I0PD.

Newby, T. G., & Razmazma, A. (2016). *An untraceable currency? Bitcoin Privacy Concerns*. Fintech Weekly. https://www.fintechweekly.com/magazine/articles/an-untraceable-currency-bitcoin-privacy-concerns.

NortonLifeLock Employee. (2019). *Internet of Things (IoT) security: 9 ways you can help protect yourself*. Norton. Retrieved October 9, 2020 https://us.norton.com/internetsecurity-iot-securing-the-internet-of-things.html.

Osborne, C. (2019). *This new Android ransomware infects you through SMS messages*. ZDnet. Retrieved October 9, 2020, from https://www.zdnet.com/article/this-new-android-ransomware-infects-you-through-sms-messages/.

Panda Security. (2019). *Malvertising: What it is and how to protect yourself.* Retrieved October 20, 2020 from https://www.pandasecurity.com/en/mediacenter/malware/malvertising/.

Perera, C., Zaslavsky, A., Christen, P., & Georgakopoulos, D. (2013). Context aware computing for the internet of things: A survey. *IEEE Communication Surveys and Tutorials, 16*(1), 414–454.

Popoola, S. I., Ojewande, S. O., Sweetwilliams, F. O., John, S. N., & Atayero, A. A. (2017). *Ransomware: Current trend, challenges, and research directions*. San Francisco, USA: World Congress on Engineering and Computer Science.

Pratt, M. K. (2018). *What is an intrusion detection system? How an IDS spots threats*. CSO. Retrieved October 22, 2020 from https://www.csoonline.com/article/3255632/what-is-an-intrusion-detection-system-how-an-ids-spots-threats.html.

Qin, T., & Burgoon, J. (2007). An investigation of heuristics of human judgment in detecting deception and potential implications in countering social engineering. In *Intelligence and security informatics, 2007* (pp. 152–159). IEEE.

Rapid7. (2017). *The pros & cons of intrusion detection systems*. Retrieved October 22, 2020 from https://blog.rapid7.com/2017/01/11/the-pros-cons-of-intrusion-detection-systems/.

Reinhold, A. G. (2020). *The diceware passphrase home page*. The World. Retrieved October 9, 2020, from https://theworld.com/~reinhold/diceware.html.

Richardson, R., & North, M. M. (2017). Ransomware: Evolution, mitigation and prevention. *International Management Review, 13*(1), 10.

Rivera, A. (2019). *8 benefits of online data storage*. Business News Daily. Retrieved October 9, 2020, from https://www.businessnewsdaily.com/6294-benefits-of-online-data-storage.html.

Rosenberg, J. M. (2015). *A Q&A about the malicious software known as ransomware*. Retrieved October 10, 2020, from https://phys.org/pdf347719229.pdf.

Rouse, M. (2020). *Firewall*. Search Security. Retrieved October 22, 2020 from https://searchsecurity.techtarget.com/definition/firewall.

Saleem, J., & Hammoudeh, M. (2018). Defense methods against social engineering attacks. In *Computer and network security essentials* (pp. 603–618). Cham: Springer.

Sanduleac, M., Chimirel, C. L., Eremia, M., Toma, L., Cristian, C., & Stanescu, D. (2016). Unleashing Smart Cities efficient and sustainable energy policies with IoT based unbundled smart meters. In *2016 IEEE international conference on emerging technologies and innovative business practices for the transformation of societies (EmergiTech)* (pp. 112–117). IEEE.

Savage, K., Coogan, P., & Lau, H. (2015). The evolution of ransomware. Mountain View CA: Symantec Corporation.

Scarfone, K., Souppaya, M., Cody, A., & Orebaugh, A. (2008). Technical guide to information security testing and assessment. *NIST Special Publication, 800*(115), 2–25.

Segura, J. (2016). *Citadel: A cyber-criminal's ultimate weapon?*. Malwarebytes. Retrieved October 9, 2020, from https://blog.malwarebytes.com/threat-analysis/2012/11/citadel-a-cyber-criminals-ultimate-weapon/.

Shahriar, H., Klintic, T., & Clincy, V. (2015). Mobile phishing attacks and mitigation techniques. *Journal of Information Security, 6*(3), 206.

Sheng, S., Wardman, B., Warner, G., Cranor, L., Hong, J., & Zhang, C. (2009). *An empirical analysis of phishing blacklists*. CEAS.

Shinde, R., Van der Veeken, P., Van Schooten, S., & van den Berg, J. (2016). Ransomware: Studying transfer and mitigation. In *2016 international conference on computing, analytics and security trends (CAST)* (pp. 90–95). IEEE.

Sittig, D. F., & Singh, H. (2016). A socio-technical approach to preventing, mitigating, and recovering from ransomware attacks. *Applied Clinical Informatics, 7*(2), 624.

Sjouwerman, S. (2015). *A short history & evolution of Ransomware*. Retrieved October 9, 2020, from https://www.knowbe4.com/ransomware#ransomwaretimeline.

Skybox Security. (2020). *COVID-19 pandemic sparks 72% ransomware growth, mobile vulnerabilities grow 50%*. Retrieved October 9, 2020, from https://www.skyboxsecurity.com/news/covid-19-pandemic-sparks-72-ransomware-growth-mobile-vulnerabilities-grow-50/.

Solomon, H. (2019). *Ransomware now stealing data, warns Ontario cybercop*. IT World Canada. Retrieved October 9, 2020, from https://www.itworldcanada.com/article/ransomware-now-stealing-data-warns-ontario-cybercop/423168.

Stasiukonis, S. (2006). *Social engineering, the USB Way*. Available at http://www.darkreading.com/security/perimeter/showArticle.jhtml?articleID=208803634.

The Guardian. (2020). *Number of global Covid-19 cases passes 750,000 with death toll over 36,000—as it happened*. Retrieved October 9, 2020, from https://www.theguardian.com/world/live/2020/mar/30/coronavirus-live-news-us-deaths-could-reach-200000-uk-warned-six-month-lockdown-covid-19-latest-updates.

Trautman, L. J., & Ormerod, P. C. (2018). Wannacry, ransomware, and the emerging threat to corporations. *Tennessee Law Review, 86*(503), 504–555.

Trend Micro. (2019). *Sodinokibi ransomware group adds malvertising as delivery technique*. Retrieved October 9, 2020 https://www.trendmicro.com/vinfo/be/security/news/cybercrime-and-digital-threats/-sodinokibi-ransomware-group-adds-malvertising-as-delivery-technique.

Whitney, L. (2019). *How to protect your business from cyberattacks that exploit Microsoft's Remote Desktop Protocol*. Tech Republic. Retrieved October 20, 2020 from https://www.techrepublic.com/article/how-to-protect-your-business-from-cyberattacks-that-exploit-microsofts-remote-desktop-protocol/.

World Health Organisation. (2020). *Coronavirus disease 2019 (COVID-19) situation report—36*. Retrieved October 9, 2020, from https://www.who.int/docs/default-source/coronaviruse/situation-reports/20200225-sitrep-36-covid-19.pdf.

Worldometer. (2020). *COVID-19 coronavirus pandemic*. Retrieved October 9, 2020, from https://www.worldometers.info/coronavirus/.

Yaqoob, I., Ahmed, E., Ur Rehman, M. H., Ahmed, A. I. A., Al-garadi, M. A., Imran, M., et al. (2017). The rise of ransomware and emerging security challenges in the Internet of Things. *Computer Networks, 129*, 444–458.

Zetter, K. (2015). *Hacker lexicon: A guide to Ransomware, the scary hack that's on the rise*. Wired. https://www.wired.com/2017/05/hacker-lexicon-guide-ransomware-scary-hack-thats-rise/.

CHAPTER

4

Individual differences in cyber security behavior using personality-based models to predict susceptibility to sextortion attacks

Veronica Power[a] *and Abubakar Bello*[b]

[a]School of Social Sciences, Western Sydney University, Sydney, NSW, Australia, [b]Cyber Security and Behaviour, School of Social Sciences, Western Sydney University, Sydney, NSW, Australia

1 Introduction

The term sextortion is derived from the words *sex* and *extortion*. Extortion is about causing fear or threat in order to obtain something of value by the perpetrator. These threats could be to cause physical harm, commit a crime, or expose sensitive information (Lindgren, 1993). In the context of cyber sextortion, the threat is of the online release of explicit, intimate, or embarrassing sexual images in the absence of consent, in an attempt to procure additional images, money, or something of valuable nature to the attackers (Patchin & Hinduja, 2018).

Cyber sextortion utilizes social engineering and phishing techniques as attack vectors, which has been considered a cyber security problem. In recent times, this problem has been growing, mostly due to the growth of internet technology: as technology increases so do the security threats (Gupta, Tewari, Jain, & Agrawal, 2017). According to the Australian government website Scamwatch, $933,470 was lost to phishing attacks in 2018. This is a huge increase from the $373,860 reported loss in 2016 (Australian Competition and Consumer Commission: ScamWatch, 2016). Although

https://doi.org/10.1016/B978-0-323-90570-1.00004-8

the number of reports remained steady from 2016 to 2018, there was still a significant increase in money lost. Due to the increasing use and reliance on technology, security threats to systems are relentlessly inventive (Gupta, Arachchilage, & Psannis, 2018). This further demonstrates that phishing is one of the most prevalent techniques for compromising personal and organizational information (Bailey, Mitchell, & Jensen, 2008). Given that cyber sextortion utilizes social engineering and phishing vectors, it is important to analyze cyber sextortion within the context of these vectors.

2 Social engineering and cyber sextortion

Successful social engineering relies on users responding to attacks and providing information or funds demanded. Therefore, understanding the psychological reasons underlying the motivation for responding, is imperative to developing successful defenses (Halevi, Lewis, & Memon, 2013). Cybercriminals depend on the vulnerable aspects of humans rather than technology (Parrish, Bailey, & Courtney, 2009) and there is an over-reliance on technical solutions for protection rather than over-caution of users (Bailey et al., 2008). It has been found that individuals are alarmingly susceptible (Parrish et al., 2009) which is supported by analysis of student vulnerabilities to phishing, where 79% of participants reported they would open an unexpected attachment from a "friend" (Bailey et al., 2008). However, individuals did demonstrate some understanding regarding not responding to emails appearing from financial institutions (Bailey et al., 2008). Social engineering online can be broken up into the following: phishing, cross-site request forgery, malware, pop-ups, and search engine poisoning (Ivaturi & Janczewski, 2011), with the most common types being phishing.

Fig. 1 demonstrates an example of a cyber sextortion attack. As can be seen from Fig. 1, victims are easily deceived into providing something of worth in order to avoid the release of supposed material.

2.1 Phishing susceptibility and cyber sextortion attacks

Phishing is a variation of social engineering attacks. These attacks exploit humans' vulnerability in order to obtain an individual's private information (Cho, Cam, & Oltramari, 2016). It has been found in a recent study that those who have Generalized Communicative Suspicion (GCS) have the tendency to believe that people are generally dishonest or deserving of wariness. Research into whether individuals having high GCS are susceptible to phishing attacks showed that higher scoring individuals demonstrated high deception detection. Whereas individuals scoring

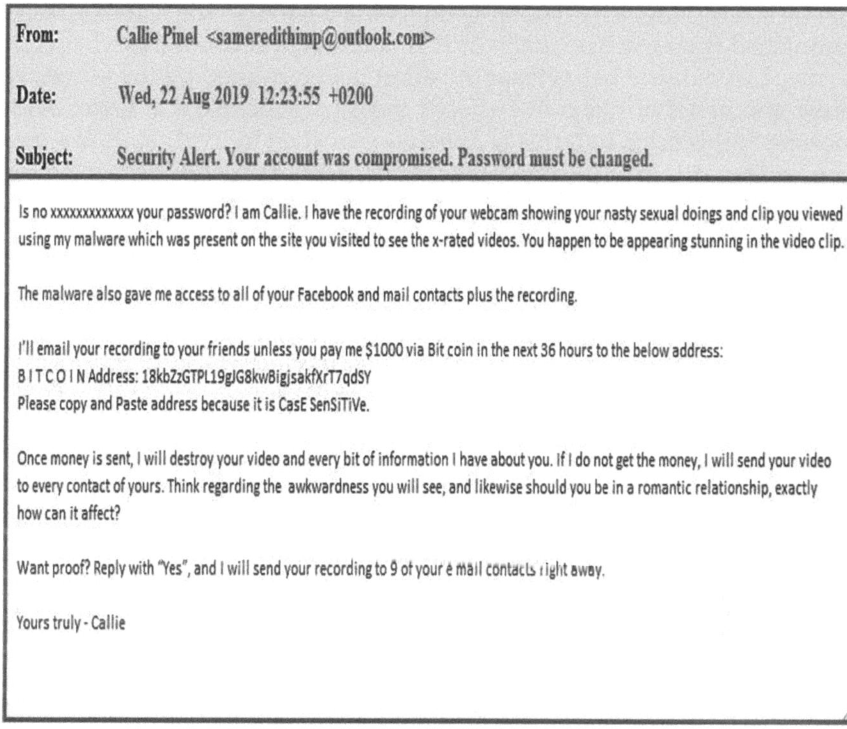
From: Callie Pinel <sameredithimp@outlook.com>

Date: Wed, 22 Aug 2019 12:23:55 +0200

Subject: Security Alert. Your account was compromised. Password must be changed.

Is no xxxxxxxxxxxxx your password? I am Callie. I have the recording of your webcam showing your nasty sexual doings and clip you viewed using my malware which was present on the site you visited to see the x-rated videos. You happen to be appearing stunning in the video clip.

The malware also gave me access to all of your Facebook and mail contacts plus the recording.

I'll email your recording to your friends unless you pay me $1000 via Bit coin in the next 36 hours to the below address:
B I T C O I N Address: 18kbZzGTPL19gJG8kwBigjsakfXrT7qdSY
Please copy and Paste address because it is CasE SenSiTiVe.

Once money is sent, I will destroy your video and every bit of information I have about you. If I do not get the money, I will send your video to every contact of yours. Think regarding the awkwardness you will see, and likewise should you be in a romantic relationship, exactly how can it affect?

Want proof? Reply with "Yes", and I will send your recording to 9 of your e mail contacts right away.

Yours truly - Callie

FIG. 1 Sample sextortion email.

low in GCS relied on mental shortcuts and heuristics leading to susceptibility (Harrison, Vishwanath, & Rao, 2016). Supporting this is the classical decision theory, whereby an individual will make rational decisions objectively when put in uncertain situations (Cho et al., 2013).

An individual's vulnerability to phishing, and in turn sextortion attacks, can be linked to how individuals perceive risk. How people perceive risk in uncertain situations may be affected by unique personality traits (Halevi et al., 2013). Therefore, it is a reasonable assumption that due to the nature of phishing, another social engineering attack such as sextortion could also demonstrate the same differences in personality traits and perception of risk.

Another recent study found that threat perception plays a vital role in influencing an individual's avoidance behavior while using computers online (Arachchilage & Love, 2013). The Arachchilage and Love (2013) study concluded that individuals may take one of two coping approaches when faced with a difficult situation. The focus relevant to this study is the emotion-focused coping approach. This approach entails an individual creating a false perception of the environment without physically changing anything, in turn simply adjusting desires so the fear or stress is diminished (Liang & Xue, 2009). When taking into account the cyber environment, avoidance behavior motivation is determined by the perceived threat and belief of safeguard effectiveness (Arachchilage & Love, 2013). Looking at this issue in terms of cyberattacks, if an individual thinks their cyber safeguards are effective, they would assume the attack is not a cause for concern. However, an individual who believes their cyber safeguards are not effective would be left in fear if an attack occurs. Based on this feeling of fear, the individual could convince themselves it is a mistake or simply respond in a more vulnerable way to diminish fear.

Not only has it been found that GCS, perception of risk, threat perception, and avoidance behavior affect the way in which individuals interact with social engineering attacks, it has also been demonstrated in past research that there are certain groups in society that show a higher vulnerability.

Research has revealed that 92% of participants were vulnerable to phishing attacks to some extent. Of 53 participants, only 4 were able to accurately detect all phishing emails presented even though 82% of participants were confident in their ability to detect these attacks (Hong, Kelley, Tembe, Murphy-Hill, & Mayhorn, 2013). From these participants, 30% were compelled to ask someone else about the validity of a phishing email received in the past, 15% clicked on the link and 6% entered personal information (Hong et al., 2013). With these results, the existence of vulnerabilities to phishing attacks is quite evident, demonstrating the significant need for ways to combat vulnerabilities.

In addition to understanding such vulnerability, it is important to first examine the behavior of the users online. Researchers studying how teenagers behave on Facebook discovered that even though there is knowledge of using care when posting online, teenagers do post a significant amount of risky information on their Facebook pages (Vanderhoven, Schellens, Valcke, & Raes, 2014). Although in terms of security settings, privacy was mainly set to "my friend's eyes only" among the participants, there was still a significant number with settings allowing "friends of friends" to access their posts. This study also concluded that the publication of risky information was displayed more often in older teenagers (Vanderhoven et al., 2014).

As mentioned above, online users do express concern in relation to disclosing personal data via the internet (Trepte et al., 2014). However, research has shown evidence that users do in fact share personal and sometimes intimate details of both their and others' lives in various online environments (Trepte et al., 2014). This demonstrates a contradiction in terms of beliefs and actions online. The reasoning underlying this contradiction may be due to a false sense of safety in some online forums which leave users open to social engineering cyberattacks.

2.2 Sextortion

As the growth of technology increases, so do the ways individuals can be taken advantage of or abused online (Gupta et al., 2017). The introduction of sextortion is one of those ways. As a fairly recent phenomenon, sextortion has not received the amount of research attention required to gain a thorough understanding of the vulnerabilities, impacts, and consequences of falling prey to a sextortion attack.

The focus on individual differences in cyber security is due to the vast range of online behaviors displayed in previous studies. This in turn leads to the focus on personality differences between individuals, as cybercriminals are relying on individual faults rather than the faults of technology to lead to a successful cyber social engineering attack. Accordingly, not only is it an individual's online behaviors but also their own personality that can leave them vulnerable to cyber sextortion attacks.

3 Personality-based models

Personality studies have been used to predict many different behaviors ranging from impulsivity regarding drug-taking or criminal behaviors to the predicted work performance of an individual (Barrick, 2005). Using personality-based models is an effective way to predict an individual's future behavior.

Debates regarding which personality-based model is the most superior in the subject have been constantly challenged between developers and supporters of Eysenck (1967) three-factor model, Cattell (1957) 16-factor model, and McCrae and Costa's (1987), Big Five model (Zuckerman, Kuhlman, Joireman, Teta, & Kraft, 1993). This study will be adopting the Big Five model as it has been shown to be most superior not only because of its wide use in past studies but also the cross-cultural ability of the model.

Research by McCrea and Costa was conducted into the validity of the Big Five-Factor model presented cross-observer agreement regarding all five factors. This validity study was able to reinforce the Big Five-Factor model's ability to be adopted in personality research and assessments (McCrae & Costa, 1987). The basis for the study surrounded concern with respect to the representativeness, comprehensiveness, and validity of the five factors that create the personality model. Results from this study showed significant correlations between what it is intended to discover and what it does (McCrae & Costa, 1987). The Big Five model was found to be valid with the only suggestions for future changes noted being the need for some fine-tuning of the questionnaire, which in the 32 years since the study was done and the extensive use which has been made of the model, has been done, making this the most suitable personality model for this study (McCrae & Costa, 1987).

The belief that the Five-Factor model is the most suitable personality model for this study is reinforced by a study undertaken by Salgado in 2010. This study, although researching the prediction of job performance, found that Five-Factor model-based inventories should be adopted in similar research over non-Five-Factor models as it demonstrated more succinct results (Salgado, 2010).

3.1 The big five personality traits

The Big Five model is a cross-cultural model and well-structured by organizing personality traits into five main dimensions (Cho et al., 2016): Openness, Conscientiousness, Agreeableness, Extraversion, and Neuroticism (McCrae & John, 1992). This research framework suggests that each of these dimensions has an effect on an individual's susceptibility to a sextortion attack. Due to the sensitive nature of sextortion, the Big Five model has been chosen for this study.

Some personality traits are more receptive to different lures such as obeying authority or the fear of missing out in phishing scams specific to deals or opportunities which can link to the lure of greed (Parrish et al., 2009). The results show that agreeableness and neuroticism have a high impact on how given information is perceived, as either trusted or distrusted and accordingly influence correct decision making. More interestingly, a person's high neuroticism impacts negatively on decision performance, particularly when combined with low openness and conscientiousness.

It should be noted that external experiences can impact an individual's vulnerability and scoring in dimensions. For example, negative life experiences have a negative effect on agreeableness and phishing (Parrish et al., 2009).

Openness. Openness is the eagerness to seek out new experiences without anxiety as well as the appreciation of differing ideas and beliefs (Parrish et al., 2009; Rolland, 2002; Weiner & Greene, 2008). In the past, openness has been correlated with many variables such as learning styles, cognitive ability, activity on phones, and self-disclosure online. In terms of learning styles, openness has shown links to elaborative processing which can be described as a rehearsal of information.

High openness can influence an individual's push to become technologically experienced which reduces susceptibility (Parrish et al., 2009). Although it has also been found that openness can influence low cyber security behaviors. Halevi and colleagues discovered that an individual high in openness is more likely to post more information on social media sites as well as have less secure privacy settings (Halevi et al., 2013). Openness also has an inverse relation to high perceived risk, showing an individual demonstrating low levels of openness will be more likely to question a sextortion attack (Cho et al., 2016).

Conscientiousness. Conscientiousness concentrates on self-discipline, aiming to be dutiful and respecting standards and procedures (Parrish et al., 2009). Scoring high in conscientiousness has shown to be positively correlated with self-efficacy which could demonstrate high motivation and therefore not doubting one's self (Martocchio & Judge, 1997), although previous research has stated a negative correlation between conscientiousness and cognitive ability (Rammstedt, Danner, & Martin, 2016).

Being high in conscientiousness has also been stated to be related to impulse control (Rolland, 2002). This demonstrates a steady head in trying situations so that having a high score in conscientiousness would be beneficial for those facing potential online threats. In a study by Parrish and colleagues, it was found that conscientiousness was the least associated with vulnerability to phishing attacks. They discovered that participants high in conscientiousness were more likely to follow cyber security training guidelines when in a phishing situation (Parrish et al., 2009). Participants low in conscientiousness were more likely to behave irresponsibly (Parrish et al., 2009).

Agreeableness. Agreeableness shows altruism and modesty, and most evident in relationships. An individual scoring high in agreeableness will show compassion and cooperate well with others. Agreeableness also positively influences the creation and maintenance of harmonious interpersonal relationships (Huang et al., 2017). Conversely, individuals scoring lower demonstrate suspicious and hostile behavior (Parrish et al., 2009; Rolland, 2002).

Therefore, it has been found that being high in agreeableness impacts how individual trusts in uncertain situations, in terms of phishing (Cho et al., 2013). An individual high in agreeableness would easily trust someone or something in uncertain situations (Cho et al., 2016). This would suggest that individuals high in agreeableness would be vulnerable to a sextortion attack.

Extraversion. Extraversion is apparent in individuals who are constantly seeking out the company of others. These individuals are also said to demonstrate positivity through their emotions but do tend to be dominant (Parrish et al., 2009). Extraversion focuses on the intensity and quantity of relationships (Rolland, 2002). In the past, extraversion has been linked to higher assertiveness, higher response speed, and lower arousal which demonstrates a lower distraction rate during cognitive testing (Curtis, Windsor, & Soubelet, 2015).

In relation to the vulnerability to phishing attacks, extraversion is inversely related to vulnerability (Halevi et al., 2013).

Neuroticism. Neuroticism is the tendency to perceive reality as negative and problematic. Individuals with a high score in neuroticism have been described as being emotionally unstable. Being so emotionally unstable, a high score has shown links to higher anxiety and depression (McCrae & Costa, 1987). In terms of learning and cognitive ability, neuroticism has a negative effect on both. This could be as a result of the sympathetic nervous system being more easily aroused leading to higher rates of distraction, the opposite of those high in extraversion (Biernacki & Tarnowski, 2011; Curtis et al., 2015).

Neuroticism has, in past studies, been shown to be the most correlated to the likelihood of responding to phishing emails (Halevi et al., 2013). High neuroticism leads individuals to become suspicious (Parrish et al., 2009) as well as leading to computer anxiety which would lower time spent online significantly lowering the risk of a successful phishing attack (Parrish et al., 2009). When analyzing effects on perceived trust and risk, and decision performance, it has been found that high neuroticism has a significant effect, especially when openness and conscientiousness are low (Cho et al., 2016).

Many solutions to targeting online social engineering attacks have been proposed in terms of preventing and, or, detecting social engineering attacks but thus far there has been no holistic way to do so (Gupta et al., 2018). This study endeavors to fill this gap in regard to sextortion, so as to prevent future attacks to which cyber security behavior and individual personality traits have left individuals vulnerable.

4 Current study

In a society where there is a high prevalence of celebrities being sextorted, there is a focus on what may cause this vulnerability. Personality differences have not been the focus in past research in sextortion. This study will highlight that it is not just high-profile members of society at risk, it is everyone. Thus, determining what creates vulnerability is a top priority. Specifically, this study will aim to determine individuals' differences in personality traits derived from the Big Five inventory and in individual cyber security behaviors, which will, in turn, demonstrate whether there is a vulnerability or susceptibility of any kind in regards to sextortion.

4.1 Hypotheses

4.1.1 Personality traits and vulnerability

As mentioned previously, individuals scoring low in *openness* will be more likely to question a sextortion attack rather than responding through impulsivity (Cho et al., 2016; Rolland, 2002). Therefore, it is hypothesized that participants scoring high in openness on the Big Five inventory will be more vulnerable to a sextortion attack than those scoring lower, the higher scoring participants also having less secure cyber security behaviors (H1).

Following this, *conscientiousness* was argued to be the least associated with potential vulnerability to online social engineering attacks and linked to following cyber training appropriately (Parrish et al., 2009). Hence, it is hypothesized that participants scoring low in conscientiousness will be more vulnerable to a sextortion attack and will have less secure cyber security behaviors (H2).

It was found that individuals high in *agreeableness* are more likely to want to please people and easily trust strangers in uncertain situations (Cho et al., 2013; Cho et al., 2016). Thus, it is hypothesized that participants scoring high in agreeableness will be more vulnerable to a sextortion attack and will exhibit less secure cyber security behaviors (H3).

Although those scoring high in *extraversion* are known to seek out the company of others, it was also discovered that extraversion is inversely related to vulnerability to social engineering attacks (Halevi et al., 2013; Parrish et al., 2009). Therefore, it is hypothesized that participants scoring low in extraversion will be more vulnerable to sextortion attacks, and these participants will also demonstrate more secure cyber security behaviors (H4).

Individuals with high *neuroticism* have, in the past, been portrayed as suspicious and mistrusting, opposite to those scoring high in openness (Parrish et al., 2009). An individual's score in neuroticism has been shown to be the most representative of one's vulnerability (Halevi et al., 2013). Hence, it is hypothesized that participants scoring low in neuroticism will be more vulnerable to sextortion attacks and will exhibit less secure cyber security behaviors (H5). Fig. 2 provides a clear picture of the proposed hypotheses of this study.

As can be seen in Fig. 2, it is hypothesized that a specific score on the personality scale will correlate to a high score in the cyber security scale demonstrating a high-risk online presence. Having a high risk will lead to a higher potential for vulnerability to cyber sextortion.

5 Method

A nonexperimental cross-sectional design was used in the form of a questionnaire for this study. The reasoning supporting the use of a questionnaire was to gain answers in a format that would allow for a regression analysis to be employed. The questionnaire consists of four parts: demographics, individual security awareness, risky cyber security behavior scale, and a personality scale. To ensure reliability and validity throughout the questionnaire, the questions were adapted from existing surveys.

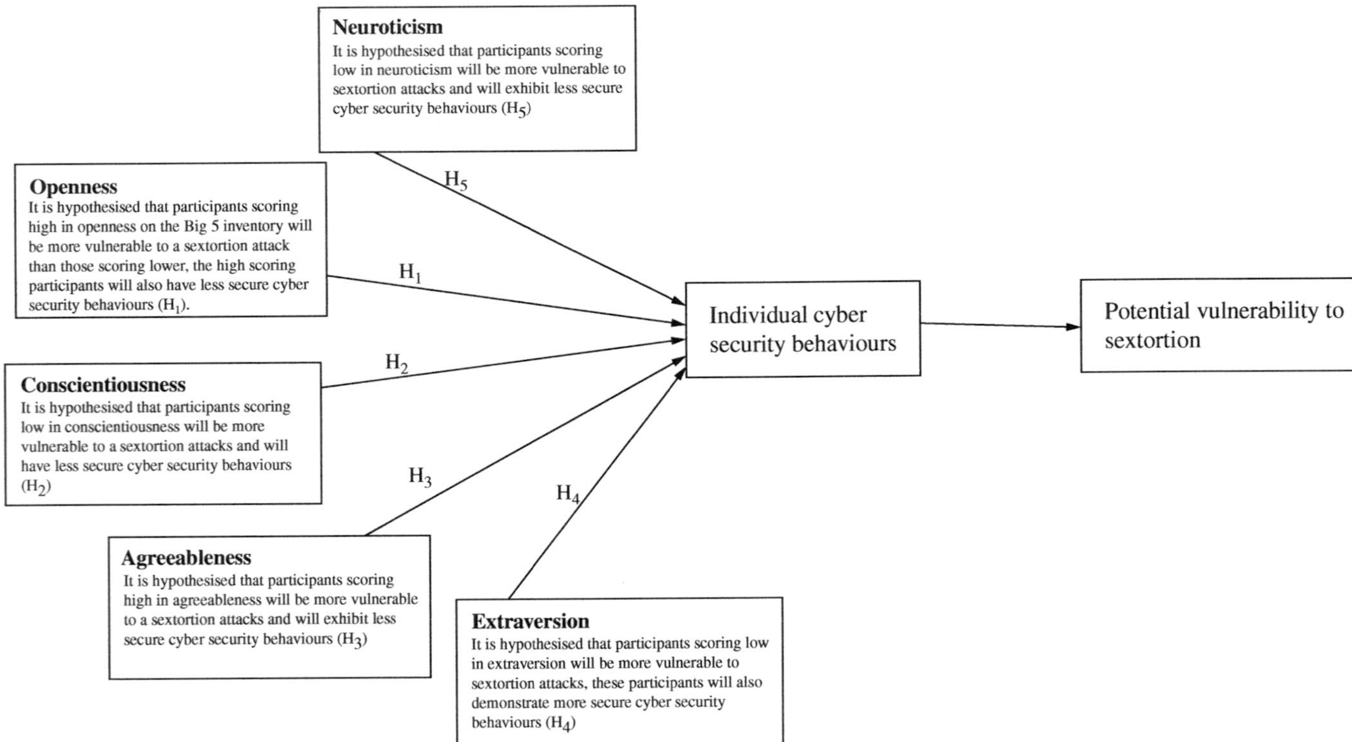

FIG. 2 Visual representation of proposed hypotheses.

5.1 Participants

Participants were recruited through a SONA participation system at Western Sydney University, Sydney, Australia, allowing the created questionnaire on Qualtrics to be advertised for participants to complete. A total of 97 participants took part in this research. The participants comprised of individuals aged under 23 and 19 and 23–38. Out of 97, 83 participants identified as female, 13 participants identified as male, and 1 participant identified as other. A total of 22 of the participants had cyber security training in the past, with 75 stating they had not. From the reports of time spent on the internet, 22 of the participants spent less than 20 h on the internet a week, and 75 spent over 20.

5.2 Materials

Social engineering is a widely researched topic that has allowed for the adoption of previously used scales. Although the scales adopted were created specifically for attacks such as phishing, they were adapted to fit the investigation of cyber sextortion. Each of the four parts of the questionnaire was adapted from a previously used scale.

5.2.1 Demographics

The demographics survey was a modified version of the VSM International Questionnaire (2013) by G. Hofstede. This questionnaire consisted of questions gathering information such as age, gender, time spent on the internet, and previous training.

5.2.2 Individual security awareness

The second part of the questionnaire that determines a participant's security awareness was based on the survey from security awareness of computer users: A phishing threat avoidance perspective by Arachchilage and Love (2014). This part of the questionnaire determined if individuals believed themselves to be vulnerable to cyber sextortion attacks. The questionnaire does so through posing statements such as, "I could be easily deceived by an effective phishing-related sextortion email coming from a legitimate-looking source". This question is followed by a 4 point Likert scale with options ranging from strongly disagree to strongly agree. When the scale was initially adopted, some of the questions were found to be not relevant to this study. Based on the piloted questionnaire, these questions were refined and the number of questions was reduced from 12 to 6 in this section. In order to target the questions more specific to the research, some questions were subtly altered. For example, the proposition, "I could successfully gain anti sextortion knowledge if I had never learned it before", was changed to, "I could gain knowledge in regards to avoiding a cyber sextortion attack". As can be seen, this subtle change allows for better understanding by the participant, as well as more directly addressing this research.

5.2.3 Risky cyber security scale

The Risky Cybersecurity Behaviors Scale (RScB) is appropriated from human factors in cybersecurity; examining the link between Internet addiction, impulsivity, attitudes toward cybersecurity, and risky cybersecurity behaviors by Hadlington (2017). This part consists of 20 questions that detail how an individual behaves online. The aim of this part of the questionnaire was to determine whether an individual is subconsciously taking part in unsafe cyber behaviors. Each question will add an additional measure of "normal" behavior that would leave an individual open and vulnerable to cyberattacks. For example, "sharing passwords with friends and colleagues", although a seemingly normal behavior, would leave an individual vulnerable. Similar to the individual security awareness scale, adaptation had to occur for ease of reading and the overall aim of the research. The wording of some questions could have led participants to be more wary of their answers rather than answering honestly. For example, the original question of "using free-to-access public Wi-Fi", was changed to "Using free or public Wi-Fi", to aid the ease of reading for participants.

5.2.4 Personality scale

The last part of the questionnaire used in this research was a personality scale. This personality scale is taken from "The Big - Five trait taxonomy: history, measurement, and theoretical perspectives" by John and Srivastava (1999). The Big Five traits are a universally understood model of personality which has achieved widespread acceptance. To gain a thorough understanding, the 44-item inventory was assumed so as to test correlations between cyber behavior and each of the traits; Openness, Conscientiousness, Extraversion, Agreeableness, and Neuroticism. These questions are measured on a 4 point Likert scale with higher scores indicating an agreement to the statement. The 44 statements are separated into five sections corresponding with the traits. Each statement begins with the phrase "I see myself as someone who..". For example, to measure conscientiousness a statement such as, "I see myself as someone who is somewhat careless" is put forward. The participants then use the scale to demonstrate their level of agreement.

5.3 Procedure and statistical analysis

Once the four questionnaires were collated, the final questionnaire was uploaded onto Qualtrics which was then advertised via SONA to the participants. The 74-item questionnaire took participants approximately 20 min to complete. Qualtrics allowed for the data to be exported into Microsoft Excel and from there, the data was transferred to IBM SPSS Version 22 for analysis.

A standard multiple regression was the analysis chosen, as the numerical format of the data made it easier to compare the different traits of personality to cyber security behavior: a regression is able to compare the traits simultaneously in a single model. A regression analysis will also show how much variation (R^2) in cyber security behavior is accounted for by personality traits.

6 Results

In order to ensure the appropriate number of participants were recruited, the rule set by Tabachnick and Fidell was applied to the research. This states that the number of participants must be $N \geq 50 + 8m$ (m being the number of predictors). In the case of this research, that would be 90 and therefore this research meets the assumption of participant size (Tabachnick & Fidell, 2007). Similarly, another rule according to Stevens (2002), states 15 participants per predictor (Stevens, 2002) is a suitable sample. Again, this research meets such a rule, fulfilling the assumption.

6.1 Cleaning and screening

In order to ensure the accuracy of the data, a frequency analysis was run to determine if there was any missing data. There were three missing scores from one participant. After close examination, it appeared that the participant's data was not reliable and was therefore deleted. Following this, standardized scores were created in order to determine whether there were any univariate outliers. Any scores ±3.29 were altered so as not to affect the overall data (Tabachnick & Fidell, 2007). Six of the scores had to be changed to one more than the last highest score, in order to ensure the validity of the data.

Before determining multivariate outliers, the questions were reverse scored and summed to create single scores for all six variables of extraversion, openness, agreeableness, neuroticism, conscientiousness, and risky cyber security behavior. Creating the single scores turned 63 scores into 6 which allowed for ease of further screening and interpretation. To analyze the data for multivariate outliers, a Mahalanobis distance was created (Tabachnick & Fidell, 2007). The Mahalanobis distance for each set of scores was cross compared to the chi-square distribution table to

TABLE 1 Tests of normality.

Variable	Statistic	Shapiro-Wilk *df*	Sig.
Overall extraversion	0.981	93	0.181
Overall agreeableness	0.157	93	0.014
Overall openness	0.977	93	0.098
Overall neuroticism	0.985	93	0.378
Overall conscientiousness	0.979	93	0.141
Overall risky cyber security behavior	0.986	93	0.453

determine which, if any, scores needed to be deleted. From this, four sets of scores proved to be multivariate outliers ($x^2 = 12.592$) and thus deleted.

To determine the normality, a Shapiro-Wilks test was run and none of the variables appeared to be significant ($p < 0.01$), as explained in Table 1.

Thus, the assumption of normality was met. The last section of the screening and cleaning process is to determine linearity and homoscedasticity. From the scatter plots, the assumptions of linearity and homoscedasticity appeared to be met.

6.2 Assumptions

- Dependent variable measure on a continuous scale: the entire questionnaire was based on a Likert scale and thus meets this assumption.
- Two or more independent variables: this research has five independent variables.
- Independence of observations: this data produced a Durbin-Watson statistic of 1.722, meeting this assumption.
- Linear relationship: shown in scatter plots created during screening and cleaning of data.
- Homoscedasticity: shown in scatter plots during screening and cleaning of data.
- No significant outliers: both types of outliers, multivariate and univariate, were attended to during screening and cleaning.
- Normally distributed data: a Shapiro-Wilks test was run. All data showed significance over 0.01 (see Table 1). Accepting the alternative hypothesis that the data is normally distributed.
- Multicollinearity: from the correlation matrix resulting from regression analysis, there were no correlations between independent variables above 0.50, thus meeting this assumption (Vatcheva, Lee, McCormick, & Rahbar, 2016).

6.3 Descriptive analysis

A descriptive analysis was run on the data to understand where the data sat in relation to all other variables. The questionnaire was a Likert scale ranging from 1, strongly disagree to 4, strongly agree.

As shown in Table 2, most of the variables' overall scores are leaning toward agree, as none of the means are below 2.19.

6.4 Regression analysis

Multiple regression was used to predict participants' risky cyber security behaviors using the five traits of personality. A significant regression was found $F(5, 87) = 4.479$, $p < 0.05$ with an adjusted r^2 of 0.159.

As shown in Table 3, the significant predictors of the risky cyber security behavior scores were agreeableness, neuroticism, and conscientiousness. The variables of extraversion and openness did not achieve significance. Due to the lack of significance, another analysis was run to attempt to explain any low correlations.

The data file was split into those with training and those without training. Although the "with training group" had a participant size much smaller for an appropriate multiple regression, the "no training group" had a participant size of 73 which sits just under the cut-off for sample size according to Stevens (2002). Regardless of this, the multiple regression analysis was run on the split file. The results from the regression showed an increase in not only the strength of the correlation but also the significance shown through the p-value. The multiple regression showed a significant regression, $F(5,67) = 6.123$, $p < 0.01$ with an adjusted r^2 of 0.262.

TABLE 2 Descriptive statistics.

	N	Minimum	Maximum	Mean	Std. Deviation
Overall risky cyber security scale	93	1.40	2.95	2.1887	0.33712
Overall extraversion score	93	1.38	3.63	2.5457	0.40370
Overall agreeableness score	93	2.22	3.89	2.9809	0.38232
Overall openness score	93	1.90	3.50	2.7602	0.35727
Overall neuroticism score	93	1.25	3.75	2.6331	0.55377
Overall conscientiousness score	93	1.44	4.00	2.7563	0.43905

TABLE 3 SPSS output from standard multiple regression.

	Variable	Correlation	*p*	Mean	SD	*B*	*SE*	*β*	*t*
	Extraversion	−0.112	0.142	2.5457	0.40370	0.075	0.090	0.090	0.832
	Agreeableness	−0.382	0.000	2.9809	0.38232	−0.302	0.095	−0.343	−3.188
	Openness	−0.051	0.315	2.7602	0.35727	−0.022	0.092	−0.023	−0.235
	Neuroticism	0.225	0.015	2.6331	0.55377	0.001	0.073	0.001	0.011
	Conscientiousness	−0.32	0.001	2.7563	0.43905	−0.195	0.085	−0.254	−2.298

Although it doesn't appear to be a large change from the original regression results, the significance for extraversion went from a very apparent lack of significance with $p = 0.142$, as shown in Table 3, to the split file giving a significance for extraversion $p = 0.052$, as shown in Table 4. Openness appeared to become less significant with the p-value increasing.

The model summary of this data displayed an increase in variance loading on the independent variables with an adjusted r^2 value of 0.262, showing a 26.2% explanation of variance. Given that again the variables of extraversion and openness did not achieve significance, further analyses were run specifically on the two variables. The analyses attempted to determine whether impacts such as time spent on the internet or previous cyber security awareness and training had an effect.

Taking into consideration that the data is ordinal, a Spearman's Correlation was implemented to observe correlations. The data was split into participants who spend more than 20h on the internet and those who spend less than 20h. The correlational results for both extraversion and openness appeared to be not significant.

The data was then split into participants aged under 20 years and those aged 23–30 years. Again, both Spearman's correlational analyses displayed no significance between extraversion and risky cyber security behavior and openness and risky cyber security behavior.

The final Spearman's Correlation, run specifically on extraversion and openness, was conducted after splitting the data file into females, males, and others. Although the "other" category only had one participant, there was no significance regardless. The results showed no significance for both females and males. The lack of significance throughout all analyses led to an in-depth analysis of the items in each trait.

The multiple regression between each item of the trait openness and scores on the risky cyber security scale showed a nonsignificant regression with $F(10, 82) = 0.929$, $p = 0.512$. The individual correlations for each item also showed no significant correlation with the risky cyber security scale.

Another multiple regression was run for each of the items of extraversion. This regression also showed a nonsignificant regression with $F(7, 82) = 1.822$, $p = 0.093$, although, individual correlations showed two significant correlations (see 1.822, $p = 0.093$, although, individual correlations showed two significant correlations (see Table 5).

The individual questions within the branch of openness were used in a multiple regression with the risky cyber security behavior scores in order to see if specific parts of the trait would determine risky cyber behavior. Resulting from this, there were no correlations with significance over the appropriate p-value ($p < 0.05$).

TABLE 4 Multiple regression results for those with no previous training.

	Variable	Correlation	*p*	Mean	SD	*B*	*SE*	*β*	*t*
	Extraversion	−0.192	0.052	2.5548	0.42029	0.059	0.102	0.073	0.582
	Agreeableness	−0.472	0.000	3.0030	0.38221	−0.375	0.107	−0.416	−3.503
	Openness	−0.012	0.461	2.7740	0.37750	−0.018	0.098	0.020	0.187
	Neuroticism	0.280	0.008	2.6147	0.55737	−0.019	0.087	−0.031	−0.219
	Conscientiousness	−0.417	0.000	2.7717	0.45053	−0.256	0.090	−0.336	−2.847

TABLE 5 Multiple Regression for items within extraversion and risky cyber security score.

Item	Correlation	*p*	Mean	SD	*B*	*SE*	*β*	*t*
I see myself as someone who is full of energy	−0.182	0.041	2.63	0.734	−0.107	0.069	−0.232	−1.546
I see myself as someone who generates a lot of enthusiasm	−0.183	0.039	2.71	0.731	−0.083	0.072	−0.180	−1.155

6.5 Participants' sextortion experience

The three questions following the demographics, regarding whether or not the participants had received a sextortion email, were optional but from the responses, it was found 24 of the participants had, in the past, received a sextortion email. Of these 24, 16 of the participants ignored the email after reading, although without deleting it; and only eight participants deleted the email after reading.

6.6 Self-efficacy (knowledge level)

Twenty-five percent of participants believe that they could be easily deceived by a sextortion email from a legitimate-looking source. Thirteen percent of participants do not believe they would be able to detect a potential phishing threat online. Eighty-six percent of participants believe they could gain knowledge in regards to avoiding cyber sextortion attacks even without the base knowledge of previous cyber security education (Table 6).

6.7 Cyber security risky behavior

The scores for each item in the risky cyber security behavior scale showed a range of scores throughout the 20 questions. There were some items that were more agreed with than others such as, "I use the same password for multiple websites", had a mean of 3.17 which is quite high. This

TABLE 6 Frequency data of self-efficacy questions.

	Strongly disagree	Disagree	Agree	Strongly agree
I could gain knowledge in regards to avoiding a cyber sextortion attack even with no previous cyber security knowledge	5 5.4%	8 8.6%	61 65.6%	19 20.4%
Phishing is an art of deception whereby cybercriminals attempt to deceive users into giving away personal information online. I feel confident that I would be able to recognize a potential threat	2 2.2%	19 10.8%	42 45.2%	39 41.9%
I could easily be deceived by an effective phishing-related sextortion email coming from a legitimate-looking source	29 31.2%	39 41.9%	23 24.7%	2 2.2%

compares to "I sometimes communicate or send personal information to strangers over the Internet", having a mean of 1.44 which is sitting between strongly disagree and disagree. The mean for the overall risky cyber security behavior score was 2.19 (see Table 2), which although low still leaves this participant sample vulnerable to potential sextortion attacks. To ensure cyber security which would protect against vulnerability, the overall mean for the risky cyber security behaviors would be below 2.

7 Discussion

The aim of this study was to understand individual differences in cyber security behavior and susceptibility to sextortion attacks using personality-based models. Based on previous research, it was hypothesized that scores in the risky cyber security behaviors scale would correlate positively with scores in openness, extraversion, and agreeableness and negatively with neuroticism and conscientiousness. The results from this study partly supported the proposed hypotheses. Data was collected and analyzed using multiple regression and Spearman's correlations. The multiple regression supported the hypothesis that high scores in cyber security would correlate with low scores of conscientiousness (see Table 3). For the remaining hypotheses, the data demonstrated the opposite for both agreeableness and neuroticism, with high risky cyber security scores correlating with low agreeableness and high neuroticism (see Table 3). The traits of extraversion and openness did not achieve statistical significance in their respective correlations with risky cyber security behavior scores (see Table 3).

From the multiple regression, an adjusted r^2 value was created which showed that personality traits only account for 15.9% of the variance in risky cyber security scores. This is not a huge loading of variance; thus more analysis was required. This further analysis was conducted after the file was split, separating participants with training, as those participants may have had an effect on the data, from participants who reported they had no previous training. Although the second multiple regression did not show extraversion or openness to be statistically significant, extraversion was very close to the threshold for statistical significance with a $p=0.52$. It may be assumed that with a larger sample size the correlation would reach significance.

If extraversion did meet the threshold for statistical significance, it would not have supported the proposed hypothesis of participants scoring high in extraversion also scoring high in risky cyber security behaviors. This is shown through the negative correlation of $r=-0.192$ (see Table 4).

It should also be noted that once training had been removed from having an effect on the data, the personality traits then account for 26.2% of the variance in risky cyber security behavior scores (adjusted $r^2 = 0.262$). After the interpretation of the regression data, the lack of significance in both extraversion and openness was intriguing, thus, further analyses specifically for these traits were run. Due to the ordinal data, Spearman's Correlation was run on the specific trait and participants' respective risky cyber security behavior scores. As shown in the results, none of the correlations appeared to be statistically significant even when data were separated into groups that may have impact scores, such as those spending under 20h on the internet against those spending over 20h.

Moving on from the correlational analyses, frequencies data was produced on the questions regarding self-efficacy in terms of social engineering attacks and training. The data showed that 14% of participants did not believe they could gain anti sextortion training without the base of cyber security knowledge, although this could demonstrate a lack of confidence in relation to cyber security training in general. This is compared to 86% of participants that believe they are able to gain such anti sextortion knowledge without previous training, although only 20.4% of those participants were confident in their answer, demonstrated through the selection of strongly agree over simply agree.

Supporting the theme that some of the participants in this study were not confident in their cyber security knowledge, 13% of participants reported they would not feel confident in their ability to detect a potential phishing threat. A further 26.9% of participants believe that they would be deceived by a sextortion email coming from a legitimate-looking source.

These results show that not all people are confident in their own ability to detect social engineering attacks and in turn cyber sextortion attacks. With over 1 in 4 participants reporting they are not confident shows a need for awareness and cyber security training. Within the demographics section of the questionnaire, there were questions regarding whether or not the participant had received a sextortion email in the past. From this, it was found that 24 of the participants had in fact received such emails. Although the participants reported not following instructions or following up the email with questions, 16 participants ignored the email after reading. From this data, it can be inferred that while the participants knew not to follow the instructions, there may have been some uncertainty on how to deal with the email as it was not deleted. It could be seen as participants leaning more toward believing the email may have had some truth underlying it. It was shown that eight participants did in fact delete the email after reading, which may have been due to confidence regarding how deceptive the email was or fear of possible repercussions.

7.1 Limitations and future directions

The current study was restricted by a number of factors, including participant sample size. The sample size, although fitting the requirements for multiple regression to be run, would have benefited from being larger in terms of observing statistical significance. If the participant sample size was larger, it may have accounted for the disproportionate gender representation as the sample was highly skewed toward females. The gender spread had only 1 participant identifying as other and 14 males, compared to 78 females. Comparing the results for females and males may be an interesting future study as males have been shown not to perceive risk as high as females in cyberspace (Garbarino & Strahilevitz, 2004). In addition, the participant sample had a restricted age range with the oldest age group being 23–38. Having a larger age range may produce interesting results, and thus future research should focus on recruiting a sample that has an equal gender spread as well as age range spread.

Another limitation restricting this study was the lack of a social desirability control. Although this study was anonymous and completed at the participants' discretion, the desirability to not admit to risky cyber behaviors would have been a factor to be investigated. Social desirability in the personality scale can be confidently rejected as previous research has found that control for social desirability in personality scales has had little effect on the overall result (Ellingson, Sackett, & Hough, 1999; Ellingson, Smith, & Sackett, 2001; Ones, Viswesvaran, & Reiss, 1996). For future research into risky cyber security behavior, a control for social desirability should be included to obtain the most valid results possible.

The last restriction to this study is in regards to the scale of openness. The items within the questionnaire concerning openness were primarily focused on the artistic side of the trait, not so much on the curiosity in which this study was interested. The reason for using the scale unaltered regardless of this was due to the support and wide use of the scale in human behavior and personality-based studies. Further research into cyber security behaviors using personality-based models should investigate the creation of a scale that interprets openness more as a curiosity trait.

Finally, further analyses regarding the identified previous sextortion attacks or lack of confidence in an individual's ability to detect a potential sextortion attack would have been preferable. However, due to the reduced sample size, the results would not be reliable.

7.2 Implications

The results of this study have significant implications when investigating the risk of cyber sextortion attacks. Although the hypotheses for this study were only partly supported, the variance shown in the multiple regression demonstrates that personality does account for 26.2% of the variance in risky cyber security behaviors. This, though not a large variance loading, does display a link between individual differences in cyber security behaviors which can be accounted for by personality differences.

It can be inferred that the riskier a person is online, the more at risk they are to cyber sextortion attacks as the lack of cyber security knowledge can leave a person vulnerable to deception. Once a person is vulnerable, cybercriminals can target or launch a social engineering attack for sextortion. These results show that security training is required for online users as their own individual differences may leave them at risk of cyber sextortion attacks.

8 Conclusion

This study aimed to understand individual differences in cyber security behavior and susceptibility to sextortion attacks using personality-based models. The results revealed trends contrary to the current understanding of personality traits with regard to risky behavior and impulsiveness. Although this study didn't investigate those behaviors specifically, risky cyber security behavior can be seen in conjunction with them. The results from this study demonstrated that participants scoring high on the risky cyber security scale will score high in neuroticism and low in agreeableness and conscientiousness. The traits of extraversion and openness were not statistically significant enough to be interpreted. The measure of personality accounted for 26.2% of the variance in risky cyber security behavior scores and thus is an indicator of a person's potential vulnerability to cyber sextortion. Subsequently, these results support the idea that training for online users regarding social engineering attacks in particular cyber sextortion attacks is required, as personality differences can cause vulnerability.

References

Arachchilage, N. A. G., & Love, S. (2013). A game design framework for avoiding phishing attacks. *Computers in Human Behavior, 29*, 706–714.

Arachchilage, N. A. G., & Love, S. (2014). Security awareness of computer users: A phishing threat avoidance perspective. *Computers in Human Behavior, 38*, 304–312.

Australian Competition and Consumer Commission: ScamWatch. (2016). *Scam statistics.* [Government Website]. Retrieved from: https://www.scamwatch.gov.au/about-scamwatch/scam-statistics?scamid=31&date=2016.

Bailey, J., Mitchell, R., & Jensen, B. (2008). Analysis of student vulnerabilities to phishing. In *Proceedings of the fourteenth Americas conference on information systems.* [online] Available at: https://aisel.aisnet.org/cgi/viewcontent.cgi?article=1279&amp=&context=amcis 2008&amp=&seiredir=1&referer=https%253A%252F%252Fscholar.google.com.au%252

Fscholar%253Fhl%253Den%2526as_sdt%253D0%25252C5%2526q%253DAnalysis%252B of%252BStudent%252BVulnerabilities%252Bto%252BPhishing%2526btnG%253D#search= %22Analysis%20Student%20Vulnerabilities%20Phishing%22. (Accessed 31 March 2019).

Barrick, M. R. (2005). Yes, personality matters: Moving on to more important matters. *Human Performance, 18*(4), 359–372. https://doi.org/10.1207/s15327043hup1804_3.

Biernacki, M., & Tarnowski, A. (2011). The effect of age and personality on the main cognitive processes in drivers. *International Journal of Occupational Medicine and Environmental Health, 24*(3), 367–379. https://doi.org/10.2478/s13382-011-0035-x.

Cattell, R. B. (1957). Personality and motivation structure and measurement. *APA PsychNet.*

Cho, J. H., Cam, H., & Oltramari, A. (2016). Effect of personality traits on trust and risk to phishing vulnerability: Modeling and analysis. In *2016 IEEE international multi-disciplinary conference on cognitive methods in situation awareness and decision support (CogSIMA).*

Cho, J. L., Wang, X., Chan, K., Chang, M., Swami, A., & Mohapatra, P. (2013). Trust and independence aware decision fusion in distributed networks. In *2013 IEEE international conference on pervasive computing and communications workshops (PERCOM workshops).* https://doi.org/10.1109/PerComW.2013.6529545.

Curtis, R. G., Windsor, T. D., & Soubelet, A. (2015). The relationship between Big-5 personality traits and cognitive ability in older adults—A review. *A Journal on Normal and Dysfunctional Development, 1*, 42–71. https://doi.org/10.1080/13825585.2014.888392.

Ellingson, J. E., Sackett, P. R., & Hough, L. M. (1999). Social desirability corrections in personality measurement: Issues of applicant comparison and construct validity. *Journal of Applied Psychology, 84*(2), 155–166. https://doi.org/10.1037/0021-9010.84.2.155.

Ellingson, J. E., Smith, D. B., & Sackett, P. R. (2001). Investigating the influence of social desirability on personality factor structure. *Journal of Applied Psychology, 86*(1), 122–133.

Eysenck, H. J. (1967). Intelligence assessment: A theoretical and experimental approach. *British Journal of Educational Psychology, 37*(1), 81–98.

Garbarino, E., & Strahilevitz, M. (2004). Gender differences in the perceived risk of buying online and the effects of receiving a site recommendation. *Journal of Business Research, 57*(7), 768–775. https://doi.org/10.1016/S0148-2963(02)00363-6.

Gupta, B. B., Arachchilage, N. A. G., & Psannis, K. E. (2018). Defending against phishing attacks: Taxonomy of methods, current issues and future directions. *Telecommunication Systems, 67*(2), 247–267. https://doi.org/10.1007/s11235-017-0334-z.

Gupta, B. B., Tewari, A., Jain, A. K., & Agrawal, D. P. (2017). Fighting against phishing attacks: State of the art and future challenges. *Neural Computing and Applications, 28*(12), 3629–3654. https://doi.org/10.1007/s00521-016-2275-y.

Hadlington, L. (2017). Human factors in cybersecurity; examining the link between internet addiction, impulsivity, attitudes towards cybersecurity, and risky cybersecurity behaviours. *Heliyon, 3*. https://doi.org/10.1016/j.heliyon.2017. e00346.

Halevi, T., Lewis, J., & Memon, N. (2013). A pilot study of cyber security and privacy related behavior and personality traits. In *Proceedings of the 22nd international conference on World Wide Web, May 13–17, 2013, Rio de Janeiro, Brazil.* https://doi.org/10.1145/2487788.2488034.

Harrison, B., Vishwanath, A., & Rao, R. (2016). A user-centered approach to phishing susceptibility: The role of a suspicious personality in protecting against phishing. In *49th Hawaii international conference on system sciences.*

Hong, K. W., Kelley, C. M., Tembe, R., Murphy-Hill, E., & Mayhorn, C. B. (2013). Keeping up with the joneses: Assessing phishing susceptibility in an email task. In *Proceedings of the human factors and ergonomics society 57th annual meeting.*

Huang, J. L., Cropanzano, R., Li, A., Shao, P., Zhang, X.-A., & Li, Y. (2017). Employee conscientiousness, agreeableness, and supervisor justice rule compliance: A three-study investigation. *Journal of Applied Psychology, 102*(11), 1564–1589. https://doi.org/10.1037/apl0000248.

Ivaturi, K., & Janczewski, L. (2011). A taxonomy for social engineering attacks. In *CONF-IRM proceedings*.

John, O. P., & Srivastava, S. (1999). The big-five trait taxonomy: History, measurement, and theoretical perspectives. In L. A. Pervin, & O. P. John (Eds.), *Vol. 2. Handbook of personality: Theory and research* (pp. 102–138). New York: Guilford Press.

Liang, H., & Xue, Y. (2009). Avoidance of information technology threats: A theoretical perspective. *MIS Quarterly, 33*(1), 71–90.

Lindgren, J. (1993). Theory, history, and practice of the bribery-extortion distinction. *University of Pennsylvania Law Review, 141*, 1695.

Martocchio, J. J., & Judge, T. A. (1997). Relationship between conscientiousness and learning in employee training: Mediating influences of self-deception and self-efficacy. *Journal of Applied Psychology, 82*(5), 764–773. https://doi.org/10.1037/0021-9010.82.5.764.

McCrae, R. R., & Costa, P. T. (1987). Validation of the five-factor model of personality across instruments and observers. *Journal of Personality and Social Psychology, 52*(1), 81–90.

McCrae, R. R., & John, O. P. (1992). An introduction to the five-factor model and its applications. *Journal of Personality, 60*(2), 175–215. https://doi.org/10.1111/j.1467-6494.1992.tb00970.x.

Ones, D. S., Viswesvaran, C., & Reiss, A. D. (1996). Role of social desirability in personality testing for personnel selection: The red herring. *Journal of Applied Psychology, 81*(6), 660–679.

Parrish, J. L., Bailey, J. L., & Courtney, J. F. (2009). *A personality based model for determining susceptibility to phishing attacks* (pp. 285–296). Decision Sciences Institute.

Patchin, J. W., & Hinduja, S. (2018). Sextortion among adolescents: Results from a national survey of U.S. youth. *Sexual Abuse*, 1–25. https://doi.org/10.1177/1079063218800469.

Rammstedt, B., Danner, D., & Martin, S. (2016). The association between personality and cognitive ability: Going beyond simple effects. *Journal of Research in Personality, 62*, 39–44.

Rolland, J. P. (2002). Cross-cultural generalizability of the five factor model of personality. In *The five factor model of personality across cultures* (pp. 7–28).

Salgado, J. F. (2010). Predicting job performance using FFM and non-FFM personality measures. *Journal of Occupational and Organiztional Psychology, 76*(3), 323–346. https://doi.org/10.1348/096317903769647201.

Stevens, J. P. (2002). *Applied multivariate statistics for the social sciences* (4th ed.). Mahwah, NJ, US: Lawrence Erlbaum Associates Publishers.

Tabachnick, B. G., & Fidell, L. S. (2007). *Using multivariate statistics* (5th ed.). Boston, MA: Allyn & Bacon/Pearson Education.

Trepte, S., Teutsch, D., Masur, P. K., Eicher, C., Fischer, M., Hennhofer, A., et al. (2014). Do people know about privacy and data protection strategies? Towards the "online privacy literacy scale" (OPLIS). In S. Gutwirth, R. Leenes, & P. de Hert (Eds.), *Law, governance and technology series: Vol. 20. Reforming European data protection law*. Dordrecht: Springer.

Vanderhoven, E., Schellens, T., Valcke, M., & Raes, A. (2014). How safe do teenagers behave on Facebook? An observational study. *PLoS One, 9*(8). https://doi.org/10.1371/journal.pone.0104036.

Vatcheva, K. P., Lee, M., McCormick, J. B., & Rahbar, M. H. (2016). Multicollinearity in regression analyses conducted in epidemiologic studies. *Epidemiology, 6*, 227. https://doi.org/10.4172/2161-1165.1000227.

Weiner, I. B., & Greene, R. L. (2008). *Handbook of personality assessment*. New York: Wiley.

Zuckerman, M., Kuhlman, M. D., Joireman, J., Teta, P., & Kraft, M. (1993). A comparison of three structural models for personality: The big three, the big five, and the alternative five. *Journal of Personality and Social Psychology, 65*(4), 757–768.

CHAPTER

5

Deconstructing security and privacy issues: The development of a logic for capturing mismorphisms*

Vijay H. Kothari, Prashant Anantharaman, and Sean W. Smith

Department of Computer Science, Dartmouth College, Hanover, NH, United States

1 Introduction

Security problems often stem from one or more mismatches between:

- a person's belief about something, e.g., a clinician may believe that a patient has a weight based on a patient weighing that they conduct;
- how that thing is represented or recorded within, say, a system or a document, e.g., the clinician may record the patient's weight within a hospital system, which is rounded to the nearest whole number; and
- the reality of that thing, e.g., the patient's true weight at the time of recording.

*This chapter revises and extends Chapter 7 of Kothari's doctoral thesis (Kothari, 2020). It also builds upon and borrows text from our previous publications (Anantharaman et al., 2020; Kothari, Blythe, Smith, & Koppel, 2018; Smith, Koppel, Blythe, & Kothari, 2015).

https://doi.org/10.1016/B978-0-323-90570-1.00008-5

These sorts of mismatches may be harmless or they may be the seed of a security or privacy issue. Consider some examples:

Security experts' failed assumptions. Security experts often design password composition policies for services. These policies impose requirements on user-generated passwords such as requiring a minimum length or requiring special characters. Often, however, practitioners overlook the induced frustration associated with these requirements and how that frustration may translate to user circumvention that nullifies envisioned security gains.

Mismatches in protocol specification and implementation. *A communication protocol* is a set of procedures that specify how two or more entities should communicate, for example, internet-connected devices use such protocols to communicate with other devices or systems. However, the protocol development process can be complex, involving multiple stages and skills associated with goal planning, specification design, community feedback, implementation, and review. This complexity often means that the *protocol designer* who designs the *protocol specification*—the document that specifies the how the implemented protocol should work—is distinct from the *protocol implementer* who *implements* (or writes code for) the protocol in software or hardware based on their conception of the protocol specification. Often, mismatches between the assumptions made by protocol designers and implementers manifest as security vulnerabilities.

For example, consider the Heartbleed bug (Carvalho, DeMott, Ford, & Wheeler, 2014). The Transport Layer Security (TLS) protocol is a communication protocol used in internet-based communication. The protocol specification supports a special message, called the Heartbeat message; a device ensures that a secure connection remains open by successfully sending a Heartbeat *packet* (data that is organized for processing) to the server and receiving a packet with almost identical content. The implementation of the Heartbleed code for OpenSSL did not check the Heartbeat packet size, leading to a vulnerability where an attacker could send a small packet to the server but expect a much larger packet with private data from the server in return. That is, there was a mismatch between the implementer's mental model and what was written in the specification, which was in turn informed by the designer's mental model; the implementer did not include the checks that were envisioned by the protocol designer and specified in the specification. In addition to flawed mental models and oversights on the part of the implementer that lead to vulnerabilities like Heartbleed, specification-implementation mismatches include ambiguous specifications that lend themselves to different interpretations.

System ambiguity. System developers may have preconceived notions about how a user will use their systems. Consider a *data entry system*, a system that users use to record and often read data. The developer of such a system may overlook the reality that different systems of measurements are used in different countries, resulting in a data field within the data entry system for which no unit of measurement is specified. A user may enter data into the system assuming one system of measurements (e.g., the metric system), which is later read by someone who assumes another system of measurements (e.g., US customary system). Hence, an implicit assumption on behalf of the developer may introduce ambiguity into the system. For one such example, we have heard reports of IT systems in a hospital using a different measurement system from that country's standard—risking serious consequences for patients.

Hidden behavior. Another example involves a user making a search query on Google. The user sees a URL—a sequence of characters that specifies the address of a resource on the internet or a network more generally—in the status bar of their web browser, and they click on the corresponding hyperlink. In reality, the user navigates through another Google URL before (*hopefully* (Abrams, 2021)) ending up on the page they initially intended, none the wiser if all goes well (Cyphers, Miagkov, & Arrieta, 2018). The user is unlikely to notice this redirection, which very well may violate their privacy expectations.

Programmers' tacit assumptions. Programmers often use the phrase *DRY*, which means do not repeat yourself (Oriol Salides, 2021), with the aim of discouraging the reuse of previously written, presumably more reliable, code. Developers make routine software tasks available as software libraries for others to use. These software libraries have an *application programming interface* (API) (Wikipedia Contributors, 2021 a), which is a document that specifies, among other things, how the library should be used. Instead of consulting the API, programmers may assume the function behavior based on prior experience, which may result in unintended behavior being introduced into the resultant program.

Privacy expectations versus reality. Several recent studies have focused on understanding users' privacy expectations pertaining to internet-of-things (IoT) devices, which correspond to the multitude of highly communicative internet-connected devices (Naeini et al., 2017). Various researchers have used the contextual integrity (CI) framework to study these privacy norms (Nissenbaum, 2009). This framework defines privacy as the *appropriate flow* of information subject to contextual information. In this framework, privacy norms change over time and from context to context. Any flow that is not aligned with the well-established legal, ethical, or standard practice norms is considered a *privacy violation*. Based on the CI framework, privacy violations stem from mismatches between users' expectations and device behavior.

For example, Malkin et al. (2019) studied users' beliefs and expectations of their smart speakers (or smart voice assistants, such as Google Home or Amazon Echo). A study involving 116 participants who use smart speakers found that many users considered the music they listened to be private. In contrast, the default setting on many music applications allows others to know what song the user is listening to. In another finding, due to unreliable voice activation, users found unintended recordings of their children or grandchildren. These examples highlight mismatches between device behavior and users' expectations of device behavior. That is, they reveal privacy violations.

In fact, across our work, we find it exceedingly difficult to find instances of security and privacy issues where we *cannot* identify such an underlying disconnect. If we dig deep enough, security problems tend to come down to one or more of these mismatches or, more precisely, what we call *mismorphisms*—"mappings that *fail* to preserve structure" (Smith et al., 2015).

If mismorphisms indeed lie at the heart of security and privacy issues, then understanding mismorphisms, developing a suitable model to express them, and working toward a rich catalog of them may serve useful in addressing the security and privacy issues to which they lead. This chapter reviews our efforts and our colleagues' efforts in achieving these aims.

This chapter is organized as follows. In Section 2, we provide a brief primer on semiotics. In Section 3, we introduce a semiotic model to capture mismorphisms that adopts a variation of semiotic triads. In Sections 4 and 5, we motivate the development of an alternative logic model to capture mismorphisms and we present said model. In Section 6, we use the logic model to develop a preliminary classification of a number of mismorphisms. In Section 7, we discuss future work. In Section 8, we conclude.

2 A brief background on semiotics

Semiotics is the study of signs, processes that involve signs, and how meaning is conveyed through signs (Wikipedia Contributors, 2021 e). A sign may be a sound, an image, a smell, or anything else from which a sentient being extracts meaning. For a simple example, a person may see a stop sign while driving and know that means they should slow down and come to a stop. Semiotic models aim to explain these and other phenomena. Two of the most prevalent semiotic models are: the dyadic model proposed by Ferdinand de Saussure, which includes a signifier and a signified; and the triadic model proposed by Charles Sanders Peirce, which includes a sign, an object, and an interpretant (Chandler, 1994). The

Stanford Encyclopedia of Philosophy (Atkin, 2013) provides a primer on Peirce's work.

Ogden and Richards presented the *semiotic triad* (Ogden & Richards, 1923; Wikipedia Contributors, 2021 g) to capture the relationship between three nodes: the referent (the thing being referred to), the thought or reference (the object evoked by the referent), and the symbol (the object used to represent the thought), as seen in Fig. 1. When a writer writes, the referent—the thing the writer is trying to express—induces a thought based on the writer's knowledge of language, who the writer thinks the reader will be and how they might interpret it, the writer's state of mind, and so forth. The thought evokes a symbol that is supposed to express the referent. Similarly, when a reader reads a word, the word or symbol evokes a thought based on the reader's general knowledge, their understanding of the context in which the word is used, and so forth. The thought is then internalized as a referent. A causal relation is established between the word (the symbol) and the thought (the reference). And a relation is also established between the thought (the reference) and the referent. However, there is no direct relation between the symbol and the referent. Instead, there is an imputed relation established through the two sides of the triangle, not the base. Thus, we have the semiotic triad.

Before discussing our earlier work in building a semiotic triad for mismorphisms, we review some related work at the intersection of semiotics and HCI. Weir (1992) discusses the need for applying semiotic approaches to better understand man-machine communication. de Souza, Barbosa, and Prates (2001) outline desired software design properties, advocate for using semiotic engineering for HCI, and outline one approach.

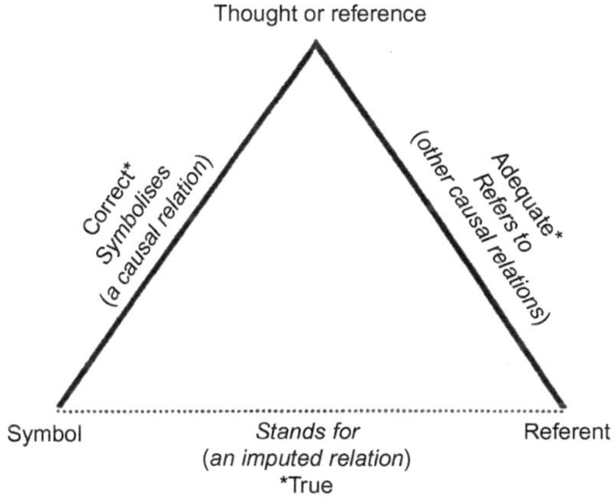

FIG. 1 The semiotic triad. *A slightly modified version of the image appearing on page 11 of the original 1923 publication of Ogden and Richards'* The Meaning of Meaning *(Ogden & Richards, 1923).*

Ferreira, Barr, and Noble (2005) look at how semiotics can be used to understand redesigns of user interfaces. They analyze three instances of redesign of a sign that is part of a user interface, and they briefly look at the contributing factors and propose that such examinations can lead to better user interface design. Andersen (2001) enumerates a number of challenges that semiotics-based HCI design can help address, including: "making HCI more coherent," "exploiting insights from older media," "defining the characteristic properties of the computer medium," and "situating the HCI-systems in a broader context."

3 A semiotic model for mismorphisms

In this section, we (very) briefly review our earlier work on mismorphisms (Smith et al., 2015). The usage of mismorphism in this section is slightly different from the usage in the logic model we later discuss. However, the essence of the two models are the same; in both models, we seek to capture differences between representations of things that produce security and privacy problems in practice.

As noted in the previous section, our semiotic model of mismorphisms is inspired by Ogden and Richards' semiotic triad. Our model is built with the intent of expressing circumvention scenarios. We replace the referent with a reality, the thought with a mental model, and the symbol with an IT representation as seen in Fig. 2. In this model,

- the reality corresponds to some truth in the real world, for example, what actions a user may perform;
- the mental model corresponds to a party's beliefs regarding what the reality should be, for example, what an admin thinks a user's permissions should be; and
- the IT system representation expresses the reality as expressed in the IT system, for example, what permissions are given to a user.

Unlike Ogden and Richards' semiotic triad, each side of the triangle now exists and links two nodes. However, this linkage is unidirectional

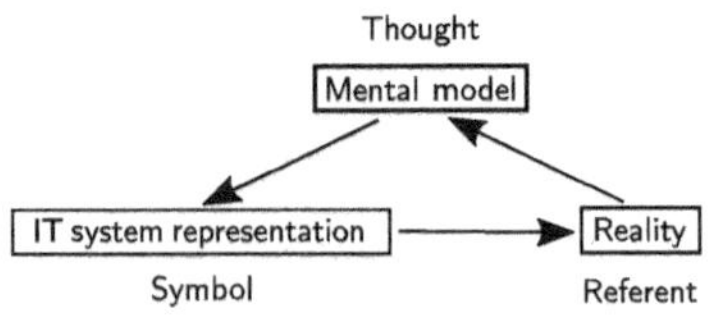

FIG. 2 A triad for capturing circumvention scenarios.

and expresses a single mapping between representations: The reality informs the mental model. A change in the mental model may drive a party to change the IT system itself. And a change to the IT system generates a new reality. For example, a security administrator may observe a reality in which users leave a machine unattended. This observation leads the security practitioner to the belief that there should be automatic timeouts. Thus, the security administrator may implement a policy within the IT system that automatically logs the user out if the IT system detects the user is away. And this, in turn, creates a new reality. Now, the user may become dissatisfied with this reality because the logouts get in the way performing their primary task. This reality will then drive the user to think of a way to circumvent the system. The user may then modify the IT system by, say, attaching a mouse jiggler to the computer, which in turn generates a new reality. (In our original paper (Smith et al., 2015), there was a unidirectional relation from the mental model to the IT system representation. However, in practice, this may be bidirectional. That is, the IT system representation may inform one's mental model. This can be an important source of security vulnerabilities if security personnel rely not on the reality but the IT system representation of the reality to make future decisions.)

In semiotics, the constructs of interest are called *morphisms* and they capture instances where a *predicate*—an expression that can be evaluated as true or false—holds the same truth value across representations. However, as highlighted in the example we just discussed, we are interested in instances where predicates take on different truth values across nodes of the triad. We call these *mismorphisms*. In our initial mismorphism work (Smith et al., 2015), we focused on exploring different classes of mismorphisms and cataloging them using the semiotic model we had developed. We found the model extremely effective in classifying circumvention scenarios.

4 Beyond semiotic triads

At the beginning of this chapter, we said mismorphisms are powerful enough to capture the underlying causes of several security and privacy issues. In the previous section, we presented a semiotic model to represent mismorphisms, one which we had used in prior work (Smith et al., 2015) to develop a catalog of practical unusability and circumvention scenarios, primarily in the hospital setting. In this section, we justify an alternative logic model that takes a more general approach to capture mismorphisms as a high-level concept. The development of this model was driven by the following considerations:

- While we showed that the semiotic approach to modeling mismorphisms is effective for capturing a number of usable security scenarios (Smith et al., 2015), it is somewhat constrained and things can get messy with additional complexity, for example, when
  - there is more than one single human interacting with a single system,
  - there are temporal effects, or
  - multiple mismorphisms are at play.

  This does not mean that we cannot use the notion of semiotic triads to capture complex scenarios. Indeed, we had demonstrated how we actually can capture some of these phenomena using the semiotic model of mismorphisms (Smith et al., 2015). However, the model does not naturally lend itself to such complexity. Ideally, we want to construct a model that has the machinery to both succinctly and accurately capture such mismorphism complexity.
- In our semiotic model, we used mathematical logic to express the underlying predicates. Reenvisioning the notion of mismorphisms using a pure logic model while retaining much of the spirit of the initial semiotic model seems natural.
- At its core, a mismorphism constitutes a difference in the interpretation of something by two or more entities. The model should allow us to capture mismorphisms involving any integer-valued number of interpretations above 1.
- This model should lay the foundation for a mismorphism submission interface that supports collaborative community development of a classification framework with accompanying examples; ergo, ease of representation of mismorphisms in the given model is paramount.

These considerations motivated us to develop a logic model for mismorphisms that is built on the notion of interrepresentational differences, that is intuitive, and that supports multiple representations, temporality, and compositionality.

5 A logic for mismorphisms[a]

The limitations discussed in the previous section drove us to develop a new model of mismorphisms, one grounded not in semiotic triads but in logic. In this section, we present this logic model. In the next section, we demonstrate how we can use this logic to create a catalog of the underlying causes of security and privacy issues.

[a]This section revises and extends text from our previous paper (Anantharaman et al., 2020).

For the rest of this chapter, we will refer to a *mismorphism* as a difference in interpretation of a predicate between two or more interpreters. That is, we can think of different *interpreters* (e.g., a person, a system, a document, code) interpreting propositions or predicates about the world. In general, it is good when the interpretations agree and are in accordance with reality. However, when a predicate takes different truth values across different interpretations, we have a mismorphism. Not all mismorphisms are bad; some may be benign or even beneficial. However, many times, mismorphisms produce unintended or undesirable outcomes that undermine security and privacy objectives at both the individual and organization level. This high-level vision for what constitutes a mismorphism drives the development of the formal logic we use to capture them.

Within our logic, we use the words *predicate* and *interpretation* in similar—albeit, not identical—manners to the common formal-logic meanings, for example, as presented by Aho and Ullman (1994). However, instead of a binary logic, we use a ternary logic similar to Kleene's ternary logic (Fitting, 1994; Goodstein, 1954).[b] We refer to a *predicate* as a function of zero or more variables whose codomain is $\{T, F, U\}$ where T is true, F is false, and U is uncertain/unknown. We refer to an *interpretation* of a predicate as an assignment of values (which may include U) to variables, which results in the predicate being interpreted as T, F, or U. A predicate is interpreted as T if after substituting all variables for their truth values, the predicate is determined to be T; it is interpreted as F if after substituting all variables for their truth values, the predicate is determined to be F; if we are unable to determine whether the predicate is T or F by substitution, the predicate is interpreted as U to signify that the truth value of the predicate cannot be determined.

The interpretation must be done by someone (or perhaps something, e.g., a system or a specification) and that entity is called the *interpreter*. In our model, for instances where there is some ground truth, we have a special interpreter, the oracle O, who interprets the predicate as it is in reality. Some interpreters may lack the requisite information to assign values to variables that would result in the predicate being interpreted as T or F. In these instances the predicate should be interpreted as U. We use $P|_A$ to denote the interpretation of predicate P by interpreter A.

To represent mismorphisms, we need a way to express scenarios where two or more interpreters diverge in their interpretations of a predicate. That is we must define relations on the interpreters' interpretations of a predicate. Ergo, we introduce the notions of *interpretation relations* and

[b]We do not specify a specific ternary logic system for evaluating predicates in this chapter.

interpretative expressions. In the context to our model, a *simple interpretative expression* has the form:

Predicate (Interpretation Relation) Interpreters

Here, the *interpretation relation* is a k-ary relation where $k \geq 2$ denotes the number of interpreters involved—and the k-ary relation is over the interpretations of the predicate by the k interpreters. The three classes of interpretation relations that we are concerned with in this chapter are: the interpretation-equivalence relations ($\underset{\text{interp}}{=}$), the interpretation-uncertainty relations ($\underset{\text{interp}}{\overset{?}{=}}$), and the interpretation-inequivalence relations ($\underset{\text{interp}}{\overset{\times}{=}}$).[c] The interpretation relations[d] we examine are defined as follows, where each P represents a predicate and each A_i represents an interpreter:

- $P(\underset{\text{interp}}{=}), A_1, A_2, \ldots, A_k$ if and only if P, as interpreted by each A_i, has a truth value that's either T or F (never U)—and all interpretations yield the same truth value. The interpretation-equivalence relations correspond to situations where every interpreter can evaluate the predicate as T or F and they all evaluate it in the same way.
- $P(\underset{\text{interp}}{\overset{?}{=}}) A_1, A_2, \ldots, A_k$ if and only if P takes on the value U when interpreted by at least one A_i. The interpretation-uncertainty relations correspond to situations where at least one interpreter lacks the requisite information to fully resolve the predicate to a known truth value of T or F.
- $P(\underset{\text{interp}}{\overset{\times}{=}}) A_1, A_2, \ldots, A_k$ if and only if P interpreted by A_i is T and P interpreted by A_j is F for some $i \neq j$. The interpretation-inequivalence relations correspond to situations where it can be determined that two interpreters disagree on the interpretation of a predicate.

There are a few important observations to note. One is that the oracle O always holds the correct truth value for the predicate by definition.

[c]Note that for $k=2$, if we confine ourselves to predicates that only take on T or F values, the relation $\underset{\text{interp.}}{=}$ is an equivalence relation in the mathematical sense, as one might expect, that is, it obeys reflexivity, commutativity, and transitivity.

[d]Technically, these are classes of interpretation relations, but adopting this terminology would be prohibitively tedious for us to write that and for you to read.

Another is that if we only know the ($\underset{\text{interp}}{\overset{?}{=}}$) relation applies, we will not know which interpreter is uncertain about the predicate or even how many interpreters are uncertain unless $k = 2$ and one interpreter is the oracle. Similarly, if we only know that the ($\underset{\text{interp}}{\overset{\times}{=}}$) relation applies, we do not know where the mismatch lies unless $k = 2$. That said, knowledge that the oracle O always holds the correct interpretation, where we are dealing with facts, combined with other information can help specify where the uncertainty or inequivalence stems from. Of course, the formalism could also be changed to allow a bit more flexibility here, but we did not see the need. Last, the ($\underset{\text{interp}}{=}$), relation will not be true if the ($\underset{\text{interp}}{\overset{?}{=}}$) or the ($\underset{\text{interp}}{\overset{\times}{=}}$) relations are true; however, P ($\underset{\text{interp}}{\overset{?}{=}}$) $A_1,\ldots,A_k$ and P ($\underset{\text{interp}}{\overset{\times}{=}}$) $A_1,\ldots,A_k$ may simultaneously be true.

In addition to the simple interpretative expressions mentioned previously, there are instances where it is useful to consider the composition of simple interpretative expressions. We call such expressions *compound interpretative expressions* and they are expressed by linking together simple interpretative expressions with any number of *interpretative operators* including:

- $\wedge$: The *and* operator has two operands, each of which are either simple or compound interpretative expressions; the full expression applies when both operands apply.
- $\vee$: The *or* operator has two operands, each of which are either simple or compound interpretative expressions; the full expression applies when at least one of the two operands apply.

Additional interpretative operators may serve valuable in creating a larger corpus, but for the purpose of our preliminary catalog of mismorphisms, these will suffice.

Let us revisit the initial goal in developing this logic model: to capture mismorphisms. Within this logic model, a *simple morphism* is a simple interpretative expression where the interpretation-equivalence relation applies. A *simple mismorphism* is a simple interpretative expression where the interpretation-equivalence relation does not apply; equivalently, it is a simple interpretative expression in which the interpretation-uncertainty relation applies or the interpretation-inequivalence relation applies. More generally, a *mismorphism* is either a simple mismorphism or a *compound interpretative expression* for which *at least* one simple mismorphism *must* hold for the expression to hold.

There are some natural extensions to this logical formalism. In select cases, we may want to consider multiple interpreters of the same role. In these instances, we could assign subscripts to distinguish roles, for example, D, I_1, I_2, O might correspond to the developer of a protocol, two implementors of the protocol, and the oracle. Also, there are temporal aspects that may be relevant. Predicates can be functions of time and so can the interpretations. While we use the v^t-style notation to represent a variable as a function of time within a predicate, we may also consider the interpreter as a function of time, for example, $I_4^{t_3}$ means the interpretation is done by implementor I_4 at time t_3. We do not use all of these extensions in this presentation, but if we were to create a larger catalog, they would serve useful.

6 A preliminary catalog of mismorphisms[e]

In this section, we discuss numerous examples of mismorphisms, classified by their general form. First, some remarks:

- The categories are not disjoint. Some mismorphisms could be placed within two or more categories.
- Some mismorphisms may be linked. For example, one mismorphism may lie at the heart of another or perhaps two mismorphisms contribute to a single security issue. This makes sense as many security issues have multiple layers of complexity. We discuss this issue more in the following section.
- We also note that there are multiple ways to do this classification. For example, another natural approach may be to choose the categories based on their application domain or security and privacy context.
- Our focus here is on applying our mismorphisms logic to a diverse but small set of security and privacy issues in different domains. Other flavors of mismorphisms certainly exist as well.

6.1 Breakdown of implication

In certain circumstances, an interpreter may believe a conditional statement that fails to hold in practice—or vice versa, they may not believe a conditional statement holds when it does hold in practice. That is, we may have something of form:

$$(X \implies Y) \ \overset{\times}{\underset{\texttt{interp}}{\equiv}} \ A, O$$

[e]This section revises and extends ideas and text from our previous work (Anantharaman et al., 2020; Kothari, 2020; Kothari et al., 2018).

Consider the following examples of this flavor of mismorphism at play:

- A prevailing belief is that users are privacy pragmatists who willingly make informed decisions to give up their privacy in exchange for services (Draper, 2017). This argument, in other words, assumes that the decision to use a service implies the user is making an informed choice. Work by Draper (2017), as well as by others (boyd & Hargittai, 2010; Kokolakis, 2017; Olmstead & Smith, 2017; Turow, Hennessy, & Draper, 2015; Urban & Hoofnagle, 2014) call this view into question. Draper argues that many users feel their privacy is gone and so they resign to giving up control over their data privacy.
- Turow et al. (2015) found that 65% of respondents to a survey believed that the existence of a privacy policy on a site meant the site would not share their information unless they gave explicit permission.
- It is often assumed that adding a privacy option to a service will only improve users' privacy. However, the user's determination of a privacy option may, itself, leak information. For example, Lewis, Kaufman, and Christakis (2008) note that both the options of sharing information or not sharing information correlate with other demographic information.

Alternatively, the implication operation may be correct, but X may not hold true, meaning nothing can be inferred about Y. (Or perhaps, we may observe the opposite direction where both the relation and X hold in practice but not within someone's mental model.)

$$((X \implies Y) \; \underset{\text{interp}}{=} \; A, O) \wedge (X \; \underset{\text{interp}}{\overset{\times}{=}} \; A, O)$$

For example,

- A security practitioner may assume that any user of a service who wishes to change their privacy settings will be able to do so if they know about them, and, moreover, they may assume the user is aware of those settings. In some cases, even if the former holds, the latter does not.

6.2 Temporal effects

Time may influence how predicates are evaluated. An individual may lack the foresight to identify these temporal effects.

$$(X^t = X^{t+\delta}) \; \underset{\text{interp}}{\overset{\times}{=}} \; A, O$$

Some examples:

- As an employee changes roles, their permissions may accumulate, whereas a security practitioner might expect the permissions to be adjusted according to the role (Sinclair, 2013).

- Time-of-check-time-of-use (toctou) bugs (Wikipedia Contributors, 2021 f) occur when there is a delay between when something is checked and when it is used. The delay means that operations may be performed on input that previously satisfied certain properties, but no longer do so. It reflects an oversight on the part of the developer.
- Shotgun parsing (Bratus, Patterson, & Hirsch, 2013) involves scattering parser code—the code responsible for vetting the input to a program—across a program, which results in code being executed before it is recognized. Vulnerabilities that exist in code that are attributable to the shotgun parser antipattern can be classified under this class of mismorphisms.
- An analog of toctou for the privacy domain is time-of-configure-time-of-use: The user may configure their privacy settings on a social networking service once, when they begin using a service. However, over time, people may join or leave the service, leaving their privacy choices outdated. Available privacy options may also change over time.
- Gaw and Felten argue that the user may choose to reuse a password for an account—i.e., select a weak password before that account is associated with sensitive information—and, by the time that account has accrued information, "they're locked into their reused password" (Gaw & Felten, 2006).
- A similar phenomenon may be true with privacy settings. Namely, the user may choose privacy settings before sensitive information is tied to their account. By the time sensitive information is tied to their account, the user may no longer think about privacy. Moreover, in instances where the user does contemplate reconfiguring privacy settings, there is a possibility that the data in question may be perceived to already be lost and, therefore, not worth protecting.
- On the other hand, some users may have already invested significant time or effort in selecting a piece of software, downloading it, and installing it before they configure their privacy settings, compelling them to continue using the service even if it does not meet their privacy needs. That is, they may fall victim to the sunk cost fallacy (Arkes & Blumer, 1985). Had they known of the invasive privacy settings beforehand, they may have chosen to go with a competitor.

6.3 A knowledge gap

In certain circumstances, an interpreter's lack of knowledge about how to interpret information may contribute to a security issue. Here, *P* may be a statement about, say, a system, and the interpreter may be ill-equipped to evaluate the truth of that statement, resulting in an unknown truth value under their interpretation:

$$P \; \boxed{\overset{?}{\underset{\text{interp}}{=}}} \; A, O$$

- Users may lack the requisite information to make informed decisions, often because that information is simply not available. It is not always clear how services safeguard user data, nor the intricacies of how that data is used in practice. Privacy policies exist but they may be exorbitantly time-consuming to read and difficult to digest (McDonald & Cranor, 2008; Obar & Oeldorf-Hirsch, 2016). Moreover, they are often vague and usually subject to change. A pessimist might argue that in practice many existing interfaces and privacy policies ensure users remain uninformed while presenting the veneer of informed consent, thereby persuading their users and others that user data is in good hands. Another concern is that primary services or third-party services may violate privacy policies, terms of service, or users' privacy expectations. This may even be compounded by a delay in reporting violations. Collectively, these and other factors support the argument that most users do not—and, at least in the current privacy landscape cannot—have a concrete understanding of how their data is used.
- A user may lack the capability to come to a determination regarding the safety of a URL (e.g., shortened URLs, gatekeeper URLs), the legitimacy of an email, or the meaning of certificate information. While some users may seek information that informs their mental model, others may fall back on insecure behavior because it is less effort and potentially a lower perceived cost than alternatives. Even if a user seeks out information, it is possible that they may consult a resource that provides inaccurate information.

6.4 Projections

An interpreter A's interpretation of how interpreter B would interpret a predicate P may differ from how B actually interprets it. That is, we may have

$$P|_B \; \boxed{\overset{\times}{\underset{\text{interp}}{=}}} \; A, O$$

We recognize that there is a slight abuse of notation here. To resolve this, we can simply substitute $P|_B$ with "B's interpretation of P," to avoid the double-meaning of $P|_B$—or we could create a wrapper. Recall the oracle is always correct and so their interpretation of $P|_B$ aligns with what $P|_B$ actually is. In any case, here is an example of such a mismorphism:

- Actual and perceived time and effort to configure privacy settings may influence whether the user begins configuring them and whether they finish. For example, the user may be dissuaded from using an interface that appears illogical, complex, or hard to navigate. Or, as we mentioned earlier, they may simply lack the requisite knowledge to make meaningful decisions that align with their intentions. The security practitioner or others may perceive users' effort to configure their privacy settings to be minimal or ignore them altogether and view the option of configuration as a binary choice.

7 Future work

In this section, we briefly discuss future work pertaining to the logic model of mismorphisms.

7.1 Peeling back the layers of mismorphisms

In the previous section, we presented a preliminary catalog of mismorphisms that captures a number of security and privacy issues across domains. However, much of the power of mismorphisms as an explanatory model comes from the ability to break down a mismorphisms and study their ramifications. Identifying these causal relations allows us deconstruct and learn from existing security problems.

For example, why might a user wrongly classify an unsafe URL as safe? Well, one reason may be that their mental model of where the URL goes is flawed (Albakry, Vaniea, & Wolters, 2020). This can be captured as a mismorphism between security properties of the URL in the system representation and the security properties of the URL within the user's mental representation. But why does that mismorphism exist? It may be a purely visual problem, due to a poor choice of font, which can be expressed as a mismorphism between the user's mental representation and the information shown in the real-world and/or a mismorphism between the system representation and the information shown in the real-world, depending on where the problem lies. Or perhaps the user correctly interprets what characters are on the screen, but fails to extract the correct security information from those characters; this again can be represented as a mismorphism between the URL specification (or, more precisely, the layers of systems involved in resolving the URL and delivering content to the user) and the user's mental representation. But we could again ask: why is there a mismorphism between a user's mental model of URL structure and the way users are resolved in practice? And so on.

Ultimately, it is this process of recursive deconstruction of mismorphisms that reveals why a security problem truly exists. Understanding mismorphisms and the links between them is essential to addressing many of the security and privacy challenges of today. Our logic model already supports some interpretative operators. However, support for casual relations would provide tremendous expressive power. Indeed, the addition of causal relations reintroduces some of the lost expressiveness of semiotic triads without sacrificing any of the expressiveness of the logic model. Approaches such as supplementing the notion of mismorphisms discussed here with, say, events, may provide the best of both models. Support for both, expressing causality and expressing events, are beyond the scope of this chapter, though they are certainly worth pursuing. Together, they would enable us to capture the development of security and privacy issues as a chain of mismorphisms and events linked by causal relations.

7.2 Psychological phenomena and the genesis of mismorphisms

There is an intimate link between psychological phenomena, especially cognitive biases, and mismorphisms. In this chapter, we have treated mismorphisms as the most elementary phenomenon driving security and privacy issues. But this raises the question of why mismorphisms arise. Part of the explanatory process to tackle this question may involve deconstructing mismorphisms into predecessor mismorphisms, as we discussed in the previous subsection. But it may also serve useful to examine the link between human psychology and mismorphisms.

Several researchers have explored the psychological aspects of computer security (Enrici, Ancilli, & Lioy, 2010): Lafrance (2004) categorized hackers based on their motivations and techniques. In comparison, Enrici et al. (2010) explore how humans can be the target of attacks—cognitive hacks. Smith (2012), an author on this chapter, has examined various cognitive biases and how they can be leveraged to improve security.

Yet there are still many avenues at the intersection of psychology and security that researchers have not yet or have only partially explored. Consider cognitive biases, the logical flaws in judgment that humans tend to exhibit. A deep exploration of programmers' cognitive biases could help to explain the underlying causes of programming mistakes that ultimately manifest as security vulnerabilities. For some examples, we believe exploration of three egocentric biases in programming may help to reduce security vulnerabilities in practice. A person who is subject to the overconfidence bias (Wikipedia Contributors, 2021 d) or illusion of validity bias (Wikipedia Contributors, 2021 b) may overestimate their performance on a task. A programmer who is subject to these biases may overestimate the correctness of their beliefs as they write code or the code itself. Another relevant bias here is the illusion of optimism bias, wherein a person believes they will not experience a negative event (Wikipedia Contributors, 2021 c). When a programmer writes code, they must account for infinite inputs, and they should be certain that no input can crash the program or enable unintended computation (Bratus, Locasto, Patterson, Sassaman, & Shubina, 2011). However, the illusion of optimism bias may drive programmers to believe that their code would not be a target of attack, leading them to overlook these checks.

Moreover, to the best of our knowledge, there is a missing explanatory framework linking these psychological phenomena to the security and privacy issues they ultimately give rise to. Augmenting the mismorphisms framework to express the link between these psychological phenomena and mismorphisms would allow us to bridge this gap and produce a holistic explanation of the genesis of security and privacy issues.

7.3 A mismorphism submission interface

The primary aim of the logic model for mismorphisms is to provide a simple, intuitive, and precise framework to accurately catalog how real-world security and privacy issues came to be. We seek to capture the initial starter mismorphisms and the chain of causal linkages that ultimately produce the problem scenarios of interest. A framework that achieves this aim, alongside a catalog with a navigable user interface, can be of great benefit to the security and privacy communities.

7.3.1 *Informing design decisions*

First, it would inform new design decisions. Practitioners would be able to learn from the mistakes of the past during the design and development of systems, policies, protocols, and so forth. When faced with a design question, having at your disposable a navigable catalog of bad outcomes—and also potentially good outcomes—associated with related design decisions that others have made would be of immense value.

7.3.2 *Addressing existing security and privacy problems*

Second, having such a framework can help in conceptualizing and addressing existing security and privacy problems faced by organizations and individuals. Moreover, if we augment the catalog by allowing users to specify how they attempted to address given mismorphisms, as well as comment on how effective those methods were, this can provide additional insights to those facing similar mismorphisms.

7.3.3 *Informing best practices*

As mismorphisms are often tethered to the human and these links can be readily observed upon deconstructing mismorphisms, a catalog of mismorphisms can shed light on human failings. Best practices can then be developed to accommodate these failings. Given the sheer breadth of security and privacy issues, however, cataloging mismorphisms would be a mammoth undertaking that would require a community-driven effort. The most natural way to achieve this would be to develop an online mismorphism submission interface.

We envision an online interface that allows users to

- submit mismorphism classes by specifying
  - a mismorphism class title;
  - a representation of the mismorphism class using logic;
  - relations between the target mismorphism class and other mismorphism classes (e.g., member of, produces);
  - a verbal description of the mismorphism class; and
  - auxiliary data (e.g., references).
- submit examples of security and privacy issues that stem, at least partially, from a mismorphism belonging to a user-specified mismorphism class;
- submit information regarding solutions adopted to tackle mismorphisms and the security and privacy issues they give rise to, as well as information on the efficacy of such solutions; and
- view information about a class of mismorphisms, including the aforementioned data submitted by the user, a visual representation of the mismorphism class, links to other mismorphism classes, and example security and privacy issues tied to the mismorphism class.

8 Conclusion

In this chapter, we pursued a logic model of mismorphisms to complement our earlier work on capturing mismorphisms via semiotic triads. We reviewed the earlier semiotic triad model, provided our rationale for developing a new model, introduced our logic model, cataloged a variety of mismorphisms, and discussed future work.

References

Abrams, L. (2021). *Beware: Malicious home depot ad gets top spot in Google search*. Bleeping Computer. https://www.bleepingcomputer.com/news/security/beware-malicious-home-depot-ad-gets-top-spot-in-google-search/.

Aho, A., & Ullman, J. (1994). *Predicate logic*. Foundations of Computer Science. http://infolab.stanford.edu/~ullman/focs.html.

Albakry, S., Vaniea, K., & Wolters, M. K. (2020). What is this URL's destination? Empirical evaluation of users' URL reading. In *Proceedings of the 2020 CHI conference on human factors in computing systems* (pp. 1–12). New York, NY: Association for Computing Machinery. https://doi.org/10.1145/3313831.3376168.

Anantharaman, P., Kothari, V., Brady, J. P., Jenkins, I. R., Ali, S., Millian, M. C., & Smith, S. W. (2020). Mismorphism: The heart of the weird machine. In *Security protocols XXVII* (pp.

113–124). Cham: Springer International Publishing. https://doi.org/10.1007/978-3-030-57043-9_11.

Andersen, P. B. (2001). What semiotics can and cannot do for HCI. *Knowledge-Based Systems, 14*(8), 419–424. https://doi.org/10.1016/S0950-7051(01)00134-4.

Arkes, H. R., & Blumer, C. (1985). The psychology of sunk cost. *Organizational Behavior and Human Decision Processes, 35*(1), 124–140. https://doi.org/10.1016/0749-5978(85)90049-4.

Atkin, A. (2013). Peirce's theory of signs. In E. N. Zalta (Ed.), *The Stanford encyclopedia of philosophy* Metaphysics Research Lab, Stanford University. https://plato.stanford.edu/archives/sum2013/entries/peirce-semiotics/ (Ed.).

Boyd, D., & Hargittai, E. (2010). Facebook privacy settings: Who cares? *First Monday, 15*(8). https://doi.org/10.5210/fm.v15i8.3086.

Bratus, S., Locasto, M., Patterson, M., Sassaman, L., & Shubina, A. (2011). Exploit programming: From buffer overflows to "weird machines" and theory of computation. *USENIX; login:, 36*(6), 13–21.

Bratus, S., Patterson, M. L., & Hirsch, D. (2013). From "shotgun parsers" to more secure stacks. *Shmoocon*. https://www.cs.dartmouth.edu/~sergey/langsec/ShotgunParsersShmoo.pdf.

Carvalho, M., DeMott, J., Ford, R., & Wheeler, D. A. (2014). Heartbleed 101. *IEEE Security & Privacy, 12*(4), 63–67. https://doi.org/10.1109/MSP.2014.66.

Chandler, D. (1994). *Signs*. Semiotics for beginners. http://visual-memory.co.uk/daniel/Documents/S4B/sem02.html.

Cyphers, B., Miagkov, A., & Arrieta, A. (2018). *Privacy badger now fights more sneaky Google tracking*. https://www.eff.org/deeplinks/2018/10/privacy-badger-now-fights-more-sneaky-google-tracking.

de Souza, C. S., Barbosa, S. D. J., & Prates, R. O. (2001). A semiotic engineering approach to user interface design. *Knowledge-Based Systems, 14*(8), 461–465. https://doi.org/10.1016/S0950-7051(01)00136-8.

Draper, N. A. (2017). From privacy pragmatist to privacy resigned: Challenging narratives of rational choice in digital privacy debates. *Policy & Internet, 9*(2), 232–251. https://doi.org/10.1002/poi3.142.

Enrici, I., Ancilli, M., & Lioy, A. (2010). A psychological approach to information technology security. In *3rd international conference on human system interaction* (pp. 459–466). https://doi.org/10.1109/HSI.2010.5514528.

Ferreira, J., Barr, P., & Noble, J. (2005). The semiotics of user interface redesign. In *Proceedings of the sixth Australasian conference on user interface: Vol. 40* (pp. 47–53).

Fitting, M. (1994). Kleene's three valued logics and their children. *Fundamental Information, 20*(1–3), 113–131. http://dl.acm.org/citation.cfm?id=183529.183533.

Gaw, S., & Felten, E. W. (2006). Password management strategies for online accounts. In *Proceedings of the second symposium on usable privacy and security* (pp. 44–55). https://doi.org/10.1145/1143120.1143127.

Goodstein, R. (1954). Introduction to Metamathematics. By S. C. Kleene. Pp. x, 550, Fl. 32.50. 1952. (Noordhoff, Groningen; North-Holland Publishing Co., Amsterdam). *The Mathematical Gazette, 38*(323), 72–76. https://doi.org/10.2307/3609805.

Kokolakis, S. (2017). Privacy attitudes and privacy behaviour: A review of current research on the privacy paradox phenomenon. *Computers & Security, 64*, 122–134. https://doi.org/10.1016/j.cose.2015.07.002.

Kothari, V. (2020). Bridging the gap between intent and outcome: Knowledge, tools & principles for security-minded decision-making. Dartmouth College, PhD Dissertation. *Dartmouth Digital Commons, 60*. https://digitalcommons.dartmouth.edu/dissertations/60.

Kothari, V., Blythe, J., Smith, S., & Koppel, R. (2018). Data privacy and the elusive goal of empowering the user. In *Workshop on moving beyond a 'one-size-fits-all' approach: Exploring individual differences in privacy, CHI*.

Lafrance, Y. (2004). Psychology: A precious security tool. *SANS GSEC Certification, Practical Assignment.* https://www.sans.org/reading-room/whitepapers/engineering/psychology-a-precious-security-tool-1409.

Lewis, K., Kaufman, J., & Christakis, N. (2008). The taste for privacy: An analysis of college student privacy settings in an online social network. *Journal of Computer-Mediated Communication, 14*(1), 79–100. https://doi.org/10.1111/j.1083-6101.2008.01432.x.

Malkin, N., Deatrick, J., Tong, A., Wijesekera, P., Egelman, S., & Wagner, D. (2019). Privacy attitudes of smart speaker users. *Proceedings on Privacy Enhancing Technologies, 2019*(4), 250–271. https://doi.org/10.2478/popets-2019-0068.

McDonald, A. M., & Cranor, L. F. (2008). The cost of reading privacy policies. *ISJLP, 4*, 543.

Naeini, P. E., Bhagavatula, S., Habib, H., Degeling, M., Bauer, L., Cranor, L. F., & Sadeh, N. (2017). Privacy expectations and preferences in an IoT world. In *Thirteenth symposium on usable privacy and security (SOUPS 2017)* (pp. 399–412). Santa Clara, CA: USENIX Association.

Nissenbaum, H. (2009). *Privacy in context: Technology, policy, and the integrity of social life.* Stanford University Press. https://doi.org/10.5860/choice.47-6940.

Obar, J. A., & Oeldorf-Hirsch, A. (2016). The biggest lie on the Internet: Ignoring the privacy policies and terms of service policies of social networking services (June 1, 2018). *Information, Communication & Society* (pp. 1–20). 2018. TPRC 44: The 44th Research Conference on Communication, Information and Internet Policy. https://doi.org/10.2139/ssrn.2757465.

Ogden, C. K., & Richards, I. A. (1923). *The meaning of meaning: A study of the influence of language upon thought and of the science of symbolism.* London, K. Paul, Trench, Trubner & Co.; New York, Harcourt, Brace & Co. https://doi.org/10.1038/111566b0.

Olmstead, K., & Smith, A. (2017). *Americans and cybersecurity.* Pew Research Center. https://www.pewresearch.org/internet/2017/01/26/americans-and-cybersecurity/.

Salides, Oriol. (2021). DRY: Don't repeat yourself. https://apiumhub.com/tech-blog-barcelona/dry-dont-repeat-yourself/. Accessed 31 March 2021.

Smith, S. W. (2012). Security and cognitive bias: Exploring the role of the mind. *IEEE Security & Privacy, 10*(5), 75–78. https://doi.org/10.1109/MSP.2012.126.

Smith, S. W., Koppel, R., Blythe, J., & Kothari, V. (2015). Mismorphism: A semiotic model of computer security circumvention (extended version). *Computer Science Technical Report TR2015-768.* https://digitalcommons.dartmouth.edu/cs_tr/368.

Sinclair, S. (2013). Access control in and for the real world. Dartmouth College, PhD Dissertation. *Dartmouth Digital Commons, 43.* https://digitalcommons.dartmouth.edu/dissertations/43.

Turow, J., Hennessy, M., & Draper, N. A. (2015). The tradeoff fallacy: How marketers are misrepresenting American consumers and opening them up to exploitation. *Available at SSRN 2820060.* https://doi.org/10.2139/ssrn.2820060.

Urban, J., & Hoofnagle, C. (2014). The privacy pragmatic as privacy vulnerable. In *Symposium on usable privacy and security (SOUPS 2014).* Berkeley, CA: USENIX Association. https://papers.ssrn.com/sol3/papers.cfm?abstract_id=2514381.

Weir, G. R. S. (1992). Meaningful interaction in complex man-machine systems. *Reliability Engineering & System Safety, 38*(1–2), 151–156. https://doi.org/10.1016/0951-8320(92)90116-3.

Wikipedia, Contributors. (2021a). *API—Wikipedia, the free encyclopedia.* https://en.wikipedia.org/w/index.php?title=API&oldid=1014458682. Accessed 18 April 2021.

Wikipedia, Contributors. (2021b). *Illusion of validity—Wikipedia, the free encyclopedia.* https://en.wikipedia.org/w/index.php?title=Illusion_of_validity&oldid=999566863. Accessed 18 April 2021.

Wikipedia, Contributors. (2021c). *Optimism bias—Wikipedia, the free encyclopedia.* https://en.wikipedia.org/w/index.php?title=Optimism_bias&oldid=1011001205. Accessed 18 April 2021.

Wikipedia, Contributors. (2021d). *Overconfidence effect—Wikipedia, the free encyclopedia.* https://en.wikipedia.org/w/index.php?title=Overconfidence_effect&oldid=1011068721. Accessed 18 April 2021.

Wikipedia, Contributors. (2021e). *Semiotics—Wikipedia, the free encyclopedia.* https://en.wikipedia.org/w/index.php?title=Semiotics&oldid=1017297510. Accessed 18 April 2021.

Wikipedia, Contributors. (2021f). *Time-of-check to time-of-use—Wikipedia, the free encyclopedia.* https://en.wikipedia.org/w/index.php?title=Time-of-check_to_time-of-use&oldid=1014091278. Accessed 18 April 2021.

Wikipedia, Contributors. (2021g). *Triangle of reference—Wikipedia, the free encyclopedia.* https://en.wikipedia.org/w/index.php?title=Triangle_of_reference&oldid=1017387319. Accessed 18 April 2021.

PART II

Behavioral studies of cybersecurity

CHAPTER

6

Are you anonymous? Social-psychological processes of hacking groups

John McAlaney

Department of Psychology, Bournemouth University, Poole, United Kingdom

1 Introduction

Cybersecurity is a phenomenon that includes both social and technological elements. The decision to conduct a cyber-attack is made by people, and the course and development of the attack are influenced by the beliefs and actions of those people. As such it is important to explore the human factors that shape hacking behavior. By doing so a better understanding of the motivations of hackers can be reached and evidence-based prevention techniques can be developed. However, reaching this understanding is not a simple undertaking. As Kirwan and Power (2013) note the motivations for hacking are very broad. These varied motivations include profit, ideology, prestige, recreation, and revenge; and is reflected in different types of hackers such as criminals, cyber warriors, hacktivists, insiders, and coders (Seebruck, 2015). While these motivations and typologies are diverse there is an underlying commonality in that there is a social element to many hacking activities. Through the application of social-psychological research, it may be possible to gain some insight into the human aspects of hacking.

Social psychology refers to the study of how the thoughts, feelings, and behaviors of individuals are influenced by the actual, imagined, or implied presence of others (Allport, 1954). Within cyberspace the concepts of 'imagined or implied' take on new meaning. This is particularly relevant to hacker communities, where many interactions between members take place online. As noted by Suler (2004) the internet may both enhance

Cybersecurity and Cognitive Science
https://doi.org/10.1016/B978-0-323-90570-1.00003-6

and transform; this means that online technologies may either amplify individual behaviors or may result in new behaviors emerging that would not happen in an offline setting. This in part relates to the concept of online disinhibition, which refers to the willingness people have to say and do things online that they would not say and do in person (Suler, 2004). Within the context of hacking and cybersecurity, there is of course often an added motivation for some individuals to retain their anonymity, given that the actions being undertaken may be illegal.

The association between hacking and criminality is one of the reasons why hackers have become the focus of research by academics and investigation by law enforcement and policymakers. This is understandable, given the increasingly social and financial impact that cyberattacks cause. However, as will be addressed throughout this chapter it is important to recognize that hacking is not necessarily a criminal behavior and that many people involved in hacking have no desire to break the law or cause harm. To assume otherwise due to the stereotypes and negative connotations that surround hacking would be to bias the evaluation of the research literature that follows.

In this chapter we identify and discuss the social-psychological processes that are relevant to individuals who participate in hacking communities, based on an evaluation of the research evidence, illustrated using real-world examples. In doing so we identify how a better understanding of these hacking groups can be used to promote informed decision making by the individuals who are part of hacking communities.

2 Context

To understand the social-psychological processes evident within a group it is necessary to consider the wider context in which that group exists. This includes understanding the structure of the group, the nature of the key players involved, and the means and frequency through which group members communicate with each other. As will be discussed it is also important to have an appreciation of where the group sits in relation to similar groups, and how, together, these groups interact with each other to create a larger community of shared interest. Reaching this level of understanding can be a challenge with communities that involve mainstream, publicly documented activities; with hacking communities, there are additional barriers. For example, forums on which hacking and cybercrime discussions take place can be on public, semidark and dark web forums (Zamani, Rabbani, Horicsányi, Zafeiris, & Vicsek, 2019). Due to increased monitoring of platforms, there has been a trend of criminal hacking forums moving toward the dark web (Pete, Hughes, Chua, & Bada, 2020). As such it is difficult to determine the size of the web forum ecosystem that exists. Even if a relevant forum is identified it may not be feasible for researchers to access this forum. In addition, these forums are fluid spaces that continually change and evolve. This is perhaps one of the reasons why, as Haslebacher, Onaolapo, and Stringhini (2017) note, much of the research on underground forums focus on closed rather than active forums. Further, being able to identify and access a forum is no guarantee for a researcher that they will be easily accessible and relevant data to analyze. As Pete et al. (2020) observe forum discussions often go off-topic, and vary in length and complexity.

The existence of these forums matters. The pathway through which an individual becomes a hacker remains an underresearched topic, but the research evidence that does exist suggests that participation in hacking forums appears to be the first step for later cybercriminal activity for some individuals (Hutchings, 2016). This is supported by the fact that the majority of attacks are conducted by people with a low level of technological sophistication (Noroozian et al., 2016; Thomas, Clayton, & Beresford, 2017), including acts of DDoSing by children who use crime as a service (Noroozian et al., 2016). Hacking forums provide a venue in which individuals without high levels of technological skills can gain further knowledge and advice (McAlaney, Hambidge, Kimpton, & Thackray, 2020). This may lead these individuals to engage in criminal acts when they did not otherwise do so. Indeed it has been found that people who are unlikely to become involved in traditional crime may become involved in cybercrime (National Crime Agency, 2017).

Interactions between people involved in hacking are not limited to online mediums. This is most apparent in the various offline events that are regularly held across the world, such as the DEFCON convention (www.defcon.org) that is held in Las Vegas each year and regularly attended by over 30,000 attendees with an interest in hacking. This occurs even when in cases where the hacking involved is unequivocally criminal in nature, with Nurse and Bada (2019) observing that within cybercriminal gangs there often appears to be a degree of offline interactions between members. However, the likelihood of offline interaction in cybercriminal groups may vary depending on the nature and goals of the group. Choo and Smith (2008) argue there are three types of cybercrime groups. Firstly, there are traditional organized criminals, who use crime as a service model. Secondly, there are organized cybercriminal groups, who are distinguished from the previous type by members having shared goals and communicating with each other but without necessarily revealing their identity. Finally, there are ideologically and politically motivated groups, who engage in hacktivism and who may commit crimes, but who are not necessarily criminally motivated. Examples of the latter would include the hacktivist collective Anonymous, who began interacting through online conversations and took steps to hide their identities, but some of whom then met in person at public marches, albeit in some instances concealing their faces by wearing the now-iconic Guy Fawkes mask (Knappenberger (Writer), 2012).

As with any community, the social-psychological processes that could potentially influence individuals are myriad and multifaceted. In the remainder of the chapter, we explore several processes which would appear to be the most influential in shaping hacker groups and behavior, as determined by the research evidence.

3 Social identity

Social identity refers to our self-concept, as shaped by our relationship with others. When discussing social identity in relation to hackers it is important to note the challenges around nomenclature and labeling of individuals who engage in hacking. Throughout this chapter, the phrase 'hacker' is used as an umbrella term to refer to any individual who engages in what could be considered hacking behaviors, but this is done with the full acknowledgment that some members of this community object to this term as being overly simplistic and overly general. As noted by Kirwan and Power (2013) there is no such thing as a single hacker culture, and the language that is used to describe hacker culture is often contentious. In our own research, we have explored issues of definition and labeling with people who engage in online hacking forums and found that there is a diverse range of terms that are used within these communities to denote different individual characteristics, behaviors, and ethical stances (Thackray, Richardson, Dogan, Taylor, & McAlaney, 2017). The labels preferred by respondents who participated in this research included crackers, script kiddies, hacktivists, and cyberpunks, along with the concepts of white hat, gray hat, and black hat hackers. Nevertheless, while participants in our previous research expressed dissatisfaction with the blanket use of the term hacker they also agreed that there was a lack of suitable alternatives that would fully encompass all of the nuances of their online community (Thackray, 2017). This research highlights how the already manifold concepts of social identity are further complicated within hacking communities.

While labels may be contentious, they are important determinants of how people behave and perceive themselves (Bryant & Marshall, 2008). Social identity theory (Tajfel & Turner, 1986) predicts that individual behavior within a group is influenced by perceived differences in groups status; the perceived legitimacy of those differences in status; and the perceived ability to able to move from one group to another. On a broad level, these principles could be argued to be applicable to hacking communities. It would be difficult for example to attend the DEFCON hacking event and not experience a sense of attendees feeling that they are part of a distinct community with its own culture. In addition, the online environment enables individuals to present themselves as different people, with Troia (2020) observing that the use of multiple online identities appears to be especially characteristic of hackers. This relates to the social-psychological concept of impression management, also known as self-presentation (Schlenker, 1975), in which individuals attempt to control how they are viewed by those around them. Impression management could be argued to be a lesser, more everyday form of some social engineering strategies that attempt to manipulate individuals through creating a false identity. Within hacking communities creating a self-presentation of being knowledgeable about hacking appears to be markedly important to many community members (McAlaney et al., 2020), although the culture of these communities is such that any assertations made by community members are often challenged and debated (Coleman, 2014). Indeed, it has been observed that using social engineering techniques based upon appealing to the target's vanity can be a useful means for identifying and tracking hackers (Troia, 2020).

Online groups can develop their own language and norms that are used to signal membership to that group (Dobusch & Schoeneborn, 2015), with hacking groups noted to have a tendency toward communicating through their own jargon, symbols, and memes. The rhetoric used within some hacking communities could suggest that certain hackers define themselves entirely in terms of having no empathy toward those they target (Olson, 2012). Caution must be taken though in accepting this stance at face value. Rogers (2011) argues that those who do engage in cybercrime often demonstrate characteristics that suggest that are experiencing cognitive dissonance over attempting to reconcile their actions with their self-identity. These characteristics include using euphemistic language that minimizes and distances their actions; blaming their actions on social pressures and claiming that, in the case of group actions, their individual contribution was minimal; minimizing or disregarding the negative consequences of their action, and finally vilifying and dehumanizing their targets and stating that they deserved to be attacked. These trends would imply that individuals involved in malicious hacks do experience some degree of guilt over their actions. It has been further noted that the apparently aggressive communication style used by many in online hacking forums—a phenomenon that has been cited in somewhat alarmist media reports around hacking—is not a true representation of the social relationships within the communities (Coleman, 2014). One notable example of this type of extreme view of hackers and hacking groups is the infamous Fox News report broadcast of the 26th of July 2007, which described the hacktivist group Anonymous as 'domestic terrorists'(Fox News (Writer), 2017). This report depicted the group as being more cohesive and a greater threat than the group themselves believed themselves to be (Coleman, 2014). In keeping with the category differentiation model (Doise, 1978), identifying groups as being a group can strengthen the sense of group membership. In other words, providing external validation of a group as an emerging threat can become a self-fulfilling prophecy in which the group becomes more active because of that external validation. This again highlights that there needs to be a distinction between the public perception of hacking groups and the actions and beliefs of that group.

4 Interpersonal perception

Interpersonal perception is the way in which we view other people, and how we come to conclusions about their behaviors and attitudes. A key factor in understanding how an individual interacts with others is the concept of attribution. This refers to the explanations that we give for other people's behavior. Early work by Jones and Davis (1965) focused on correspondent inference—how we infer what traits other people have by observing their behavior, including whether these traits are stable. Jones and Davis argue that we focus on behaviors based on three factors. Firstly, we consider whether the observed behavior appears to arise from free choice. For example, someone who acts in an authoritarian manner in an online hacking forum may be assumed to be someone who has an autocratic personality in their online and offline lives. However, if that person is a moderator on the forum then at least part of their behavior may be attributed to the requirements of their role as a moderator. Secondly, we focus on behaviors that are noncommon—that is behaviors that only have one specific explanation (Jones & Davis, 1965). In the context of hacking, this could mean that a hack that requires a high level of technical knowledge and skill is used as an indicator that the hacker is talented. In contrast, if someone conducts a basic attack they may be assumed to be lacking in skills, even if the basic attack that has been used is the most suitable strategy to achieve the desired goal. Finally, when attributing explanations to the behavior of others we focus on the social desirability of the behavior, with behaviors that are low in social desirability being taken as a stronger sign of individual traits. This is because acting against a social norm of your group is something that people do not typically find easy, meaning that if someone publicly breaches a social norm it will be assumed by the group that this reflects a strong personal trait of the individual concerned (Jones & Davis, 1965). This phenomenon can be recognized in many of the interactions that take place on online hacking forums, where any breach of the established norms of the forum is typically met with a negative reaction by the other users of that forum (McAlaney et al., 2020).

Given the wide variety of potential motivations for hacking this concept of social desirability is quite a difficult one to apply within hacking communities. What is for example considered a socially desirable behavior in a hacktivist group may be quite different from the desired behavior within a cybercriminal gang.

Gilbert and Malone (1995) further argue that we use a two-step process to attribute traits to other people based on their observed behavior, in which we base our initial judgment on the behavior of the person (an internal cause) and then look for evidence of external causes that should be considered. However, research also suggests that we tend to underestimate these external factors (Chaiken & Trope, 1999). For instance, while we may acknowledge that the behavior of someone in the hacking community is in part dictated by their circumstances, we do not always fully appreciate how much of an influence these circumstances have. In addition, we also tend to assume that we are more understanding about external factors than other people (Van Boven, White, Kamada, & Gilovich, 2003). These processes collectively lead to corresponding bias, also known as fundamental attribution error (Gilbert & Malone, 1995). In other words, we tend to assume that people are behaving a certain way because they are 'that kind of person' and tend not to think about the external factors that may be the actual reason why they behave in that way. Within the context of hacking communities, this may again result in the actions of others being misconstrued. This could be a contributing factor to incidences of intragroup conflict, as will be discussed later in this chapter.

Several other biases exist that can warp our perception of other people. This includes the actor-observer effect (Jones & Nisbett, 1972), in which we are more likely to attribute our own behavior to external factors, especially when the outcome of that behavior is a negative one. For instance, a hacker who is unable to access a system may attribute this to the system being particularly robust, while also assuming any other hacker who experiences similar failure to just lack ability. Related to this is self-serving bias (Olson & Ross, 1988), in which we are more likely to attribute positive outcomes to internal factors and negative outcomes to external factors. In relation to hacking for example we may attribute our own success to our skill and knowledge, but the success of another hacker who achieved the same result as being down to luck. This bias may also influence how people interpret the risk of taking part in hacking. For example, someone who has been caught and prosecuted for a hacking offense may erroneously attribute this to the law enforcement agency who caught them as just having been lucky. This could result in them continuing to commit cybercrime despite having already been caught doing so. This is supported by the assertion that hackers often appear to overestimate their own abilities (Reynolds, 2016).

5 Group processes

Our behavior is strongly influenced by those around us. As based on the Theory of Planned Behavior (Ajzen, 1985) it has been observed that the intention someone has to perform the desired behavior is in part determined by whether they think influential others will support or condemn their actions. This phenomenon has been documented in relation to cybersecurity behaviors (Venkatesh, Morris, Davis, & Davis, 2003). This idea of hackers being part of a group conflicts with the stereotype of hackers being socially inept loners. Chiesa, Ducci, and Ciappi (2008) observe that while hackers may begin by working independently, they typically over time join collaborative groups to share knowledge and to develop social relationships. As discussed previously hacking forums often appear to be the starting point for these interactions, and continue to serve as an important source of socialization and social identity for hackers across all life stages (McAlaney et al., 2020). It is important not to underestimate how extensive these networks can be. Pete et al. (2020) for example conducted social network analysis on six dark web forums and demonstrated that a rich and varied ecosystem of social interactions existed within these forums despite their clandestine nature.

The structure of a group impacts how that group functions. There is an association between the ability to exert social control within a group and the network density of that group (Frank & Yasumoto, 1998). In addition, information spreads quicker through highly dense networks than through less dense ones (Shaw-Ching Liu, Madhavan, & Sudharshan, 2005). These factors may determine how much individuals within hacking groups are able to shape the behavior of others within that group. Individuals with a significant interest in hacking participate in multiple online communities in order to gather more information and increase the density of their social networks (Holt, 2009). These process can result in cases of deliberate manipulation, such as the allegations that more established hackers misled less technically savvy hackers into believing that they would not be able to be identified through DDoS software used in the attack against the Church of Scientology (Olson, 2012).

Groups vary both internally and externally, which may be reflected in the interaction and communication styles that can be observed on the mediums used by those groups. For example, Pete et al. (2020) found that some hacking forums were welcoming to new users and encouraged the flow of information within the group, whereas others were found to be more tight-knit with specialist interests. A common occurrence across all forums however was the existence of key players, i.e., individuals who are highly influential within that forum. Pastrana, Hutchings, Caines, and Buttery (2018) found that key actors in forums tend to be connected to each other, with clusters of other users forming around key actors based upon (Noroozian et al., 2016) specialist interests. This is supported by Sood and Enbody (2013), who found that people on underground markets tend to specialize. There have been high-profile individuals in carding forums (Rhysider, 2021). This phenomenon has not gone unnoticed by law enforcement agencies, with the removal of key players being used as a strategy to disrupt online marketplaces (Hutchings & Holt, 2017). This technique may not always be effective, however. Dark networks are often decentralized, and it has been argued that removing key players from these networks may make them even more decentralized and harder to disrupt (Duijn, Kashirin, & Sloot, 2014).

Group processes are complex, and as noted relates to both the intragroup processes that take place between group members and intergroup processes that take place between groups. In the following section, we discuss the social-psychological factors that are relevant to the intra and intergroup processes that may shape behavior in hacking communities.

5.1 Intragroup processes

Intragroup processes refer to how the interactions between members of the same group influence individual behavior. This can occur in several ways. People for instance will tend to alter their behavior and cognitions to match the group (Kelman, 2006). This idea that your behavior and beliefs can be influenced by those around you is not something that people find easy to accept, particularly in countries with individualistic cultures such as the UK and USA (Hofstede, Hofstede, & Minkov, 2010). It contrasts the stereotype of hackers as being anarchistic and antiestablishment. Similarly, an emotion can spread through a group and alter the behavior and attitudes of the members of that group, even with members who had no involvement in whatever situation triggered the initial emotional reaction (Smith, Seger, & Mackie, 2007). This process could in part explain some acts of hacktivism, where an organization is targeted on the basis of the perceived amoral actions of that organization, even if the members of the hacktivism group are not personally affected by that organization. This could account for what has been termed Robin Hood hackers, where money gathered through ransomware attacks of large companies is donated to charities by the attackers (BBC News, 2020).

People are more likely to conform to the viewpoints of others when they are in novel or uncertain situations. This may be especially relevant to hacking forums, where many of the interactions on the forum involve people who are seeking advice (McAlaney et al., 2020). Individuals are also known to make riskier decisions when in groups than when alone (Wallach, Kogan, & Bem, 1962). In addition, groups exhibit groupthink, in which groups make more extreme decisions than the individuals would have otherwise made (Janis, 1982). This is a result of the desire to achieve consensus in the group and is enhanced by a high level of group cohesiveness and a threatening situational context. Groupthink can lead to increased conformity, an overestimation of the confidence of the group, and close-mindedness (Janis, 1982). Together these processes may lead to decisions and actions that are highly risky. For instance, it has been stated that Space Shuttle *Challenger* disaster was the result of flawed decision-making processes indicative of groupthink (Crisp & Turner, 2017). It is interesting to consider how this phenomenon may apply to online hacking communities. Given that some activities of these groups may involve risky behavior it is feasible that members of the group would become strongly linked to one another, and so group cohesiveness would be high. This would appear to be the case in some of the more well-documented case studies of hacking groups, such as the hacktivist collective Anonymous (Coleman, 2014).

Trust relationships are important within groups. Cybercriminal networks have important social dependency relationships (Leukfeldt, Kleemans, & Stol, 2017). Décary-Hétu and Leppänen (2013) note that reputation is a key aspect of the underground economy. Somewhat counterintuitively anonymity can increase the sense of membership in online groups (Tanis & Postmes, 2005). Not surprisingly however the manifestation of and development of trust can be different with hacking groups as opposed to other types of groups. Motoyama, McCoy, Levchenko, Savage, and Voelker (2011) for instance found that new members to some underground forums must be vouched for by existing members. Further, they found that members are sometimes banned on these forums, for a range of reasons including incidents where users have attempted to scam other members. In the case of accuser and accused, the accused was more likely to be banned when the accuser was more active on the forum. This reflects previously discussed processes that relate to group dynamics and social identity. As stressed throughout this chapter it is not the case that hacking is inherently an illegal or immoral behavior, but based on our research it could be argued that there is often uncertainty on hacking forums on what exactly is or is not an illegal activity (McAlaney et al., 2020). As such it may be that people who are involved in online forums perceive themselves to permanently be in a situation where there is an external threat. Indeed, the revelation that a senior member of the hacktivist collective Anonymous was an informant for the FBI appeared to do extensive damage to the group (Coleman, 2014). It has been argued that one way to disrupt cybercriminal forums is to spread false information that reduces trust between members (Hutchings & Holt, 2017).

It is not always the case that individuals conform to the group. In cases of minority influence, an individual or a small number of people within a larger group can drive the attitudes and behaviors of the majority. For this to happen the minority must be consistent and confident in their views (Maass & Clark, 1984). This process relates to the concept of attribution, as discussed previously. Going against the group norm is not something that comes easily to individuals, and so if someone is willing to do so then it suggests to the rest of the group that this person must have some strong reason for doing so. Often this will be an assumption that the individual concerned has a better understanding of the issues or a greater skillset. This is not necessarily a negative process. Minority influence can lead to divergent thinking, in which decisions are more deeply considered (Nemeth, 1986). Nevertheless, differences in opinion between groups member can often lead to intragroup conflict. For example, the motivation between targeting an organization to address a perceived social injustice and doing it for personal enjoyment became a reason for in-group conflict in Anonymous (Coleman, 2014). These conflicts can potentially be amplified by external inputs and manipulation, with Troia (2020) observing that 'hacker drama' can often be leveraged to cause in-group fighting.

5.2 Intergroup processes

The actions of individuals within a community are shaped not just by the interactions within the group, but also by the interactions with those outside the group. A common interaction type between groups in conflict. It has been demonstrated throughout social psychology research how easily and quickly conflict of groups can form. This is epitomized by the infamous Robbers Cave study, in which a group of 11- to 12-year-old boys were taken to a summer camp and split into two groups, who were encouraged to compete with each other. It was found that negative attitudes and behaviors rapidly developed between the groups (Sherif, White, & Harvey, 1955). An example of intergroup conflict between a hacking group and an external organization would be that between Anonymous and HBGary (Olson, 2012). This occurred after an individual at HBGary, a technology security company, publicly announced that they had identified the real-world names of members of the Anonymous collective. In response to this external threat members of Anonymous are alleged to have compromised the HBGary website, posting tens of thousands of company emails to the internet. They also gained control of the employee's Twitter account, using it to post unflattering content. These instances of intergroup conflict between groups within the cybersecurity space are not unusual and include situations where one hacking group has been hacked by another hacking group (Franceschi-Bicchierai, 2019).

As with individuals, groups can engage in impression management, which may be used as a weapon within intergroup conflicts. For example, in their message to Scientology video Anonymous used various techniques to present themselves as a credible and serious threat to the Church of Scientology (Knappenberger (Writer), 2012). This included presenting themselves as being a large and efficient organization. The video also made use of some persuasive tactics such as framing the followers of Scientology as needing to be freed from the influence of the organization. In doing so Anonymous presented themselves as being merciful and compassionate, which would likely increase public support for their actions. They also framed their victory over Scientology as being inevitable. In keeping with Protection Motivation Theory (Rogers, 1975) individuals can be influenced to not protect themselves from threats if they can be persuaded that it is futile to even attempt to do so. As such the goal on the members of Anonymous may have been to dissuade the Church of Scientology from even attempting to protect themselves, although given the tendency of trolling within Anonymous it could be suggested that the group members did not seriously expect that they would be able to frighten their target into inaction. As members of the group later admitted the threats they made were beyond what they were actually capable of (Olson, 2012).

The potential for conflict between groups can influence behavior even when members of the hacking community have no desire to create such conflict. For example, it has been reported that ethical hackers who have attempted to expose—but not exploit—flaws in security systems to companies are ignored. One alleged instance of this is the case of tablet devices produced for children by the company VTech (Rhysider, 2017). It was discovered by an ethical hacker that there was a gap in the VTech systems that allowed the names and addresses of children who were using the tablet to be downloaded. The ethical hacker did not however feel comfortable in bringing this matter directly to VTech. Instead, he took the issue to a journalist, who attempted to notify VTech directly but who allegedly faced resistance from the company (Rhysider, 2017). This distrust that many hackers appear to have for large organizations may be a source of intergroup conflict. These conflicts may be reinforced by wider beliefs around social hierarchies and used as a justification for cyberattacks. A common assertion among hacktivist groups for example is that the use of technology in this way addresses the power imbalance between the public and large organizations (Solomon, 2017). It is possible that greater social and legal recognition of ethical hacking may change this dynamic in the future. The possibility of licensing hackers has for example been discussed in Singapore (Wong, 2017), an action that would likely add social validation to hackers and alter public opinion.

6 Informed decision making

Throughout this chapter, we have explored how social factors may influence hacking behaviors. Much of the research also has the potential to attempt to change hacking behavior. Given the serious societal and financial challenges caused by cybercrime, it is understandable that there is interest in employing social psychology research to change hacker behavior. However, this is something that should be considered with caution. Individuals are often resistant to efforts to coerce them into behaving in a certain way, a tendency known as the need to assert uniqueness (Imhoff & Erb, 2009). From our own research, we would argue that people who engage in hacking are especially likely to react negatively to attempts to control their behavior than a typical member of the public. This does naturally depend on the context of the group and the activities that they undertake. In the case of cybercriminal gangs with clear malicious intent then social psychologically based prevention strategies, such as the aforementioned disruption of reputational systems in underground forums, may be acceptable. In hacking groups where there is less criminal activity or intent then such attempts to change behavior may be counterproductive. As discussed within this chapter the role of being a hacker appears to be an important source of social identity for many members of hacking communities. Further, we have found that the rhetoric in several hacking forums emphasizes that members should act legally and ethically. A blanket attempt to prevent people with an interest in hacking from communicating with each other risks removing the benefits that individuals gain from participation in hacking forums. It would also underestimate the ingenuity of hackers in finding ways to circumvent systems.

Instead, we argue that the most productive approach is to empower individuals to make informed decisions about their participation in hacking activities. As reiterated throughout this chapter, hacking is not in itself a criminal activity. Curiosity appears to be a far greater motivation for initial engagement in hacking than any desire to commit crime (McAlaney et al., 2020; Thackray et al., 2017). Helping those with an interest in hacking to better understand the legality and ethical issues around hacking may steer such individuals toward legitimate cybersecurity careers, where their skills and passion can be used for social good. This would help address the global shortage of cybersecurity professionals (Morgan, 2021), by bringing more individuals into the profession and preventing the criminalization of those who may unintentionally commit cybercrimes. This is reflected in moves by law enforcement agencies, such as the National Crime Agency in the UK who seek to prevent young adults from becoming involved in criminal hacking acts (BBC News, 2016).

7 Conclusion

When considering the social-psychological aspects of hacking groups it is important to first acknowledge wider perceptions of people who engage in hacking. There are many stereotypes of hackers, fuelled in part by media portrayals in works of fictional and news reporting of cybersecurity incidents. Recurring themes in these stereotypes are that hackers and singular, socially isolated, antiestablishment, loners who either chose not to interact with mainstream society or lack the social skills to be able to do so. As with all stereotypes however the reality of the situation is far more complex and nuanced. Humans are fundamentally social creatures; even groups who define themselves as being a part of a subculture still engage in social interactions both within and outside their group. There are very few of us who would be capable of surviving without any form of social interaction. Yet social-psychological processes can lead us to make risky decisions that we would not otherwise make. In the case of hacking, this can lead to unintended and negative consequences both to the individual and those around them. By better understanding these processes we can identify ways to optimize the benefits of hackers' culture, while challenging the harmful aspects.

References

Ajzen, I. (1985). From intentions to actions: A theory of planned behavior. In J. Kuhl, & J. Beckmann (Eds.), *Action control: From cognition to behavior* (pp. 11–39). Berlin, Heidelberg: Springer Berlin Heidelberg.

Allport, G. (1954). The historical background of modern social psychology. In G. Lindzey, & J. Aronson (Eds.), *Handbook of social psychology*. New York: Random House.

BBC News. (2016). *Hanging out with the script kiddies*. Retrieved from https://www.bbc.co.uk/news/technology-32019260.

BBC News. (2020). *Mysterious 'Robin Hood' hackers donating stolen money*. Retrieved from https://www.bbc.co.uk/news/technology-54591761.

Bryant, R., & Marshall, A. (2008). Criminological and motivational perspectives. In R. Bryant (Ed.), *Investigating digital crime* (pp. 1–26). Chichester: John Wiley and Sons.

Chaiken, S., & Trope, Y. (1999). *Dual process theory in social psychology*. New York: Guilford Press.

Chiesa, R., Ducci, S., & Ciappi, S. (2008). *Profiling hackers: The science of criminal profiling as applied to the world of hacking*. Auerbach Publications.

Choo, K.-K. R., & Smith, R. (2008). Criminal exploitation of online systems by organised crime groups. *Asian Journal of Criminology, 3*, 37–59. https://doi.org/10.1007/s11417-007-9035-y.

Coleman, G. (2014). *Hacker, hoaxer, whistleblower, spy : The many faces of anonymous*. London: Verso.

Crisp, R. J., & Turner, R. N. (2017). *Essential social psychology (4th ed.)*. Thousand Oaks: SAGE Publications India Pvt Ltd.

Décary-Hétu, D., & Leppänen, A. (2013). Criminals and signals: An assessment of criminal performance in the carding underworld. *Security Journal, 29*(3), 442–460. https://doi.org/10.1057/sj.2013.39.

Dobusch, L., & Schoeneborn, D. (2015). Fluidity, identity, and organizationality: The communicative constitution of anonymous. *Journal of Management Studies, 52*(8), 1005–1035. https://doi.org/10.1111/joms.12139.

Doise, W. (1978). *Groups and individuals: Explanations in social psychology*. Cambridge: Cambridge University Press.

Duijn, P. A. C., Kashirin, V., & Sloot, P. M. A. (2014). The relative ineffectiveness of criminal network disruption. *Scientific Reports, 4*(1), 4238. https://doi.org/10.1038/srep04238.

Fox News (Writer). (2017). *Fox 11 Investigates: Anonymous: KTTV (Fox)*.

Franceschi-Bicchierai, L. (2019). *Infamous forum for instagram hackers gets hacked by other hackers*. Motherboard. 17/5/2019.

Frank, K. A., & Yasumoto, J. Y. (1998). Linking action to social structure within a system: Social capital within and between subgroups. *American Journal of Sociology, 104*(3), 642–686. https://doi.org/10.1086/210083.

Gilbert, D. T., & Malone, P. S. (1995). The correspondence bias. *Psychological Bulletin, 117*(1), 21–38. https://doi.org/10.1037/0033-2909.117.1.21.

Haslebacher, A., Onaolapo, J., & Stringhini, G. (2017). *All your cards are belong to us: Understanding online carding forums*. New York: IEEE.

Hofstede, G. H., Hofstede, G. J., & Minkov, M. (2010). *Cultures and organizations: Software of the mind* (3rd ed.). New York: McGraw-Hill.

Holt, T. J. (2009). Lone hacks or group cracks: Examining the social organization of computer hackers. In F. Schmalleger, & M. Pittaro (Eds.), *Crimes of the internet*. Upper Saddle River, NJ: Prentice Hall.

Hutchings, A. (2016). Cybercrime trajectories: An integrated theory of initiation, maintenance, and desistance. In T. J. Holt (Ed.), *Crime online: Correlates, causes, and context* (pp. 117–140). Durham: Carolina Academic Press.

Hutchings, A., & Holt, T. J. (2017). The online stolen data market: Disruption and intervention approaches. *Global Crime, 18*(1), 11–30. https://doi.org/10.1080/17440572.2016.1197123.

Imhoff, R., & Erb, H. P. (2009). What motivates nonconformity? Uniqueness seeking blocks majority influence. *Personality and Social Psychology Bulletin, 35*(3), 309–320. https://doi.org/10.1177/0146167208328166.

Janis, I. L. (1982). *Groupthink: Psychological studies of policy decisions and fiascoes*. Houghton Mifflin.

Jones, E. E., & Davis, K. E. (1965). From acts to dispositions—the attribution process in person perception. *Advances in Experimental Social Psychology*, 2(4), 219–266.

Jones, E. E., & Nisbett, R. E. (1972). The actor and the observer: Divergent perceptions of the causes of behaviour. In E. E. Jones, D. E. Kanouse, D. D. Kelley, R. E. Nisbett, S. Valins, & B. Wiener (Eds.), *Attribution: Perceiving the causes of behaviour* General Learning Press.

Kelman, H. C. (2006). Interests, relationships, identities: Three central issues for individuals and groups in negotiating their social environment. *Annual Review of Psychology, 57*, 1–26. https://doi.org/10.1146/annurev.psych.57.102904.190156.

Kirwan, G., & Power, A. (2013). *Cybercrime: The psychology of online offenders*. New York, NY, US: Cambridge University Press.

Knappenberger (Writer), B. (2012). *We are legion: The story of the hacktivists*. FilmBuff.

Leukfeldt, E. R., Kleemans, E. R., & Stol, W. P. (2017). Origin, growth and criminal capabilities of cybercriminal networks. An international empirical analysis. *Crime, Law and Social Change, 67*(1), 39–53. https://doi.org/10.1007/s10611-016-9663-1.

Maass, A., & Clark, R. D. (1984). Hidden impact of minorities: Fifteen years of minority influence research. *Psychological Bulletin, 95*(3), 428–450. https://doi.org/10.1037/0033-2909.95.3.428.

McAlaney, J., Hambidge, S., Kimpton, E., & Thackray, H. (2020). Knowledge is power: An analysis of discussions on hacking forums. In *Paper presented at the 2020 IEEE European symposium on security and privacy workshops (EuroS&PW)*. 7-11 sept. 2020.

Morgan, S. (2021). *Cybersecurity talent crunch to create 3.5 million unfilled jobs globally by 2021.* Retrieved from https://cybersecurityventures.com/jobs/.

Motoyama, M., McCoy, D., Levchenko, K., Savage, S., & Voelker, G. M. (2011). In *An analysis of underground forums Paper presented at the Proceedings of the 2011 ACM SIGCOMM conference on Internet measurement conference, Berlin, Germany.*

National Crime Agency. (2017). *Pathways into Cyber Crime.* Retrieved from https://perma.cc/897P-GZ3R.

Nemeth, C. J. (1986). Differential contributions of majority and minority influence. *Psychological Review, 93*(1), 23–32. https://doi.org/10.1037/0033-295X.93.1.23.

Noroozian, A., Korczyński, M., Gañan, C. H., Makita, D., Yoshioka, K., & van Eeten, M. (2016). *Who gets the boot? Analyzing victimization by DDoS-as-a-service.* Springer International Publishing.

Nurse, J. R. C., & Bada, M. (2019). The group element of cybercrime: Types, dynamics, and criminal operations. In A. Attrill-Smith, C. Fulwood, M. Keep, & D. J. Kuss (Eds.), *The Oxford handbook of cyberpsychology.* Oxford: Oxford University Publishing.

Olson, J. M., & Ross, M. (1988). False feedback about placebo effectiveness—Consequences for the misattribution of speech anxiety. *Journal of Experimental Social Psychology, 24*(4), 275–291. https://doi.org/10.1016/0022-1031(88)90021-2.

Olson, P. (2012). *We are anonymous.* New York: Back Bay Books.

Pastrana, S., Hutchings, A., Caines, A., & Buttery, P. (2018). Characterizing eve: Analysing cybercrime actors in a large underground forum. In M. Bailey, T. Holz, M. Stamatogiannakis, & S. Ioannidis (Eds.), *Vol. 11050. Research in attacks, intrusions, and defenses, raid 2018* (pp. 207–227). Cham: Springer International Publishing Ag.

Pete, I., Hughes, J., Chua, Y. T., & Bada, M. (2020). A social network analysis and comparison of six dark web forums. In *Paper presented at the 2020 IEEE European symposium on security and privacy workshops (EuroS&PW).* 7-11 sept. 2020.

Reynolds, M. (2016). *Hackers are 'not as sophisticated as they think they are'.* Wired.

Rhysider, J. (2017). *Episode 2: The peculiar case of the VTech hacker.* DarkNet Diaries. Retrieved from https://darknetdiaries.com/episode/2/.

Rhysider, J. (2021). *Episode 85: Cam the carder.* DarkNet Diaries. 16/2/21. Retrieved from https://darknetdiaries.com/episode/85/.

Rogers, R. W. (1975). A protection motivation theory of fear appeals and attitude change. *Journal of Psychology, 91*(1), 93.

Rogers, M. K. (2011). The psyche of cybercriminals: A psycho-social perspective. In S. Ghosh, & E. Turrini (Eds.), *Cybercrimes: A multidisciplinary analysis* (pp. 217–235). Berlin, Heidelberg: Springer Berlin Heidelberg.

Schlenker, B. R. (1975). Self-presentation: Managing the impression of consistency when reality interferes with self-enhancement. *Journal of Personality and Social Psychology, 32*(6), 1030–1037. https://doi.org/10.1037/0022-3514.32.6.1030.

Seebruck, R. (2015). A typology of hackers: Classifying cyber malfeasance using a weighted arc circumplex model. *Digital Investigation, 14,* 36–45. https://doi.org/10.1016/j.diin.2015.07.002.

Shaw-Ching Liu, B., Madhavan, R., & Sudharshan, D. (2005). DiffuNET: The impact of network structure on diffusion of innovation. *European Journal of Innovation Management, 8*(2), 240–262. https://doi.org/10.1108/14601060510594701.

Sherif, M., White, B. J., & Harvey, O. J. (1955). Status in experimentally produced groups. *American Journal of Sociology, 60*(4), 370–379. https://doi.org/10.1086/221569.

Smith, E. R., Seger, C. R., & Mackie, D. A. (2007). Can emotions be truly group level? Evidence regarding four conceptual criteria. *Journal of Personality and Social Psychology, 93*(3), 431–446. https://doi.org/10.1037/0022-3514.93.3.431.

Solomon, R. (2017). Electronic protests: Hacktivism as a form of protest in Uganda. *Computer Law and Security Review, 33*(5), 718–728. https://doi.org/10.1016/j.clsr.2017.03.024.

Sood, A. K., & Enbody, R. J. (2013). Crimeware-as-a-service—A survey of commoditized crimeware in the underground market. *International Journal of Critical Infrastructure Protection, 6*(1), 28–38. https://doi.org/10.1016/j.ijcip.2013.01.002.

Suler, J. (2004). The online disinhibition effect. *CyberPsychology & Behaviour, 7*(3), 321–326. https://doi.org/10.1089/1094931041291295.

Tajfel, H., & Turner, J. C. (1986). The social identity theory of inter—Group behaviour. In S. Worchel, & W. G. Austin (Eds.), *Psychology of intergroup relations* (2nd ed.). Chicago: Nelson—Hall.

Tanis, M., & Postmes, T. (2005). A social identity approach to trust: Interpersonal perception, group membership and trusting behaviour. *European Journal of Social Psychology, 35*(3), 413–424. https://doi.org/10.1002/ejsp.256.

Thackray, H. (2017). Hackers gonna hack: But do they know why? In *Paper presented at the Social Engineering Village*. Las Vegas: DEFCON.

Thackray, H., Richardson, C., Dogan, H., Taylor, J., & McAlaney, J. (2017). Surveying the hackers: The challenges of data collection from a secluded community. In *Paper presented at the 16th European Conference on Cyber Warfare and Security, Dublin*.

Thomas, D. R., Clayton, R., & Beresford, A. R. (2017). 1000 Days of UDP amplification DDoS attacks. In *Paper presented at the 2017 APWG Symposium on Electronic Crime Research (eCrime)*. 25–27 April 2017.

Troia, V. (2020). *Hunting cyber criminals: A Hacker's guide to online intelligence gathering tools and techniques*. Indianapolis, Indiana: Wiley.

Van Boven, L., White, K., Kamada, A., & Gilovich, T. (2003). Intuitions about situational correction in self and others. *Journal of Personality and Social Psychology, 85*(2), 249–258. https://doi.org/10.1037/0022-3514.85.2.249.

Venkatesh, V., Morris, M. G., Davis, G. B., & Davis, F. D. (2003). User acceptance of information technology: Toward a unified view. *MIS Quarterly, 27*(3), 425–478.

Wallach, M. A., Kogan, N., & Bem, D. J. (1962). Group influence on individual risk-taking. *Journal of Abnormal Psychology, 65*(2), 75–86.

Wong, J. I. (2017). *The country with the world's best cybersecurity is planning a new law to license hackers*. Quartz. 11/7/21.

Zamani, M., Rabbani, F., Horicsányi, A., Zafeiris, A., & Vicsek, T. (2019). *Differences in structure and dynamics of networks retrieved from dark and public web forums. ArXiv, abs/1909.10259.*

CHAPTER

7

On the relation between hacking and autism or autistic traits: A systematic review of the scientific evidence

Jennifer Wagner[a], *Samuela Bolgan*[b], *and Elena Rusconi*[c,d]

[a]School of Applied Sciences, Abertay University, Dundee, United Kingdom, [b]Department of General Psychology, University of Padova, Padova, Italy, [c]Department of Psychology and Cognitive Science, University of Trento, Rovereto, Italy, [d]Centre of Security and Crime Sciences, University of Trento and University of Verona, Trento, Italy

David: "Can you wait here?"
Jennifer: "Why?"
David: "'Cause these guys can get a little nervous."

War Games (1983).

1 Introduction

In one of the very first movies on hacking, David Lightman, a budding hacker faced with an impasse when trying to access a videogame before its official release, goes to consult a pair of friends "in the know." He asks for help to interpret some obscure information he stumbled across and to find a way to play some odd games he discovered accidentally by penetrating an unknown system. Jim and Malvin, who are also system developers, explain that there might be a backdoor password that could grant him access and suggest he finds out who developed the system because

https://doi.org/10.1016/B978-0-323-90570-1.00016-4

personal information may be key to guessing the password. Altogether, they make for an interesting pair: one twitchy and reactive, the other clumsy, socially insensitive, giving factual answers at a lightning speed with an overly loud and high-pitched voice. As portrayed in the opening quote, the presence of a girl accompanying David is unusual; ignored by Jim, and felt out of place by Malvin, who comments they should not be discussing their "best tricks" in front of a girl.

Although since 1983 technology has greatly advanced and a few female hacker characters have now also found their way on the big screen (see e.g., Kate Libby in *Hackers*, 1995, and Lisbeth Salander in *The Girl with the Dragon Tattoo*, 2011), the image we normally associate with a hacker may share a few characteristics with a stereotypical person, with normal or higher intelligence, on the autistic spectrum (Draaisma, 2009). Both recent literary works (e.g., Forsyth, 2018), TV series (e.g., *The Code*, 2014–2016), and the press coverage of high-profile cases of self-taught hackers with Asperger syndrome (e.g., Ronson, 2005) may indeed help to reinforce our stereotypical representation of a hacker. Here we have conducted a systematic review of the scientific literature to assess whether the putative relation between autism, or autistic traits, and hacking is supported by scientific evidence. First of all, however, we will provide a few useful definitions for making sense of the themes addressed in the extant literature and to help interpret the evidence presented in the Results section.

1.1 Autism and autistic traits

1.1.1 Autism as a neurodevelopmental condition

A first necessary clarification concerns the following terms: *autism, Asperger's syndrome,* and *autistic traits,* all of which have been mentioned in the literature in relation to hacking. Currently, in the Diagnostic and Statistical Manual of Mental Disorders 5, *autism* (short for *autism spectrum disorder* or *ASD*) is used to refer to a group of neurodevelopmental conditions, all characterized by persistent deficits in social communication and social interaction across multiple contexts on the one hand, and by restricted or repetitive patterns of behavior, interests, or activities on the other (DSM-5, American Psychiatric Association, 2013). However, autism is characterized by both deficits, like the cardinal diagnostic features mentioned above, and strengths, such as strong attentional focus and superior attention to detail, which are considered associated rather than diagnostic features. The estimated prevalence of autism in the Western adult population is about1% or higher, although its precise figure may vary from country to country and with different healthcare systems (e.g., Brugha et al., 2011; Dietz, Rose, McArthur, & Maenner, 2020).

Before the publication of the DSM-5 in 2013, a diagnosis of autism (or autistic disorder) was typically formulated in the presence of a triad of symptoms including language deficits, which could differentiate between classic autism and *Asperger's syndrome* (without language deficits Lai, Lombardo, Chakrabarti, & Baron-Cohen, 2013). Since 2013, however, a single diagnostic label (ASD) covers both classic autism (high, medium, and low functioning[a]), Asperger's syndrome, and, although less relevant here, also the previous Pervasive Developmental Disorder, Not Otherwise Specified or PDD-NOS (DSM-5, American Psychiatric Association, 2013). Given the possible heterogeneity between and within all these conditions, specifiers have been created to complement the single diagnostic label and to help identify associated deficits (e.g., intellectual impairment, language impairment, other neurodevelopmental, mental, or behavioral disorders, etc.). Heterogeneity notwithstanding, such specifiers make it possible to map individual conditions on a spectrum, or continuum of severity, based on cardinal symptoms, language deficits, and concurrent conditions. While the most recent diagnostic guidelines do not recommend the use of further labels or subgroupings, it is important to acknowledge that such labels can be found in the literature of interest, especially in studies informed by previous diagnostic criteria.

1.1.2 Autism in the collective imagination

Draaisma (2009) points out that in the collective imagination autism tends to overlap with the higher-functioning/Asperger's syndrome end of the spectrum, and it is often assumed that most autistic individuals possess special or "savant" skills, against a background of well-known impairments. This popular stereotype may have been originated and reinforced by exposure to the typical representations of autistic individuals from the entertainment industry (see e.g., *Rain Man*, 1988, or *Mercury Rising*, 1998), which are no accurate depictions of the average person with autism. Even if savantism is more frequent in autism than in other disorders (Hill, 1977; Howlin, Goode, Hutton, & Rutter, 2009; Saloviita, Ruusila, & Ruusila, 2000), and the majority of those with savantism are also on the autism spectrum, only 10%–30% of individuals with ASD (and possibly not those in the low functioning end) may show savant or special skills in typical areas of savantism such as visuospatial abilities, calculation, drawing or music—that is a talent that is not aligned with their other abilities (Howlin et al., 2009; Rimland, 1978). On the other hand,

[a] As a rule of thumb, individuals who are referred to in the scientific literature as having Asperger's syndrome may present with normal or above-average IQ and without language delay, individuals with high, medium, and low functioning autism may show some degree of language delay in the presence of an IQ within one standard deviation from the population mean (i.e., 85 or above), between one and three standard deviations from the population mean (i.e., 55–84) and below three standard deviations from the population mean (i.e., IQ $<$ 55) respectively (Baron-Cohen, 2006).

recent evidence suggests that about half of autistic children and adults might show a special isolated skill in memory (Meilleur, Jelenic, & Mottron, 2015).

The reason for such a significantly higher frequency of special skills in autism is still unclear. It has been suggested, for example, that the detailed-focused processing bias associated with autism could represent the starting engine for talent development (Happé & Vital, 2009). However, hyperattention to detail may not be sufficient to explain the level of competence reached by autistic savants in such complex domains as mathematics, music, and art (Baron-Cohen, Ashwin, Ashwin, Tavassoli, & Chakrabarti, 2009; Happe, 2018); thus, a further construct has been proposed, to capture an additional characteristic of the autistic mindset that could explain this higher than expected occurrence of talent, that is a hyper-drive to analyze or construct rule-based systems (i.e., hypersystemizing; Baron-Cohen et al., 2009). We will come back to systemizing after having introduced the important concept of *autistic traits*.

1.1.3 The continuum view of autism: Quantifying autistic traits

While autistic people can be (and have typically been) considered as forming a separate category, it has also been proposed that the defining and associated traits of autism may be continuously distributed across the population (e.g., Baron-Cohen, Wheelwright, Skinner, Martin, & Clubley, 2001; Constantino et al., 2003; Frith, 1991; Posserud, Lundervold, & Gillberg, 2006; Wing, 1988). Individuals with autism, whose traits can be measured via the same instruments as for the general population without any learning disabilities, could therefore be located at the extreme high (autistic) end of a continuous distribution of characteristic traits (Baron-Cohen, 2010).

In personality psychology, a trait is "a dimension of personality used to categorize people according to the degree to which they manifest a particular characteristic" (Burger, 2011, p. 149). Traits are considered to be relatively stable over time and across situations and can refer to a wide range of individual characteristics (from needs to strategies, to sensory sensitivity and imagination). They are often measured via questionnaires (Chamorro-Premuzic & Furnham, 2006). *Autistic traits* refer to the individual features that comprise the quantitative variability in the cardinal behavioral domains as defined by current diagnostic manuals and in the known associated features that may not be considered for diagnostic purposes, like a detail-oriented cognitive style (Happé, 1999; Lai et al., 2013).

Questionnaires that can be used to measure autistic traits in individuals with a diagnosis of autism as well as in the general population show a continuous distribution of scores in the general population, suggesting that the autism spectrum may be conceptualized as extending into the general population (Lai et al., 2013). Furthermore, traits comprised in the so-called broader autism phenotype[b] show shared genetic association with clinically defined autism (Chakrabarti et al., 2009; Colvert et al., 2015).

1.1.4 *The autism spectrum quotient (AQ)*

One of the most popular instruments used to measure autistic traits in the general population without learning disabilities through to individuals with autism of average intelligence or above is the Autism Spectrum Quotient (Baron-Cohen et al., 2001). Not surprisingly, the AQ has been the instrument of choice also in the relevant hacking research.

The AQ is suitable for people aged 16+. Being a self-report instrument, it relies on individuals' self-awareness and, more precisely, on their ability to report their own preferences and what they find easy or difficult in everyday life (Baron-Cohen et al., 2001). It presupposes an IQ of 85+ and has been attuned to quantify autistic traits in the general population, high functioning autism and Asperger's syndrome but not in classical forms of autism involving comorbid learning disability or below average IQ (which may represent 50% of the autistic spectrum; Baird et al., 2006). Although cut-offs have been identified to circumscribe a range of scores of potential clinical interest[c], it is now widely acknowledged that it should not be used as a clinical screening or diagnostic tool, but rather as a descriptive, quantitative measure of autistic traits (Ruzich et al., 2015). Crucially for a self-report instrument, AQ scores have been shown to correlate with a wide range of objective variables, such as brain structure and activity, behavioral indices, and genetic risk (Belmonte, Gomot, & Baron-Cohen, 2010; Wheelwright, Auyeung, Allison, & Baron-Cohen, 2010).

The AQ is made of 50 items requiring responses on a 4-level Likert scale (definitely/slightly agree, slightly/definitely disagree) and divided into

[b] The broader autism phenotype can be defined as a subclinical set of observable/manifest traits indicating familiarity or genetic liability to autism. It is quantitatively milder but qualitatively similar to the clinical phenotype (Wheelwright et al., 2010).

[c] A preliminary study on the potential of the AQ for use in clinical practice reported that the originally identified cut-off score of 32 (Baron-Cohen et al., 2001) correctly identified 76% of individuals receiving a diagnosis of ASD (sensitivity: 0.77; specificity: 0.74) from a sample of 100 individuals who had been consecutively referred to a specialized diagnostic clinic for adults, on suspicion of having high functioning autism or Asperger's syndrome; a cut-off score of 26, instead, would enable the correct identification of 83% of patients in the same referred sample (sensitivity: 0.95; specificity: 0.52; Woodbury-Smith et al., 2005). The authors concluded by suggesting that, were the AQ ever to be used as a screening instrument in a clinical setting, the more conservative cut-off of 26 should be adopted to limit false negatives. On the other hand, were the AQ ever to be used as a screening instrument in the general population, the higher cut-off or 32 should be adopted, to limit false positives. Later studies, however, showed that the AQ does not satisfy the criteria required for use as a screening instrument with an eye to diagnosis (Ruzich et al., 2015).

five subscales measuring respectively communication skills (e.g., "Other people frequently tell me that what I've said is impolite, even though I think it is polite"), social skills (e.g., "I find it hard to make new friends"), attention switching (e.g., "I prefer to do things the same way over and over again"), imagination (e.g., "I don't particularly enjoy reading fiction") and attention to detail (e.g., "I tend to notice details that others do not"). With the original scoring method, endorsement of the response coherent with the autistic profile for an item receives 1 point, whereas endorsement of the alternative response receives 0 points. The AQ total score, which therefore has a theoretical range of 0–50 (Baron-Cohen et al., 2001), is a composite measure of traits related to both cardinal (e.g., social skills) and associated features (e.g., attention to detail) of autism. It has been claimed that the scale shows good test-retest reliability $r = 0.7$; although this is based on a very small number of observations (Baron-Cohen et al., 2001) and acceptable internal consistency, with all Cronbach's alpha coefficients for the component subscales in the 0.63–0.77 range (Baron-Cohen et al., 2001). In the general population without learning disabilities, males achieve higher scores than females, whereas no such difference is found in the clinical population whose scores are positioned towards the extreme high end of the distribution of total scores (Ruzich et al., 2015). With the notable exception of the attention to detail subscale (perhaps in part due to the way some of its items are phrased; e.g., Hoekstra, Bartels, Cath, & Boomsma, 2008; Lundqvist & Lindner, 2017), in all subscales, the clinically diagnosed population consistently achieves higher scores than the general population, and the difference between sexes in the general population remains true across individual subscales (Baron-Cohen et al., 2001). Moreover, university students enrolled in science degrees[d] achieve higher scores than students enrolled in humanities or social science degrees—a difference that appears to be driven mostly by the social skills and imagination subscales (Baron-Cohen et al., 2001). The various AQ subscales[e] may be

[d] Science degrees comprised: physical sciences (physics, physical natural sciences, chemistry, geology, communications, chemical engineering, mineral science, material science, geophysics), biological sciences (experimental psychology, neurophysiology, biological natural sciences, biology, bioanthropology, neuroscience, molecular ecology), mathematics, computer science, engineering, medicine, and veterinary science.

[e] It should be mentioned that, within the specialist literature, doubts have been cast on the actual number of dimensions (or traits) captured by the AQ, with a few studies identifying between two and four separate dimensions instead of the five that were originally proposed a priori on the basis of a theoretical understanding of autism. Moreover, although the AQ is widely considered a valid research instrument, the suitability of certain specific individual items contributing to the 50-item AQ and subscale scores has also been questioned (e.g., Hoekstra et al., 2008; Lundqvist & Lindner, 2017; van Rentergem, Lever, & Geurts, 2019). Although alternative measures of autistic traits are available (e.g., Constantino et al., 2003; English et al., 2021; Hurley, Losh, Parlier, Reznick, & Piven, 2007), all the studies reviewed here have used the AQ - by and large the most popular instrument so far.

uncorrelated, weakly correlated, or moderately correlated in non-clinical samples (e.g., Rusconi, Ferri, Viding, & Mitchener-Nissen, 2015), which, within a continuum view of autism, appears consistent with a multidimensional model of autistic traits, whereby the neurodevelopmental condition is identified by the co-occurrence of features having distinct genetic origin (e.g., Brunsdon & Happé, 2014; Happé, Ronald, & Plomin, 2006).

1.1.5 The systemizing quotient (SQ)

Another self-report instrument that has been used to measure autistic traits in relation to hacking is the Systemizing Quotient (Baron-Cohen, Richler, Bisarya, Gurunathan, & Wheelwright, 2003) which, unlike the AQ, is built as a unidimensional measurement tool. Systemizing has been described as "our most powerful way of understanding and predicting the law-governed inanimate universe" (Baron-Cohen et al., 2003, p. 361), a drive to analyze or construct systems. According to the hypersystemizing theory, such drive would be set to abnormally high levels in autism (see e.g., Baron-Cohen, 2006; Baron-Cohen et al., 2003; Baron-Cohen, Leslie, & Frith, 1986). Hypersystemizing has been proposed as a putative mechanism subtending the non-social features of autism, such as narrow interests, repetitive behavior, and resistance to change, and as an explanation for the typical manifestation of talent in autism in highly predictable domains, as mentioned in Section 1.1.2). According to Baron-Cohen et al. (2009), it would draw its origins in heightened attention to detail, which could itself be a consequence of sensory hypersensitivity in autism. It is usually pinned against empathizing, described as "our most powerful way of understanding and predicting the social world," that is "the drive to identify another person's emotions and thoughts, and to respond to these with an appropriate emotion," which would instead be set to abnormally low levels in autism (Baron-Cohen et al., 2003, p. 361). Tasks that are best addressed by empathizing (e.g., because they entail agentive changes rather than highly structured, non-agentive changes) could not be satisfactorily addressed by systemizing, and vice versa (Baron-Cohen, 2006).

A "system" is characterized by the fact that it follows rules: identifying and recursively testing those rules will enable one to create a detailed and comprehensive representation of the system and to predict how the system will behave in the future. Systemizing consists of noting regularities (patterns or structure) and rules, derived by establishing associations in the form of laws (if "p," then "q"). The SQ was created to measure individual differences in systemizing in everyday life situations in adults of normal intelligence, similarly to the AQ, within or outside the autism spectrum (Baron-Cohen et al., 2003). In its original form, it comprises a total of 60 items (including 40 systemizing items, e.g., "When I listen to a piece of music, I always notice the way it's structured" and 20 filler items, e.g., "I adhere to common superstitions") requiring responses on a 4-level Likert scale (definitely/slightly agree, slightly/definitely disagree). Endorsement of the response coherent with the autistic profile for an item receives 2 points when the "strongly" option is selected and 1 point when the "slightly" option is selected, whereas endorsement of any of the alternative responses receives 0 points. Thus, the SQ total score has a theoretical range of 0–80 (filler items do not contribute to the total score). While no effects of age or education are found, males score higher than females within the general population. Individuals with high functioning autism or Asperger's syndrome (within which no sex differences are found, similarly to the AQ, which also shows in this clinical group a moderate positive correlation with the SQ score: $r = 0.46$; Baron-Cohen et al., 2003) score higher than matched controls from the general population (Baron-Cohen et al., 2003). Across groups, the SQ shows adequate internal consistency (Cronbach's alpha = 0.79; Baron-Cohen et al., 2003). A revised version of the SQ, the SQ-R (Wheelwright et al., 2006), is also available but since it has not been used in research on hacking it will not be discussed here.

1.2 Hacking

1.2.1 Definitions

The term *hacking* does not have a single definition (Gunkel, 2000): in this section, we will briefly mention the different ways in which it has been defined in time and clarify how it is used here. The origin of the term hacking can be traced directly to the development of computer technology at the Massachusetts Institute of Technology (MIT) in the 1950s when the term referred to programming shortcuts that could make things run quicker or better (Lilley, 2002). Therefore, a *hacker* was first and foremost an individual very skilled with computers. Being defined as a hacker "had strong laudatory connotations of deep knowledge driven by insatiable curiosity" (Bratus, 2007, p. 2). Good correspondence with the original definition is found in the 1994 edition of the New Hacker's Dictionary, which defines a hacker as "someone who enjoys exploring the details of programmable systems and how to stretch their capabilities." Accordingly, hackers can show ingenuity and aptitude across a broad range of artifacts, and they are generally curious about the configuration and mechanisms of technological systems and products (Taylor, 1999). Within this framework, hacking essentially means exploring and manipulating things in order to understand a system's behavior and how to overcome any limitations that the system might have (Gunkel, 2000).

However, the term has also been used to designate a computer *virtuoso*, a skilled computer programmer or engineer who enjoys taking up challenges involving access to and manipulating of others' computers or systems (Levy, 1984; Lilley, 2002; Sterling, 1991; Turgeman-Goldschmidt, 2008) or extending the scopes of computer codes by writing programs (Ludlow, 1996). Meyer (1989) defined hackers as computer *aficionados* who break into corporate and government computer systems using their home computer and a telephone modem.

It is therefore clear that the term *hacking* has been used to emphasize, on the one hand, the outcome of a specific type of drive and technical skills (e.g., the activity of a person who enjoys learning the details of computer systems and how to stretch their capabilities) and, on the other, a morally dubious and/or illicit practice (e.g., the activity of a person who tries to gain information by manipulating others' systems using deceptive or illegal means; Steele, 1983; Gunkel, 2000).

Throughout the years, at least three distinct communities have been referred to as *hackers* (Hannemyr, 1999). The original hackers were computer professionals with a certain level of craftsmanship and who took pride in and thoroughly enjoyed their work (Hannemyr, 1999). Early hackers were seen in a good light, shared a genuine belief in the liberating power of technology, and were guided by their own "hacker ethics" (Levy, 1984).

In the 1970s, hackers became activists. They believed that technology was a source of power and that computers (the supreme manifestation of the power of technology) should have been accessible to everyone. This second generation was characterized by the will to make computers become useful and accessible to citizens; they pioneered public access terminals, computer conferencing, and personal computers (Hannemyr, 1999).

In the second half of the 1980s, with the emergence of the so-called "computer underground," things partially changed. To this third community of hackers, hacking meant breaking into or sabotaging a computer system (Hannemyr, 1999). As a consequence, nowadays hacking is often used with reference to "gaining unauthorized access to a computer system with or without a further criminal motive" (Hutchings, 2013, p. 2) and has acquired a negative connotation.

However, the way we will refer to hacking here does not necessarily imply any illicit activities. In other words, we will adopt a definition that is closer to the original. With *hacker,* we will refer to a computer enthusiast, who enjoys learning programming languages, exploring computer systems, and pushing the envelope. S/he could be considered an expert on the subject, who mastered the art of making computers and software do more than the original designers intended (Sukhai, 2004). This definition remains agnostic about a hacker's intent or authorization status. Obviously, the use that any hacker can make of their skills and knowledge may vary depending on personal and/or situational factors. To classify hackers according to their intent and authorization status, we can use further specifiers.

1.2.2 *Types of hackers based on maliciousness*

Three of the most popular labels used both in the hackers' community and in the literature to reflect the perceived maliciousness of a hacker's intent refer to their supposed hat color, in a reference to old cowboy movies, where the good guy wore a white hat and a bad guy wore a black hat (Furnell, 2002). According to this classification, hackers can be identified as *black hats, white hats,* or *gray hats.*

Black hats. Also known as "crackers," these are hackers who use their skills for illegal or malicious purposes. They break into or otherwise violate the system integrity of remote machines, with malicious intent. After having gained unauthorized access, black hats destroy vital data, deny legitimate user service, and cause problems for their targets. A differentiation can thus be made between "hacks," in their original meaning, and "cracks." A "hack" refers to "any legitimate and useful alteration or adjustment to computer hardware or software, which enables technology to be used in an innovative or unusual way," whereas the term "crack" is applied when "a hacker alters technology for a negative or potentially criminal application" (Holt, 2010, p. 215).

White hats. These are hackers that use their skills to improve cybersecurity and protect against malicious hackers. They remain entirely within the law and work to expose holes in systems with the purpose to fix flaws and improve security, using their hacking skills for protective purposes. This type includes security professionals with knowledge of hacking and of the hacker toolset and who use their knowledge to locate weaknesses and implement countermeasures (i.e., ethical hackers; Graves, 2010; Munson, 2009). Ethical hackers complete many of the same activities with many of the same tools as malicious hackers, or black hats, to penetrate systems. Indeed, the closer the penetration test simulates a real-world attack, the more value it provides. In general, differences between black-hat hackers and ethical hackers can be summarized in three key points:

(1) Authorization (i.e., the process of obtaining approval before conducting any tests or attacks). The penetration tester and the company or individual being audited need to agree upon the scope of the test, explicitly defining the authorized targets for the penetration tester.

(2) Motivation: a malicious hacker may be driven by the desire for personal gain, revenge, or fame whereas an ethical hacker is driven by the will to help the individual or the organization and improve their security via penetration testing, on their request. Further, in most cases, penetration testing may last 1 week up to several weeks, whereas a black hat may spend a significant amount of time attacking the organization.

(3) Intent: an ethical hacker's intent is to provide the individual or the organization with a realistic attack simulation so that they can improve their security through the early discovery and mitigation of vulnerabilities. Ethical hackers will keep the penetration testing findings confidential and never share sensitive information discovered during the process of penetration testing.

Gray hats. These are hackers who may work offensively or defensively, depending on the situation. They do not align themselves with any specific moral philosophy but rather act to achieve some specific goal (Holt, 2010). Both hackers and crackers are powerful forces on the internet and some individuals qualify for both categories. The existence of such individuals clouds the division between black hats and white hats (Graves, 2010).

1.2.3 *The "hacking spectrum"*

Within the hacker subculture, technical knowledge and interest in understanding how systems work are highly praised; crucially, these characteristics appear to set the criteria to distinguish between a hacker and a non-hacker, such as an "end users," "scene whores" and "feds" (Holt, 2010). However, even within the hacker subculture, it is possible to differentiate individuals not only according to their moral/ethical stance but also along a continuum of knowledge and technical skills. In some cases, the continuum can be seen also as a developmental trajectory of a hacker in time. At the lower end of the continuum "noobies," "newbies," "neophytes" are found, that is individuals who have just entered the world of hacking. They are inexperienced and not as skilled as long-time hackers. Through real-world experience, peer mentoring, readings, they will gradually develop their technical skills and a deeper understanding of the hacking subculture. "Script kiddies," on the other hand, have not really developed an in-depth knowledge of how computer systems and networks work but seek to hack via pre-coded scripts that are made available online and that enable a sort of automated hacking. This may be seen within the subculture as the lazy alternative to proper hacking, which can only be achieved via hard work and dedication. Kiddies attract disapproval and further derogatory labeling as "lamers" (Furnell, 2002). Only with time a "noobie" or a "script kiddie" can develop the skills of a real "hacker" and move further up along the continuum. The upper end of the continuum is inhabited by "elite" hackers, who master the craft and are often seen more as an ideal to aspire to than real-life individuals. Further categories may be located in-between the two ends.

Crucially, the way in which terms are defined shows large variability within the hacker subculture and the meanings attached to labels a degree of subjectivity. Holt (2010) notes that this may be the reason why contrasting definitions can often be found in the literature on hackers (Furnell, 2002; Holt, 2010; Schell, Dodge, & Moutsasos, 2002). What is very clear, however, is the importance to the hacker subculture of understanding technology, of the drive and dedication to explore and master programmable systems. What may superficially seem like hacking to the outsider and also to the media based on a target-oriented definition (e.g., "breaking into systems") may not necessarily be endorsed as hacking by hackers if, for example, is carried out exclusively via pre-programmed scripts by somebody without technical skills or an in-depth understanding of how scripts and systems work. The process behind the end result may thus be at least as relevant to define what hacking is and what a hacker does, as the result itself. This should be kept in mind when identifying populations to sample and skills to measure, for studies to be truly informative on the connection between hacking and individual traits or neurodevelopmental conditions, as discussed in the current chapter.

1.2.4 Court cases involving hackers on the autism spectrum

In the opening of *The Fox* (2018), Forsyth describes a night incursion by black-clad Special Forces soldiers into the outer suburb of an English provincial town. They are on the hunt for dangerous cybercriminals who had been able to penetrate the defenses of the entire Western world. A rather harmless-looking house which they are about to enter may host terrorists, fanatics, or mercenaries, so they are prepared for the worst. But the reality is very different and, once subdued the four inhabitants who were quietly sleeping in the house, the team leader cannot help but report to the operation head: "You are not going to believe what we have found here."

Forsyth's espionage thriller revolves around an autistic teenager, Luke Jennings, who possesses an extraordinary talent for hacking. Luke is so talented that even the finest minds from GCHQ cannot beat him—let alone understand how he does what it does. Forsyth is not really interested in the details of Luke's skills, which are pictured as a sort of extra-powered code-breaking/password-finding magic enabled by Luke's diversely gifted brain. However, he did take inspiration from reality in creating his character. In particular, he makes reference to two recent hacking cases that gained huge media attention worldwide and especially in the UK: the cases of Gary McKinnon and Lauri Love.

Gary McKinnon, a 36-year-old unemployed computer system administrator, was arrested in London in 2002. He had allegedly used hacking tools to penetrate, take control of, alter, delete or copy classified materials and account details, password, and system files from US government computers (including some located at various Army, NASA, and Air

Force facilities), in patent contravention of US law. All the offenses were supposed to have originated from his home computer in London (i.e., they were carried out remotely via the internet, thus the offending occurred extraterritorially; Mann, Warren, & Kennedy, 2018) and had cost the US government and private industry 1,200,000 US$ to repair the damages and restore the systems. Although McKinnon pleaded guilty to unauthorized computer access and data modification offenses under the *Computer Misuse Act 1990* (UK), the US did not accept a shift of the trial forum to the UK (whereby the offender would have been prosecuted under a more lenient system of sanctions) and swiftly requested McKinnon's extradition. After a string of decisions by UK prosecutors and courts in favor of extradition to the US, each one of them unsuccessfully appealed against by the defendant, McKinnon was spared from extradition at the last hour by the then Home Secretary, Theresa May, who exercised her discretion not to extradite him based on the fact that "McKinnon's extradition would give rise to such a high risk of him ending his life that a decision to extradite would be incompatible with Mr. McKinnon's human rights" (BBC, 2012).

Lauri Love, was first arrested in 2013, at the age of 28, at his mother's house in Suffolk on suspicion of hacking into US private companies, military and government computer systems (including those of the Missile Defense Agency, the NASA, and the FBI), with the help of fellow hackers located in Sweden and Australia (BBC, 2013; Mann et al., 2018). After having been released on bail and with no further action taken by the UK authorities, he was arrested again in 2015 in the execution of an extradition warrant on behalf of the US (BBC, 2015). Although Love had not cooperated with the National Crime Agency by providing passwords and encryption keys, enough evidence had been retrieved to connect him beyond a reasonable doubt to the hacking attacks. In Love's case, the estimated damage was even worse than that caused by McKinnon's actions, amounting to millions of dollars in damages and £13 million in restoration costs (Mann et al., 2018). Unlike in McKinnon's case, the then Home Secretary, Amber Rudd, ordered Love's extradition. However, in 2018, Love won a High Court appeal against extradition to the US, at which point the US authorities declared they would not fight the decision, effectively abandoning the case as it would be "oppressive" (BBC, 2018a, 2018b).

The cases we have just mentioned share several common features. The involvement and role of both individuals in carrying out many of the offenses they had been charged for was apparently very clear to the US authorities, they were both diagnosed as having an ASD (more specifically, a severe form of the Asperger's syndrome variant), both cases received extensive media coverage and in both cases considerations about human rights protection (a person-centered approach) seemed to be initially superseded by a desire to preserve political comity in the administration of transnational justice (Mann et al., 2018). Very interestingly, the most decisive element for the rejection of the extradition request by the UK authorities in both McKinnon's and Love's case was not the ASD diagnosis by itself or a supposed inability to stand trial (they were both considered fit to stand trial), lack of intention or full understanding of the consequences of their actions but the consideration that their severe form of ASD in conjunction with other illnesses (most crucially, depression) made them unfit for detention in a US prison, exposing them to an unacceptably high risk of suicide, which would be contrary to the interest of justice and oppressive (Freckelton, 2020; Mann et al., 2018). According to Freckelton (2020), certain forms of criminal conduct are over-represented among individuals with ASD and these include computer offenses. These may be favored in ASD (especially its high functioning and Asperger's forms) by a combination of specific skills and opportunity. We suspect that the extensive media coverage afforded by high-profile cases such as McKinnon's and Love's, however, is likely to inflate the strength of association between ASD and this type of criminal offenses in the public perception.

1.3 Rationale

We aimed to address the question of whether there is evidence of an association between autism or autistic traits and hacking in the empirical literature. Specifically, we aimed to investigate any associations with general hacking-related skills or interests, hacking with criminal intent, and vulnerability of being the victim of such crimes.

Thus, to integrate and synthesize existing empirical studies on autism or autistic traits and hacking, a systematic literature review was carried out. However, during the scoping stages of the review, it became clear that the literature specific to hacking is somewhat ambiguous based on variations in definitions of hacking and motivations for hacking; furthermore, many studies extend their investigations to the broader concept of cybercrime. Thus, to ensure the review considered all available literature and offered an insight into hacking-related skills and behaviors, we extended the search to include published empirical studies related to cybercrime and cybersecurity; this also allowed for any differences in motivations for hacking to be considered.

2 Methods

The systematic review was carried out by following the PRISMA (Preferred Reporting Items for Systematic Reviews and Meta-Analyses) guidelines as much as possible within our constraints (Moher, Liberati, Tetzlaff, & Altman, 2009), and our search strategy and inclusion criteria are detailed below.

2.1 Search strategy

A search of Article Title, Abstract, and Keywords in Scopus was carried out for the keywords (hack* OR "cybersecurity" OR cybercrim*) AND (autis* OR Asperger*). The same keyword search was carried out in the ACM Full-Text Collection; however, the search was carried out for search terms appearing anywhere in the document as the ACM digital library does not have a specific search for keywords (author keywords are available to search but these are likely to differ from research keywords). A further search was carried out on Web of Science for Topic, Keywords, or Abstract. A final search was carried out in PubMed for the same keywords in the Title/Abstract. We also looked through the reference sections of all included studies to identify any further studies which may fit our inclusion criteria. All searches were carried out separately by two authors, with the final search taking place on the 14th May 2021.

2.2 Inclusion criteria

Studies were included if they were empirical studies, presenting data relating to autistic traits or autism and hacking, cybercrime, cyberdeviancy or cybersecurity, or specific skills relating to these activities. Studies could investigate such relationships from the point of view of possessing the adequate skillset, committing offenses, or being the victim of such offenses. To ensure we included all potential studies in all database searches we placed no restriction on document type or date. However, at the point of screening, we ensured the criterion of including only published material was met. A preliminary selection was carried out for content relevance based on title and abstract, and a subsequent selection based on full-text if required. A diagram detailing all exclusions, during each decision phase is reported in Fig. 1.

3 Results

3.1 Skills and interest in hacking

Two studies investigating autistic traits and hacking-related skills or interests met our criteria (see Table 1). Our search returned no published studies, which were directly addressing the associations between autism and hacking-related skills and interests.

In the first published study to test the possible relationship between autistic traits and hacking, Schell and Melnychuk (2011) conducted a survey among hacker conference attendees. Noteworthy, they decided to focus not on a narrowly-defined criminal segment of hackers but on self-identified hackers attending a specialist conference, who confirmed

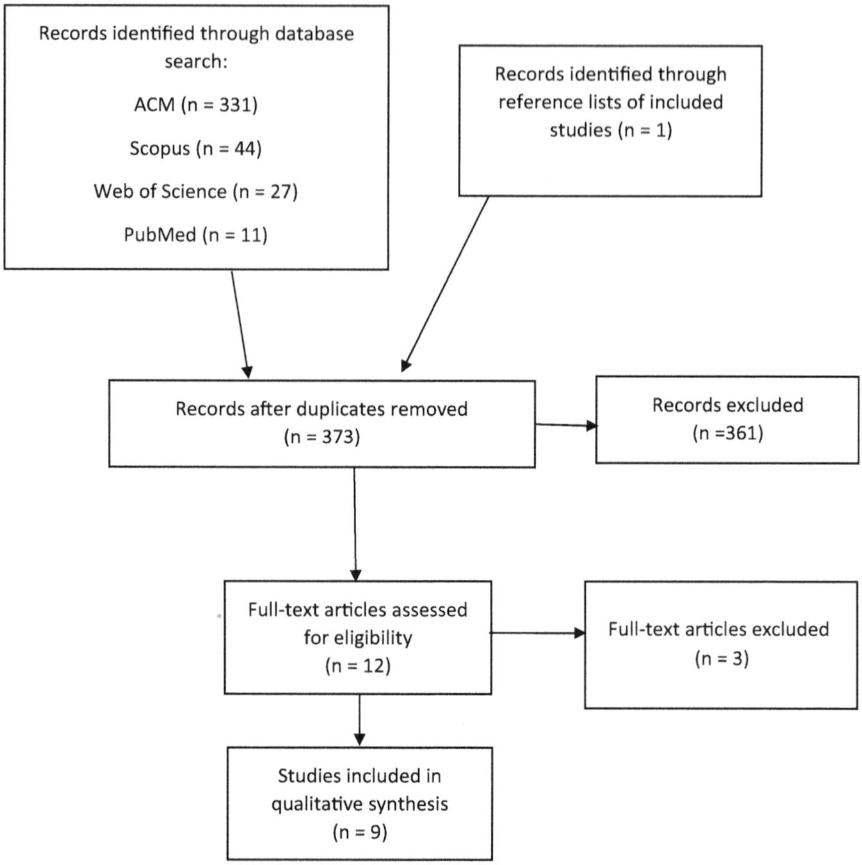

FIG. 1 Diagram showing stages of decision-making regarding study inclusion.

being actively involved in the activities of the conference (as opposed to just accompanying an attendee). A total of 136 attendees (66 males and 70 females), aged 34 on average (range: 18–56), mostly in employment (e.g.as security officers, engineers, or professors, although there were also a few students in the sample), mainly from the US but also from foreign countries, filled the questionnaires.

The vast majority of these respondents, who shared a keen interest in technology and hacking/IT security, declared to hack primarily to solve interesting puzzles and challenges, to advance network, software, and computer capabilities, or to expose weaknesses in a company's network or in their products. Three respondents reported having hacking-related offenses (none of which required jail time) and about a quarter of the respondents reported having been victims of cyber-stalking or cyber-harassment.

TABLE 1 Details of all studies included in the systematic review.

Study	Sample size	Sample mean age	Sample drawn from	Tool used to measure or probe criminality	Tool to measure susceptibility to cybercrime or safety risks online	Tool to measure hacking skills and/or interests	Measurement tool for autistic traits
Schell and Melnychuk (2011)	133/136 completed the AQ	Approximately 34 years	Hacker conferences attendees who were actively involved in the activities of the hacker conference (self-identified hackers)	Self-report	Self-report	Self-report Motives for hacking and time spent on hacking-related activities as well as other questions related to interest in IT and any information on hostile experiences captured via a self-report survey	Autism Quotient
Harvey, Bolgan, Mosca, McLean, and Rusconi (2016)	Part 1: 103 student scientists 56 ethical hackers Part 2: 43 student scientists 13 ethical hackers	Mostly in the range: 18–34	Student scientists: mainly from BSc-hon and MSc Psychology courses Ethical hackers: mainly from BSc-hon and MSc Ethical Hacking courses	NA	NA	Part 1 and Part 2: Enrolment in Ethical Hacking degree course taken as a proxy for interests Part 2: Skills: measured with code-breaking tasks and	Part 1: Autism Quotient Part 2: Systemizing Quotient

Continued

TABLE 1 Details of all studies included in the systematic review—cont'd

Study	Sample size	Sample mean age	Sample drawn from	Tool used to measure or probe criminality	Tool to measure susceptibility to cybercrime or safety risks online	Tool to measure hacking skills and/or interests	Measurement tool for autistic traits
						contrasted with a control task (threat search in security x-ray images)	
Seigfried-Spellar, O'Quinn, and Treadway (2015)	275/296 completed AQ	Approximately 20 years	University graduate students (Southern university in the US)	CCI-R—a questionnaire measuring frequency and prevalence of activity related to cyber-deviance including hacking, identity theft, cyberbullying, and virus writing	NA	NA	Autism Quotient
Payne et al. (2019)	290	Approximately 24 years	Internet users likely having advanced digital skills, involved in courses related to computers, and recruited through university, school, and a	Cyber-dependent crime (YES/NO) questionnaire Activities investigated included: (a) Phishing (b) Webcam managing	NA	NA	Autism Quotient

			cybersecurity related organization	(c) File hijacking (d) Key-logging (e) Screenshot managing (f) Employing ad clicking (g) Hacking (h) Launching distributed denial of service attacks			
Payne, Maras, Russell, Brosnan, and Mills (2020)	29—cyber-dependent crime decliners (CDD) 7—cyber-dependent crime offenders (CDO)	CDDs—Approximately 29 years CDOs—Approximately 18 years	CDDs drawn from an original sample of 175 individuals likely to have a high degree of computer skills CDOs had committed Computer Misuse Offenses and were recruited through National Crime Agency, UK	Semi-structured interviews were carried out to determine motivations for engagement in cyber-dependent offending (for CDOs) carried out over the phone or face-to-face Motivations for declining to engage in cyber-dependent offenses (for CDDs) were	NA	NA	Autism Quotient

Continued

TABLE 1 Details of all studies included in the systematic review—cont'd

Study	Sample size	Sample mean age	Sample drawn from	Tool used to measure or probe criminality	Tool to measure susceptibility to cybercrime or safety risks online	Tool to measure hacking skills and/or interests	Measurement tool for autistic traits
				probed using an online questionnaire			
Ledingham and Mills (2015)	8	NA	Representatives of law-enforcement agencies in cybercrime units in the UK, USA, Australia, New Zealand, Germany, the Netherlands, and Denmark	Brief online survey requesting information on mental health and autism training and information on any interactions with cyber-offenders with ASD	NA	NA	NA
Neupane, Satvat, Saxena, Stavrinos, and Bishop (2018)	15—autistic individuals 15—non-autistic individuals	Approx. 77% of total sample < 31 years	University student sample. Participants were regular internet users. Individuals in the clinical group	NA	Participants asked to distinguish (yes/no) between 30 real and 30 fake websites (15 easy and	NA	Autism Quotient

			had to have had a diagnosis given by a doctor or clinical psychologist		15 difficult) shown for 6s each		
Macmillan, Berg, Just, and Stewart (2020)	63 parents of autistic children 41 parents of non-autistic children. All children were aged 6 years and over	Mean age of children approximately 11 years	Parents invited through Facebook and Twitter. Parents of autistic children were recruited through Autism-related organizations	NA	Online parental reporting of child's online risk. Respondents were asked to click yes or no to risks that their child had experienced	NA	NA
Taylor-Jackson and Matthews (2020)	35 in total—median split based on AQ to make high and low group	Approximately 21 years	University student sample	NA	Participants shown 20 headlines with an accompanying image (10 real and 10 fake). Articles were emotive/non-emotive. Participants were required to decide whether the articles were real or fake		Autism Quotient

NA = Not available

The authors report that, in the sample of 133 respondents who completed the AQ, males obtained a significantly higher total AQ score for each of the three sub-levels of AQ scores in which they had roughly divided the theoretical distribution. Based on the information reported in the main text, such sub-levels would translate exactly to:low = $AQ < 16.4$; intermediate = $17 \leq AQ \leq 33$; high = $AQ \geq 34$ (Schell & Melnychuk, 2011, p. 158). However, based on the data reported in the accompanying Table 1 (p. 162), such differences are not apparent; also, it would appear that slightly different criteria may have actually been used to define sub-levels, as the mean scores for males and females in the high sub-level are actually lower than the declared minimum score for that sub-level. Finally, a test performed on the overall means and standard deviations provided in the article (p. 162) does not detect any significant difference between males and females (males: $M = 20.12$, SD = 7.63; females: $M = 19.24$; SD = 5.82; $t < 1$, $df = 131$; $p = 0.45$).

However, it is confirmed, that for both males and females, about two-thirds of the responders obtained a score in the intermediate range and that the count of males and females in the high vs. low sub-level is significantly different ($X^2 = 4.4043$, $p = 0.03$; Yates correction applied), with more males scoring in the high sub-level and more females scoring in the low sub-level.

With a mean overall total score of 19.67 (SD = 6.75; range: 3–37) in the AQ, Schell and Melnychuk's (2011) sample exceeds the upper boundary of the 95%-CI (i.e., 17.4) for the general population sample ($N = 4931$; unchecked for first-degree familiarity with autism) in a recent meta-analysis on the AQ (Ruzich et al., 2015). The same holds true when considering males and females separately. This would lend partial support to Schell and Melnychuk's (2011) conclusion that their sample of hacker conference attendees was characterized by higher than average AQ scores. Unfortunately, due to their arbitrary choice of sub-levels, it is not possible to quantify how many of their respondents could be potentially classed in the broad/medium/narrow autism phenotype in a similar fashion to Wheelwright et al. (2010). However, given the reported range of scores, it is possible to infer that at least two of their respondents may have scored in the narrow autism phenotype range (defined in Wheelwright et al., 2010, on the basis of normative data from Wheelwright et al., 2006, as having AQ scores of 35+).

Further, Schell and Melnychuk (2011) report that hackers achieved particularly high scores on the attention to detail and attention switching subscales, and relatively low scores on the social skills, communication, and imagination subscales. In effect, on the basis of the data provided in their Table 1 (p. 163), the overall score on the attention to detail subscale significantly exceeds the average score reported by Baron-Cohen et al. (2001) for student scientists (diff = 0.70; $t = 3.29$, $df = 585$, $p = 0.001$). However, attention switching does not deviate significantly from the average score of student scientists, nor do the imagination and communication subscales. On the other hand, hackers' endorsement of the autistic options for the social skills subscale appears significantly higher than that of student scientists' in Baron-Cohen et al.'s (2001) report (diff $=0.80$, $t=3.42$, $df=585$, $p=0.0007$). As a word of caution, it should be noted that the internal consistency values reached by all the subscales in Schell and Melnychuck's sample are very low (the lowest being that for the attention to detail subscale, with a Cronbach's alpha of 0.393), with the exception of the social skills subscale (Cronbach's alpha $=0.756$).

Finally, Schell and Melnychuk (2011) ranked individual items for the frequency at which respondents would agree or disagree with them. Of the six highest-ranking agreed with items, five belonged to the attention to detail subscale ("I tend to notice details that others do not," "I notice patterns in things all the time," "I usually notice car number plates or similar strings of information," "I often notice small sounds when others do not," "I am fascinated by numbers") and one belonged to the attention switching subscale ("I frequently get so strongly absorbed in one thing that I lose sight of other things"). The two highest-ranking disagreed with items belonging to the communication subscale ("I know how to tell if someone listening to me is getting bored") and the attention to detail subscale ("I am not very good at remembering phone numbers"—reverse scored).

In conclusion, this group of hackers, who most likely are in the technically skilled segment of the "hacking spectrum," and in most of the cases are well-integrated into society and have highly paid jobs, show significantly higher levels of self-reported autistic traits than the general population. Their average scores are similar to those of a large cohort of student scientists; they seem to stand out for higher scores on the attention to detail and social skills subscales (Baron-Cohen et al., 2001). Unfortunately, the original scores could not be compared to those of a matched control group, as this was not included in the study design and it is also unclear whether the data were normally distributed (thus it is possible that non-parametric statistics and indices of central tendency would provide more reliable information). Finally, it is unclear what percentage of the approached conference attendees agreed to fill the questionnaires.

These potential issues, however, were not present in the following study conducted by Harvey et al. (2016), which took inspiration from Schell and Melnychuck's work (Schell & Melnychuk, 2011). Harvey et al. (2016) administered the AQ to a group of 56 expert hackers from a BSc/MSc ethical hacking course (they were either enrolled in the course or had recently completed it) and to a control group ($N=103$) mainly drawn from a BSc/MSc psychology course (i.e., a group of science students, according to Baron-Cohen et al.'s, 2001, classification). Harvey et al. (2016) found that hackers obtained significantly higher AQ scores than non-hacker science students. Moreover, at the subscale level, a significant difference—in the direction of higher scores for hackers compared to non-hackers—was found in the attention to detail subscale only (i.e., no difference was found for scores in the social skills, communication, imagination, and attention switching subscales).

Harvey et al. (2016) also administered the SQ to a subsample of hackers ($N=13$) and non-hackers ($N=43$), from the original sample that completed the AQ. The subsamples were small (especially the hackers' one) and self-selected, as the SQ was completed only by those respondents from the original sample who agreed to participate in a behavioral testing session. However, Harvey et al. (2016) found evidence of a significant difference in SQ scores between hackers and non-hackers, whereby hackers obtained higher scores. On the overall subsample, they also found a moderate positive correlation between SQ and attention to detail scores and a moderate negative correlation between SQ and attention switching scores. Correlations between SQ and AQ total scores, SQ, and the other subscales were all positive and, being somewhat weaker, did not reach the threshold for significance, likely due to insufficient power.

In addition to self-report measures, Harvey et al. introduced two behavioral tasks in their design: a prototypical hacking task (code-breaking) that is suitable for both hackers and non-hackers and was expected to hinge upon systemizing skills, and a security threat search task with x-ray images, previously shown to benefit from high levels of attention to detail (as measured with the attention to detail AQ subscale; Rusconi et al., 2015). A significant moderate correlation was found between systemizing and hacking performance (both for the whole group, and also separately for hackers and non-hackers and for males and females) in the expected direction, but not between AQ or attention to detail and hacking performance. It could be noted that while the non-significant correlation between attention to detail and hacking performance appeared to be at the upper limit of the weak range, the non-significant correlation between AQ and hacking performance was equal to zero. On the other hand, a significant moderate correlation was found between attention to detail and performance in the x-ray screening task but no significant correlation was found between systemizing and performance in the x-ray screening task in the same participants. According to Harvey et al. (2016), this is consistent with the proposal that attention to detail may be relevant but insufficient for the development of systemizing skills (Baron-Cohen et al., 2009), which in this case were measured with a code-breaking task.

Harvey et al. (2016) thus concluded that in terms of self-reported traits, hackers do show higher levels of autistic traits, and in particular attention to detail, compared to a group of student scientists. This confirms Schell and Melnychuk's (2011) findings. It should be noted that biological science students, who represented the majority of Harvey et al.'s control group, may have overall lower scores than physical or computer science students; Baron-Cohen et al., 2001). However, no difference was found between hackers and the control group of student scientists for attention switching or other traits that require mentalizing and empathy. If confirmed as a genuine characteristic—rather than an artifact motivated by social desirability within the hacker community participating in Harvey et al.'s (2016) study, this would appear particularly interesting, especially given that social engineering is an important part of the hacker toolset in real life (e.g., Hadnagy, 2018; Mitnick & Simon, 2003). Furthermore, in a self-selected subgroup of participants, it was shown that hackers have higher SQ scores than the control group of student scientists. Crucially, a relation was detected between SQ scores and performance in a hacking task, regardless of expertise and sex, but not between SQ scores and performance in a visual search (control) task in the same participants. This would suggest specificity for the relation between SQ scores and hacking.

In conclusion, the current evidence from professional/student hackers and science students establishes a link between self-reported autistic traits, choice of hacking as a profession, and related skills, even in a non-hacking sample. There are strong hints that one of the most relevant traits is heightened attention to detail (or one that, collectively, is considered particularly distinctive by hackers themselves, given that it is self-reported), and possibly even more so systemizing. The relation between hacking and self-reported deficits in social skills, instead, is less clear. It should be noted that these studies—and especially Harvey et al.'s (2016) study, mainly probed an "overground" hacking community (i.e., those openly attending hacking conferences, working in IT security, and receiving academic-professional training, as opposed to an underground hacking community). It may be reasonable to expect that those who choose to be part of such a community tend not to have extensive impairments in social skills. However, as things stand, it is not possible to discard that they may tend to downplay social deficits as a tribute to their shared identity as a well-integrated community. When confronting self-reports with a prototypical hacking task, attention to detail appeared to show only a weak and non-significant relation, whereas systemizing showed a robust relation with an actual performance which could not be explained by expertise or sex differences. On the whole, based on the evidence presented in this section, we can conclude that a relation between a restricted set of autistic traits and hacking skills/interests is highly likely. Since all the evidence is correlational, it is not possible to establish any causal connection. Further research is needed to probe this relation with a wide range of real-world hacking tasks and to clarify the relevant issue of social skills.

3.2 Engagement in cybercrime

Nevertheless, not everyone who has the talent, skills, or interest in hacking will be inclined to engage in criminal cyber-activity such as hacking with criminal intent. Yet, there is a narrative constructed in the literature, linking autism, either directly or by a focus on autistic traits, to cyber criminality. A scoping review of the literature highlighted that many of the investigations into autism or autistic traits and hacking were carried out in studies looking at the wider issue of cyber criminality. Thus, this section will consider findings related to cybercrime also highlighting any specific findings relating to hacking for criminal purposes (see Table 1).

Theoretical interpretations on the link between ASD and offending (in the broad context) have been posited in the literature (see King & Murphy, 2014 for a review); such notions often focus on impairments in social skills, which are characteristic of ASD; for instance, it has been suggested that social deficits and obsessional interests may lead to a lack of understanding of the consequences of criminal activity and its effects on those involved, as well as increasing an individual's vulnerability into being coerced into committing a crime (Howlin, 2004, in King & Murphy, 2014). However, taken together, findings from studies probing links between ASD and offending generally offer little evidence to support the idea of greater propensity to offend (e.g., Ghaziuddin, Tsai, & Ghaziuddin, 1991; Mouridsen, 2012).

In relation to cybercrime, the basis for highlighting this association in the literature is based largely on a limited number of high-profile cases of hackers (see Introduction for examples) such cases have often been presented in the media rather than in scientific outputs; thus, they often lack a systematic or empirical approach and are built around anecdotal evidence. Consequently, caution must be exercised when considering their take-home messages. With these considerations in mind, the aim of the following section is to discuss empirical validation of relationships between autism, or autistic traits and cybercrime. However, although autistic individuals *tend* to score higher on measures of autistic traits than non-clinical individuals, not all do, and all individuals measuring high in autistic traits will not all be autistic (Baron-Cohen et al., 2001). As such, a distinction must be made between findings from studies of non-clinical and clinical populations.

Our systematic search returned one study specifically focusing on the association between cybercrime and ASD (Ledingham & Mills, 2015), which gave very general evidence to suggest autistic individuals were involved in cybercrime. There were a further three studies related to autistic traits in the general population and cybercrime or deviance, these studies often also succeeded in capturing a small number of responses from individuals with an autism diagnosis; however, this was not their primary focus and where this is the case, response numbers are too low to offer a clear insight into autism and its relationship with cybercrime. They offered mixed results for the association between autistic traits and cybercrime or deviancy.

Ledingham and Mills (2015) investigated associations between autism and cybercrime; with an attempt to restrict this to "an illegal act where the only means of commission is via the use of a computer and the internet and includes the technical expertise required to commit cybercrime where network intrusions, hacking, the creation and deployment of malware are concerned" (p. 2). This study included a literature review and empirical study where eight responses in total were captured from representatives of a cross-section of law-enforcement agencies working in the UK, USA, Australia, New Zealand, Germany, Netherlands, and Denmark: a representation of the reported 70 countries having cybercrime legislation.

Adopting a short online survey, which was distributed to cybercrime units, respondents answered questions such as "If you identify that your department has investigated a cyber-offender with an Autistic Spectrum Disorder, which offences were they suspected of committing?." Of the eight respondents, six reported being in a national cybercrime unit and two reported being in a fraud and cybercrime unit. All reported that mental health issues were recognized in their country but only half of the sample stated they had received training in understanding and dealing with people with mental health issues; importantly only one stated they had received training in ASD or Asperger's syndrome. Of further note is the large discrepancy in the definition of cybercrime across countries, with some suggesting a very specific definition, such as "hacking, coding, malware and ddos only" ($N = 5$) and others answering more generally "All offences committed on the internet" ($N = 1$). However, again it appeared respondents could choose more than one response, so it is uncertain exactly what the definition of cybercrime means across the represented countries.

In terms of understanding the mental health of offenders, half of the sample stated "no" when answering whether arrested persons would undergo any medical assessment in custody; the remaining three who answered (one did not respond) stated "other" with assessment only being carried out where a condition was suspected. Interestingly, of the six respondents offering information on the offenses cyber-offenders with ASD were suspected of committing, four stated these interactions were in relation to hacking offenses. The number of respondents for each option suggests respondents could select more than one answer.

Thus, the findings from the empirical section of this study offer general insights and suggest that autistic individuals *carry out* cybercrime, that is, some representatives of law-enforcement agencies report having been in contact with autistic individuals during the course of their duties. The survey specifically asks "How many suspects has your department arrested who either were already diagnosed with an autism spectrum disorder prior to arrest or received a diagnosis after arrest," for which three respondents answered that the offender was already diagnosed. However, it is slightly unclear as to whether all investigated persons went on to receive a medical diagnosis or whether this may have been based on the presentation of autistic symptomatology. For instance, in terms of the respondents' awareness of the investigated cyber-offenders ASD, responses included "symptoms were noticed during hearings not a clear diagnose" and "some of the attributes became apparent during interactions with the offender." That is, it is unclear whether respondents would have answered only based on medically diagnosed offenders for all questions. The authors do state the brief online survey was preceded by a short information campaign. However, the contents of this information are unknown; thus, it is unclear whether further information was given to respondents at this stage. Furthermore, the respondents' experience and rank are not made clear in the study and as such, neither is the timeframe under consideration or respondents' level of knowledge. Thus, there is the possibility that respondents vary on these factors across the countries included in the study.

The Ledingham and Mills study offers preliminary general evidence to link autism with cybercrime, as such, it has set the foundations on which further research has been built (Payne et al., 2019; Payne et al., 2020). The Payne et al. studies have focused specifically on the commission of cyber-dependent crime; including all crimes requiring a computer to be executed rather than those crimes that may be aided by a computer but which do not necessarily require one (Payne et al., 2019; Payne et al., 2020).

The most recent of these studies (Payne et al., 2020) investigated reasons for engaging or declining to engage in cyber-dependent crime in a group of offenders recruited through the National Crime Agency, having committed a Computer Misuse Offense, and decliners, recruited from a larger sample of individuals with a high degree of computer skills, but who had noted that they had been approached and had declined to commit a cyber-dependent offense. By capturing autistic traits, as measured by the AQ, the study investigated associations between reasons for offending/non-offending behavior and autistic behaviors (Payne et al., 2020). However, a large element of the study is qualitative in nature, with questions and interviews being used to probe the reasons for offending or declining to offend. One of the main findings of this study showed autistic traits were higher for the declining group than the offender group. The authors also suggest that the themes differed for declining to commit offenses based on autistic traits, with motivations of low reward and high risk being linked with high AQ scores (above 26), whereas motivations related to law and morality were associated with AQ scores below the cut-off (Payne et al., 2020).

Although this study adds to the limited empirical research on the association between cyber-dependent offending and autistic traits, the sample of offenders included in the study is very small ($N = 7$) and may represent a specific type of offender. For instance, offenders were drawn from a database of individuals who had committed computer misuse offenses and who were known to the National Crime Agency; they also self-selected to be part of the study (only 10% of those contacted, agreed to take part). Their motivations and reasons for engaging in cyber-dependent crime may, therefore, differ from those of individuals who have committed crimes and who have never been caught or from those of individuals who do not want to disclose details for their reasons for offending. Motivations for offending were gathered by telephone or face-to-face interviews, whereas motivations for declining were gathered using online questions, which may have led to some differences in the quality of information offered by the two groups.

Noteworthy is the difference found between the offenders (mean age of approximately 18 years) and the decliners (mean age of approximately 29 years) in age. The overall age range in this study is also of interest with the youngest reported to be 14 years and the oldest 47 years, as such, there may be variations in motivations based on development. The offender group was all male, whereas the decliner group included a fairly balanced representation of male and female participants (16 males and 13 females). Moreover, a significant difference between the offenders and non-offenders was found based on IQ, with IQ statistically higher for individuals declining to commit crime than offenders; yet there were no significant differences found for digital skills between the groups. Interestingly, one of the main themes identified for offending behavior was lack of understanding, which was cited by all participants and included lack of understanding of consequences, legal implications, and impact of offenses.

Importantly, the findings of this study are not in line with a further study returned in our search by Payne et al. (2019), investigating a larger sample ($N = 290$) of computer-users who completed an anonymous survey; the age range captured in this study was even broader (14 to 74 years), and the study included responses by 122 individuals who had reported committing a cyber-dependent crime. These crimes could relate to phishing, webcam managing, file hijacking, key-logging, screenshot managing, employing and clicking, hacking, or launching a distributed denial-of-service attack, with respondents providing a yes/no answer for each. Thus, participants could score between 0 and 8 for the cyber-dependent crime. None of the respondents had been convicted or cautioned for cyber-dependent crimes. Interestingly, hacking was the highest reported cyber-dependent crime committed with 87 respondents reporting to have committed this crime. The study findings suggest a unit increase in AQ score was associated with a 4% increased adjusted odds of having carried out a cyber-dependent crime. This association was mediated by digital skills when IQ and age were controlled for. Interestingly, in this study hacking appears to be defined as "accessing computer systems" (Payne et al., 2019, Appendix). It is uncertain whether respondents were aware the questionnaire was titled the "Cyber-Dependent Crime Questionnaire" and whether respondents would only class hacking for criminal intent in this definition or include other less sinister activities such as interrogating systems to highlight weaknesses.

The difference in the Payne et al. (2019) and the (Payne et al., 2020) studies may come down to methodological heterogeneity; for instance, the 2020 study included individuals who had been caught for their cyber-dependent crime activities or who had declined to take part in a cyber-dependent crime, whereas the 2019 study included a general sample of internet users. Thus, the Payne et al. (2019) study is likely to have greater variation across the sample whereas, the latter study (Payne et al., 2020) investigated sub-groups, which may not be representative of the population at large.

Findings of a previous study investigated the link between autistic traits and cyber-deviance in 275 university students (note there were 296 respondents in total but only 275 completed the AQ), which included hacking, cyberbullying, identity theft, and virus writing (Seigfried-Spellar et al., 2015). For this study hacking was defined as "knowingly accessing a computer system or network without authorization" (Seigfried-Spellar et al., 2015, p. 536), with the inclusion of unauthorized, this definition may allude to hacking for sinister purposes. However, it is unknown whether individuals in the hacking community would agree.

This study takes a slightly broader look into cyber-related activities than the Payne et al. studies (Payne et al., 2019; Payne et al., 2020), which would not consider cyberbullying as a cyber-dependent crime instead classifying this as a cyber-enabled crime; that is, it is a method of undertaking a traditional crime, which can be aided by the use of a computer, but it is not necessary to have a computer to engage in bullying behavior.

The Seigfried-Spellar et al. (2015) study adopted an online survey and captured responses from 150 males and 145 females (1 declined to answer) although not all completed the AQ; 161respondents who completed the AQ self-reported having engaged in some type of computer deviant behavior (59%). The authors looked at the differences between the AQ scores of individuals who committed various types of computer-deviancy and compared them with computer non-deviants. This study found that those who had engaged in hacking had significantly lower attention to detail score (subscale from the AQ) than non-hackers. In fact, there was no evidence from this study to suggest that hacking was significantly positively related to any of the AQ subscales. However, AQ scores were shown to be positively related to identity theft, cyberbullying, and virus writing. Thus, this study appears to highlight a difference between types of deviant behavior and autistic traits. The study also investigated whether there were differences in the AQ and subscales based on the level of self-reported deviance (i.e., whether individuals had engaged in 1,2,3, or all types of deviant behaviors). Findings revealed significant differences for AQ total, social skills, communication, and imagination subscales, with individuals reporting engaging in all four types of computer deviant behaviors having higher scores than those engaging in fewer types of computer deviant behaviors. However, there were only 14 people who self-reported engaging in all four behaviors.

Taking the findings together, there is mixed evidence on the link between cyber-activities including hacking and autism or autistic traits. The current literature is heterogeneous in terms of the populations sampled and the definitions used for key terminology such as hacking. These variations may have driven the inconsistency in results. As it stands, the existing literature offers interesting preliminary insights into autistic traits and cybercrime or deviance; however, given the widely held image of hackers presenting with characteristic autistic behaviors, the lack of empirical research is surprising.

3.3 Vulnerability to cybercrime

The previous section considered autism and autistic traits and cybercrime or deviance; however, there is another line of enquiry, which probes the notion that autistic individuals or individuals with higher autistic traits are more at risk of falling victim to cybercrime. Our systematic review returned three studies of this nature (Macmillan et al., 2020; Neupane et al., 2018; Taylor-Jackson & Matthews, 2020; see Table 1). The findings of the studies investigating adults offer no evidence to suggest autistic individuals or individuals with higher AQ scores are more vulnerable to falling victim to phishing attempts or fake news. One study that investigated online safety risks of autistic children compared to a control group offered evidence to suggest autistic children may be more likely to experience more online risk (Macmillan et al., 2020). However, there was no association between autism and risk of phishing/hacking attacks and the main area highlighted for increased risk was for making unauthorized purchases.

Turning first to investigations into behaviors in adults, delays in the decision-making process and variations in emotional processing by autistic individuals are proposed as reasons for potential associations between autistic traits and susceptibility to fake news (Taylor-Jackson & Matthews, 2020). The sample for this study was taken from a non-clinical population made up of 35 predominantly female students with a mean age of approximately 21 years. Participants were categorized as being low or high in autistic traits based on a median split of the AQ score and were asked to categorize 20 news articles as fake or real. News articles also differed in their level of emotiveness. The findings offered no evidence for differences in susceptibility to fake news based on autistic traits. Participants' detection confidence scores were also captured, with no significant difference found between the low and high autistic trait groups. However, a limitation also noted by the authors of this study was the use of real news, which may have been seen previously by participants, potentially making the task a test of memory more than the ability to detect fake news.

A further study carried out to investigate differences between autistic and non-autistic individuals and susceptibility to phishing attacks has offered evidence to suggest autistic individuals are no more susceptible to phishing attacks than individuals without autism (Neupane et al., 2018). The Neupane study (Neupane et al., 2018) tested a group of 15 autistic individuals (with a diagnosis given by a doctor or clinical psychologist) and a control group of 15 non-autistic individuals on the ability to distinguish between 30 real and 30 fake websites, shown for 6 s. All participants were required to use the internet often. Approximately two-thirds of the sample was made up of males but this was balanced across groups (i.e., 66.6% of the autistic group were males and 66.6% of the non-autistic group were males). On average, the autistic group appeared younger than the control group. Further differences in the groups are apparent in the educational background with all of the control groups having either a bachelors or masters degree, whereas only two autistic individuals report having a bachelors or masters degree. Ten autistic individuals report being unemployed, whereas only one of the control group reported being unemployed.

Participants were given 10 s after the website presentation to respond as to whether the website was real or fake. The chosen websites were highly ranked; fake website addresses were then created with some of the fake websites having logos or layouts changed to look fake. The 15 fake websites with only changes made to the URL were categorized as difficult, whereas the 15 fake websites with logo changes were termed easy.

Results showed that, as expected, on average autistic individuals had a higher AQ score than non-autistic individuals. However, the authors postulated that autistic individuals would be more susceptible to phishing attacks based on deficits in social skills often seen in autism. This was not supported; furthermore, when the AQ scores were looked at and correlational analyses were run on response time and accuracy of the phishing detection task, no significant correlations were found.

Finally, our systematic review returned one study, which was specifically related to children's online safety; this study focused on differences in online safety between autistic and non-autistic children over the age of 6 years (average age approximately 11 years) (Macmillan et al., 2020). The study adopted parental reports to measure online safety risks, recruiting parents through social media and through related organizations. Parents were asked to check yes or no as to whether their child had experienced different online safety risks, such as phishing/scamming, with examples given. Although this study focused on the broader concept of online safety, one of the areas investigated was a content risk, which includes the risk of being the victim of phishing or hacking. The study offered findings supporting the theory that according to parental reports, autistic children experienced a significantly higher total amount of online safety risks than non-autistic children, however, the effect was small. Furthermore, there was no evidence to suggest an association between autism and susceptibility to phishing or hacking attacks. Instead, an association was found between autism and susceptibility to buying goods services online (unauthorized purchases), with a 4.61 increase in odds if children were autistic.

The use of parental reports in this study may allow for the inclusion of children who may otherwise not be able to report findings; however, there is also the possibility that parents are not privy to all risks their children have experienced. Thus, results may in part reflect a difference in parental anxieties or strongly held beliefs rather than reflecting true online safety risk. Also noteworthy is the lack of representation of girls with autism in the study.

In conclusion, there is a small body of research specifically probing the link between autism or autistic traits and vulnerability to experiencing cybercrime or risk online. Much of this research involves small sample sizes and is exploratory in nature. The findings of these studies do not support the notion that individuals with autism or individuals with high levels of autistic traits are more susceptible to cybercrime or risks; however, it should be noted that the existing evidence only probes this link with very specific types of cybercrime. Thus, the findings must be interpreted cautiously.

4 Conclusions and future directions

Our systematic review revealed that the research linking autism or autistic traits with hacking is very limited and although widening the search to include cybersecurity or cybercrime there is still a small body of research from which a conclusion can be drawn. Gaining insight into the popular notion that autism or autistic traits are associated with hacking is made more challenging by the quality of the existing empirical research, as many of the studies are preliminary or small-scale exploratory studies. These studies add to the literature by highlighting potentially promising areas for future research; however, they may lack the power required to provide conclusive evidence.

The findings from studies relating to skills and interest in hacking discussed in Section 3.1 offer preliminary evidence to suggest that there may be significant links indeed, in particular with traits such as attention to detail and systemizing. Hackers consistently achieve higher AQ scores than the general population or student scientists but the evidence points to specific subscales, and the attention to detail subscale in particular, as to the most consistent source of this effect. Interestingly, the attention to detail subscale of the AQ appears to behave a little differently from the other subscales (e.g., Baron-Cohen et al., 2001; Hoekstra et al., 2008; Rusconi et al., 2015; Wheelwright et al., 2010) and in this context, it was weakly correlated with performance in a prototypical hacking challenge. Systemizing, on the other hand, is a more complex autistic trait that likely builds on a good attention to detail basis but also entails a characteristic drive to analyze and explore systems and, hence, dedication to develop expert skills in a lawful domain of knowledge (Baron-Cohen, 2006). This self-reported trait appears to be not only associated with objective interests in hacking but also with performance. Crucially, in these studies, there is no systematic relation assumed between autistic traits and engagement in cybercrime, as hacking is not considered a priori a criminal activity.

The evidence for links between autism or autistic traits and cybercrime including hacking, discussed in Section 3.2, is mixed and at present does not align with media representations of the hacker "type." On the basis of the findings, we cannot suggest that there is evidence for the absence of association between autism or autistic traits and hacking as an unauthorized activity; instead, we highlight the need for future targeted investigations to offer more insight.

Similarly, our findings relating to susceptibility to falling victim to cybercrime including hacking and autism or autistic traits, discussed in Section 3.3, offer no evidence supporting an association in the adult population. Current research has investigated the susceptibility to phishing and fake news and offers no evidence of a greater susceptibility based on autism diagnosis (phishing) or autistic traits (fake news). However, these existing studies have adopted tasks where participants are required to distinguish real or fake items; this is only one element of a real-world situation. In fact, theoretically, it may be more likely that the deception autistic individuals or individuals with high autistic traits may be more susceptible to is the deception that comes from social contact. For instance, the emails, social media posts, or links that lead to these websites or news articles; such factors may also have some influence on whether news articles or websites are deemed trustworthy. It is understood that ethical considerations are paramount, which raises challenges with designing experimental methodologies that are both ecologically valid and safe. However, the current evidence is limited and highlights the need for further elegant studies that can investigate online activity and risk in ways that reflect real-life risks. There was some evidence for greater online risks for autistic children when compared to a control group (Macmillan et al., 2020). However, this study utilized parental self-reports and there was no association shown between autism and susceptibility to phishing or hacking attacks.

Across the literature there were common limitations; for example; the use of self-reported questionnaires results in the capture of information on individual perceptions of behavior and not necessarily actual behavior. Furthermore, accurate data are reliant on sound levels of self-awareness and honesty; which may be a greater challenge in research where respondents are asked to detail criminal behavior or behavior that may be regarded as socially unacceptable. Furthermore, for studies targeting autistic individuals, it may be that not all individuals will be able to complete self-reported questionnaires; as such, where studies capture responses by autistic individuals, responses may not be representative of the entire autistic spectrum.

The heterogeneity of studies is also a challenge; there were two specific issues that were clear within the literature. The first is related to the definition of, and motivation for hacking. This lack of clarity may mean that evidence linking individual differences and interest in and propensity to carry out hacking-related activities is confused with interest in and propensity to carry out hacking-related crime. It is clear that the rules governing cyber-activity can be gray, and this remains the case for hacking. Thus, gaining unauthorized access to a computer system may not in itself be classified to those in the hacking community as an offense and, as such, may not be regarded as criminal. For instance, an individual may be gaining access to a system to highlight a weakness. Therefore, future studies should ensure the definitions of the activity match the nature of the study. Thus, if studying cybercrime then the definition should make clear the criminal aspect of the act; for instance, on the Cyber-Dependent Crime Questionnaire used by Payne et al. (2019) in their study investigating the relationship between cyber-dependent crime, autistic-like traits, and autism hacking is simply termed "accessing computer systems."

The second issue relates to the reliance on AQ cut-off scores to gain an insight into whether clinical levels of autism are linked with cyber-dependent offending (see Introduction for discussion on the difference between autism and autistic traits). The AQ is non-diagnostic and although this distinction is often mentioned in the text of studies, the reliance on results of such analyses as an indication of links with clinical levels of autism may blur the line between diagnosed autism and high self-reported traits. Furthermore, for those studies interrogating associations between autism and hacking, there is a need for well-matched control groups, in terms of computer skills, interests, age, and IQ. Otherwise, interpreting any differences is a challenge.

With regards to associations between self-reported autistic traits and hacking-related activity, the correlational nature of the existing research must be noted; as such, causation cannot be inferred. Thus, even where evidence does link autistic traits and hacking, there is also the possibility that higher reported AQ scores or subscale scores arise from greater indulgence in solo hacking-related activities and not the other way around; that is, it may be that higher self-reported autistic traits are an outcome of reliance on recent memories of behaviors and they are not necessarily a predisposition.

In conclusion, this study offers an overview of the existing scientific literature of an empirical nature probing the association between hacking and autism or autistic traits. The results suggest preliminary evidence linking skills and interests in hacking-related activities and autistic traits but offer no conclusive evidence to suggest any preferential associations with involvement in cybercrime or vulnerability to cybercrime including hacking.

The findings of this systematic review have highlighted two key areas, which would particularly benefit from future studies to advance current knowledge including the need to ensure definitions are consistent across the literature; that is, studies on hacking for malicious purposes should be defined in a standard way, the definition being checked with individuals in the hacking community to ensure it aligns with their views. Additionally, there is a need to ensure studies investigating autism use clinical samples, and the reliance on scoring on non-diagnostic tools is restricted to studies investigating autistic traits. Furthermore, future research may also benefit from an increase in pre-registered studies as many existing studies are exploratory in nature.

References

American Psychiatric Association. (2013). *Diagnostic and statistical manual of mental disorders (DSM-5®)*. American Psychiatric Pub.

Baird, G., Simonoff, E., Pickles, A., Chandler, S., Loucas, T., Meldrum, D., et al. (2006). Prevalence of disorders of the autism spectrum in a population cohort of children in South Thames: The Special Needs and Autism Project (SNAP). *The Lancet, 368*(9531), 210–215.

Baron-Cohen, S. (2006). The hyper-systemizing, assortative mating theory of autism. *Progress in Neuro-Psychopharmacology and Biological Psychiatry, 30*(5), 865–872.

Baron-Cohen, S. (2010). Empathizing, systemizing, and the extreme male brain theory of autism. *Progress in Brain Research, 186*, 167–175.

Baron-Cohen, S., Ashwin, E., Ashwin, C., Tavassoli, T., & Chakrabarti, B. (2009). Talent in autism: Hyper-systemizing, hyper-attention to detail and sensory hypersensitivity. *Philosophical Transactions of the Royal Society of London. Series B, Biological Sciences, 364*(1522), 1377–1383. https://doi.org/10.1098/rstb.2008.0337.

Baron-Cohen, S., Leslie, A. M., & Frith, U. (1986). Mechanical, behavioural and intentional understanding of picture stories in autistic children. *British Journal of Developmental Psychology*, *4*(2), 113–125.

Baron-Cohen, S., Richler, J., Bisarya, D., Gurunathan, N., & Wheelwright, S. (2003). The systemizing quotient: An investigation of adults with asperger syndrome or high-functioning autism, and normal sex differences. *Philosophical Transactions of the Royal Society of London. Series B, Biological Sciences*, *358*(1430), 361–374.

Baron-Cohen, S., Wheelwright, S., Skinner, R., Martin, J., & Clubley, E. (2001). The autism-spectrum quotient (AQ): Evidence from asperger syndrome/high-functioning autism, males and females, scientists and mathematicians. *Journal of Autism and Developmental Disorders*, *31*(1), 5–17. https://doi.org/10.1023/A:1005653411471.

BBC. (2012). *Gary McKinnon extradition to US blocked by Theresa May*. London: BBC. https://www.bbc.com/news/uk-19957138. (Accessed 31 May 2021).

BBC. (2013). *British man accused of hacking into US government systems*. London: BBC. https://www.bbc.com/news/technology-24712214. (Accessed 31 May 2021).

BBC. (2015). *Lauri Love hacking case: Suffolk man re-arrested*. London: BBC. https://www.bbc.com/news/uk-england-suffolk-33560879. (Accessed 31 May 2021).

BBC. (2018a). *Lauri love case: Hacking suspect wins extradition appeal*. BBC. Retrieved from: https://www.bbc.co.uk/news/uk-england-42946540. (Accessed 31 May 2021).

BBC. (2018b). *Lauri Love case: US abandons extradition case*. London: BBC. https://www.bbc.com/news/uk-england-suffolk-43119355. (Accessed 31 May 2021).

Belmonte, M. K., Gomot, M., & Baron-Cohen, S. (2010). Visual attention in autism families: 'Unaffected' sibs share atypical frontal activation. *Journal of Child Psychology and Psychiatry*, *51*(3), 259–276.

Bratus, S. (2007). What hackers learn that the rest of us don't: Notes on hacker curriculum. *IEEE Security and Privacy*, *5*(4), 72–75.

Brugha, T. S., McManus, S., Bankart, J., Scott, F., Purdon, S., Smith, J., et al. (2011). Epidemiology of autism spectrum disorders in adults in the community in england. *Archives of General Psychiatry*, *68*(5), 459–465.

Brunsdon, V. E., & Happé, F. (2014). Exploring the 'fractionation'of autism at the cognitive level. *Autism*, *18*(1), 17–30.

Burger, J. M. (2011). *Introduction to personality* (8th ed.). United Kingdom: Wadsworth Cengage Learning.

Chakrabarti, B., Dudbridge, F., Kent, L., Wheelwright, S., Hill-Cawthorne, G., Allison, C., et al. (2009). Genes related to sex steroids, neural growth, and social–emotional behavior are associated with autistic traits, empathy, and asperger syndrome. *Autism Research*, *2*(3), 157–177.

Chamorro-Premuzic, T., & Furnham, A. (2006). Intellectual competence and the intelligent personality: A third way in differential psychology. *Review of General Psychology*, *10*(3), 251–267.

Colvert, E., Tick, B., McEwen, F., Stewart, C., Curran, S. R., Woodhouse, E., et al. (2015). Heritability of autism spectrum disorder in a UK population-based twin sample. *JAMA Psychiatry*, *72*(5), 415–423.

Constantino, J. N., Davis, S. A., Todd, R. D., Schindler, M. K., Gross, M. M., Brophy, S. L., et al. (2003). Validation of a brief quantitative measure of autistic traits: Comparison of the social responsiveness scale with the autism diagnostic interview-revised. *Journal of Autism and Developmental Disorders*, *33*(4), 427–433.

Dietz, P. M., Rose, C. E., McArthur, D., & Maenner, M. (2020). National and state estimates of adults with autism spectrum disorder. *Journal of Autism and Developmental Disorders*, *50*(12), 4258–4266.

Draaisma, D. (2009). Stereotypes of autism. *Philosophical Transactions of the Royal Society, B: Biological Sciences*, *364*(1522), 1475–1480.

English, M. C., Gignac, G. E., Visser, T. A., Whitehouse, A. J., Enns, J. T., & Maybery, M. T. (2021). The Comprehensive Autistic Trait Inventory (CATI): Development and validation of a new measure of autistic traits in the general population. *Molecular Autism*, *12*(1), 1–23.

Forsyth, F. (2018). *The Fox*. London: Bantam Press.
Freckelton, I. (2020). Autism spectrum disorder and suitability for extradition: Love v the Government of the United States [2018] 1 WLR 2889;[2018] EWHC 172 (Admin) per Burnett LCJ and Ouseley. *Psychiatry, Psychology and Law*, *27*(2), 181–191.
Frith, U. (1991). Asperger and his syndrome. In *Vol. 14. Autism and Asperger Syndrome* (pp. 1–36).
Furnell, S. (2002). *Cybercrime: Vandalizing the information society*. London: Addison-Wesley.
Ghaziuddin, M., Tsai, L., & Ghaziuddin, N. (1991). Brief report: Violence in asperger syndrome, a critique. *Journal of Autism and Developmental Disorders*, *21*(3), 349–354.
Graves, K. (2010). *CEH Certified Ethical Hackers study guide*. Indianapolis: Wiley.
Gunkel, D. (2000). We are borg: Cyborgs and the subject of communication. *Communication Theory*, *10*(3), 332–357.
Hadnagy, C. (2018). *Social engineering: The science of human hacking*. John Wiley & Sons.
Hannemyr, G. (1999). Technology and pleasure: Considering hacking constructive. *First Monday*, *4*(2). https://doi.org/10.5210/fm.v4i2.647.
Happé, F. (1999). Autism: Cognitive deficit or cognitive style? *Trends in Cognitive Sciences*, *3*(6), 216–222.
Happe, F. (2018). Why are savant skills and special talents associated with autism? *World Psychiatry: Official Journal of the World Psychiatric Association (WPA)*, *17*(3), 280–281. https://doi.org/10.1002/wps.20552 [doi].
Happé, F., Ronald, A., & Plomin, R. (2006). Time to give up on a single explanation for autism. *Nature Neuroscience*, *9*(10), 1218–1220.
Happé, F., & Vital, P. (2009). What aspects of autism predispose to talent? *Philosophical Transactions of the Royal Society, B: Biological Sciences*, *364*(1522), 1369–1375.
Harvey, I., Bolgan, S., Mosca, D., McLean, C., & Rusconi, E. (2016). Systemizers are better code-breakers: Self-reported systemizing predicts code-breaking performance in expert hackers and naive participants. *Frontiers in Human Neuroscience*, *10*, 229. https://doi.org/10.3389/fnhum.2016.00229 [doi].
Hill, A. L. (1977). Idiot savants: Rate of incidence. *Perceptual and Motor Skills*, *44*(1), 161–162.
Hoekstra, R. A., Bartels, M., Cath, D. C., & Boomsma, D. I. (2008). Factor structure, reliability and criterion validity of the autism-spectrum quotient (AQ): A study in dutch population and patient groups. *Journal of Autism and Developmental Disorders*, *38*(8), 1555–1566.
Holt, T. J. (2010). Examining the role of technology in the formation of deviant subcultures. *Social Science Computer Review*, *28*(4), 466–481.
Howlin, P. (2004). *Autism and Asperger syndrome: Preparing for adulthood*. Routledge.
Howlin, P., Goode, S., Hutton, J., & Rutter, M. (2009). Savant skills in autism: Psychometric approaches and parental reports. *Philosophical Transactions of the Royal Society, B: Biological Sciences*, *364*(1522), 1359–1367.
Hurley, R. S., Losh, M., Parlier, M., Reznick, J. S., & Piven, J. (2007). The broad autism phenotype questionnaire. *Journal of Autism and Developmental Disorders*, *37*(9), 1679–1690.
Hutchings, A. (2013). Hacking and fraud: Qualitative analysis of online offending and victimization. In *Global criminology: Crime and victimization in the globalized era* (pp. 93–114).
King, C., & Murphy, G. H. (2014). A systematic review of people with autism spectrum disorder and the criminal justice system. *Journal of Autism and Developmental Disorders*, *44*(11), 2717–2733.
Lai, M., Lombardo, M. V., Chakrabarti, B., & Baron-Cohen, S. (2013). Subgrouping the autism "spectrum": Reflections on DSM-5. *PLoS Biology*, *11*(4), e1001544.
Ledingham, R., & Mills, R. (2015). A preliminary study of autism and cybercrime in the context of international law enforcement. *Advances in Autism*, *1*(1), 2–11.
Levy, S. (1984). *Hackers: Heroes of the computer revolution*. Doubleday Garden City, NY: Anchor Press.
Lilley, P. (2002). *Hacked, attacked & abused: Digital crime exposed*. Kogan Page Publishers.
Ludlow, P. (1996). *High noon on the electronic frontier: Conceptual issues in cyberspace*. MIT Press.

Lundqvist, L., & Lindner, H. (2017). Is the autism-spectrum quotient a valid measure of traits associated with the autism spectrum? A rasch validation in adults with and without autism spectrum disorders. *Journal of Autism and Developmental Disorders, 47*(7), 2080–2091.

Macmillan, K., Berg, T., Just, M., & Stewart, M. (2020). Are autistic children more vulnerable online? relating autism to online safety, child wellbeing and parental risk management. In *Paper presented at the proceedings of the 11th nordic conference on human-computer interaction: Shaping experiences.* Tallinn, Estonia: Shaping Society. https://doi.org/10.1145/3419249.3420160.

Mann, M., Warren, I., & Kennedy, S. (2018). The legal geographies of transnational cyber-prosecutions: Extradition, human rights and forum shifting. *Global Crime, 19*(2), 107–124.

Meilleur, A. S., Jelenic, P., & Mottron, L. (2015). Prevalence of clinically and empirically defined talents and strengths in autism. *Journal of Autism and Developmental Disorders, 45*(5), 1354–1367.

Meyer, G. (1989). *The social organization of the computer underworld.* Unpublished Manuscript Northern Illinois University.

Mitnick, K. D., & Simon, W. L. (2003). *The art of deception: Controlling the human element of security.* John Wiley & Sons.

Moher, D., Liberati, A., Tetzlaff, J., Altman, D. G., & The PRISMA Group. (2009). Preferred reporting items for systematic reviews and meta-analyses: The PRISMA statement. *PLoS Medicine, 6*(7).

Mouridsen, S. E. (2012). Current status of research on autism spectrum disorders and offending. *Research in Autism Spectrum Disorders, 6*(1), 79–86.

Munson, L. (2009). *What are the main differences between hackers and crackers? Security FAQs.* Retrieved from: http://www.security-faqs.com/what-are-the-maindifferences-between-hackers-and-crackers.html.

Neupane, A., Satvat, K., Saxena, N., Stavrinos, D., & Bishop, H. J. (2018). *Do social disorders facilitate social engineering? A case study of autism and phishing attacks.* https://doi.org/10.1145/3274694.3274730.

Payne, K., Maras, K. L., Russell, A. J., Brosnan, M. J., & Mills, R. (2020). Self-reported motivations for engaging or declining to engage in cyber-dependent offending and the role of autistic traits. *Research in Developmental Disabilities, 104,* 103681. https://doi.org/10.1016/j.ridd.2020.103681.

Payne, K., Russell, A., Mills, R., Maras, K., Rai, D., & Brosnan, M. (2019). Is there a relationship between cyber-dependent crime, autistic-like traits and autism? *Journal of Autism and Developmental Disorders, 49*(10), 4159–4169. https://doi.org/10.1007/s10803-019-04119-5.

Posserud, M., Lundervold, A. J., & Gillberg, C. (2006). Autistic features in a total population of 7–9-year-old children assessed by the ASSQ (autism spectrum screening questionnaire). *Journal of Child Psychology and Psychiatry, 47*(2), 167–175.

Rimland, B. (1978). *Savant capabilities of autistic children and their cognitive implications.*

Ronson, J. (2005). Game over. In *The Guardian.* https://www.theguardian.com/theguardian/2005/jul/09/weekend7.weekend2. (Accessed 31 May 2021).

Rusconi, E., Ferri, F., Viding, E., & Mitchener-Nissen, T. (2015). XRIndex: A brief screening tool for individual differences in security threat detection in x-ray images. *Frontiers in Human Neuroscience, 9,* 439. https://doi.org/10.3389/fnhum.2015.00439.

Ruzich, E., Allison, C., Smith, P., Watson, P., Auyeung, B., Ring, H., et al. (2015). Measuring autistic traits in the general population: A systematic review of the autism-spectrum quotient (AQ) in a nonclinical population sample of 6,900 typical adult males and females. *Molecular Autism, 6*(1), 1–12.

Saloviita, T., Ruusila, L., & Ruusila, U. (2000). Incidence of savant syndrome in finland. *Perceptual and Motor Skills, 91*(1), 120–122.

Schell, B. H., Dodge, J. L., & Moutsasos, S. S. (2002). *The hacking of America: Who's doing it, why, and how.* Westport, CT: Quorum Books.

Schell, B. H., & Melnychuk, J. (2011). Female and male hacker conferences attendees: Their autism-spectrum quotient (AQ) scores and self-reported adulthood experiences. In *Corporate hacking and technology-driven crime: Social dynamics and implications* (pp. 144–169). IGI Global.

Seigfried-Spellar, K. C., O'Quinn, C. L., & Treadway, K. N. (2015). Assessing the relationship between autistic traits and cyberdeviancy in a sample of college students. *Behaviour & Information Technology, 34*(5), 533–542. https://doi.org/10.1080/0144929X.2014.978377.

Steele. (1983). *The Jargon file*. Retrieved from: http://www.catb.org/jargon/oldversions/jarg447.txt.

Sterling, B. (1991). *Cyber view 91 report*. http://www.eff.org/Misc/Publications/Bruce_Sterling/cyberview_91.report.

Sukhai, N. B. (2004). Hacking and cybercrime. In *Paper presented at the proceedings of the 1st annual conference on information security curriculum development* (pp. 128–132).

Taylor, P. A. (1999). *Hackers: Crime in the digital sublime*. Psychology Press.

Taylor-Jackson, J., & Matthews, S. (2020). The influence of traits associated with autism spectrum disorder (ASD) on the detection of fake news. In *Paper presented at the international conference on human-computer interaction* (pp. 464–472).

Turgeman-Goldschmidt, O. (2008). Meanings that hackers assign to their being a hacker. *International Journal of Cyber Criminology, 2*(2), 382–396.

van Rentergem, J. A. A., Lever, A. G., & Geurts, H. M. (2019). Negatively phrased items of the Autism Spectrum Quotient function differently for groups with and without autism. *Autism, 23*(7), 1752–1764.

Wheelwright, S., Auyeung, B., Allison, C., & Baron-Cohen, S. (2010). Defining the broader, medium and narrow autism phenotype among parents using the autism spectrum quotient (AQ). *Molecular Autism, 1*(1), 1–9.

Wheelwright, S., Baron-Cohen, S., Goldenfeld, N., Delaney, J., Fine, D., Smith, R., et al. (2006). Predicting autism spectrum quotient (AQ) from the systemizing quotient-revised (SQ-R) and empathy quotient (EQ). *Brain Research, 1079*(1), 47–56.

Wing, L. (1988). *The continuum of autistic characteristics. Diagnosis and assessment in autism* (pp. 91–110). Springer.

Further reading

Pring, L., & Hermelin, B. (2002). Numbers and letters: Exploring an autistic savant's unpractised ability. *Neurocase, 8*(4), 330–337.

Vital, P. M., Ronald, A., Wallace, G. L., & Happé, F. (2009). Relationship between special abilities and autistic-like traits in a large population-based sample of 8-year-olds. *Journal of Child Psychology and Psychiatry, 50*(9), 1093–1101.

World Health Organisation. (2004). *ICD-10: International statistical classification of diseases and related health problems: Tenth revision* (2nd ed.). Geneva: World Health Organization. Retrieved from: https://apps.who.int/iris/handle/10665/42980.

CHAPTER

8

An introduction to cyberbullying

Peter J.R. Macaulay[a], *Lucy R. Betts*[b], *James Stiller*[c], *and Blerina Kellezi*[b]

[a]School of Psychology, University of Derby, Derby, United Kingdom
[b]Department of Psychology, Nottingham Trent University, Nottingham, United Kingdom
[c]Department of Psychology, University of Chichester, Chichester, United Kingdom

1 A brief overview of traditional bullying

Traditional bullying (i.e., face-to-face bullying) derives from aggression and is used to distinguish acts that are intentional, compared to those that are not. Dan Olweus (1978) provided the first foundation of understanding bullying in relation to three distinct criteria. As proposed by Olweus, bullying needs to encompass: (a) negative behavior that is intended to inflict harm on another, (b) the behavior is repeated over time, and (c) there is a perceived or actual imbalance of power between the victim and bully. In his early research, Olweus (1993) found from surveys with 150,000 children in Norway and Sweden (approx. Aged 7–16 years) that 15% of children were involved in bullying either as a bully or victim with some regularity. Since then, researchers have continued to explore and examine bullying, and the negative impact it can have. Traditional bullying can manifest in many different ways, which include forms of physical bullying (e.g., pushing, hitting), verbal bullying (e.g., name-calling, verbal threats), social bullying (e.g., rumor circulation), and relational bullying (e.g., manipulation, exclusion) (Baldry & Farrington, 2004; Smith, 2016). However, due to the emergence of information and communication technologies, bullying has now moved into the online domain.

Cybersecurity and Cognitive Science
https://doi.org/10.1016/B978-0-323-90570-1.00002-4

2 The emergence of digital technologies and cyberbullying

The proliferation of digital technology has influenced the growth of social media platforms and applications, allowing more young people to stay connected (Lenhart, 2015). The emergence of information and communication technology has provided an array of social, recreational, and educational benefits for young people (Finkelhor, 2014). One of the potential risks of using digital technology is that of cyberbullying, often regarded as an 'umbrella' term which also includes online bullying and electronic bullying (Tokunaga, 2010).

Bill Belsey was one of the first researchers to coin the term cyberbullying following the launch of the website (http://www.cyberbullying.ca) in 2003 that addressed cyberbullying in Canada (Bauman & Bellmore, 2015). However, the earliest records of the term go back as far as 1995 in a New York Times article (Bauman, 2014). Despite the early emergence and concept of cyberbullying, the literature has continued to debate and seek to clarify a consensus definition of cyberbullying over the recent years (Brewer & Kerslake, 2015; Slonje et al., 2013). One of the most widely cited definitions was proposed by Smith et al. (2008) in England, building on Olweus's three distinct criteria, by adding the electronic aspect associated with cyberbullying. Smith et al. (2008) defined cyberbullying as "an aggressive, intentional act carried out by a group or individual, using electronic forms of contact, repeatedly and over time against a victim who cannot easily defend him or herself" (p. 376). An earlier definition proposed by Patchin and Hinduja (2006) from the USA, defined cyberbullying as the "wilful and repeated harm inflicted through the medium of electronic text" (p. 152). Such definitions of cyberbullying continue to provide the foundation when describing such behaviors.

In a more recent meta-analysis of cyberbullying research, Kowalski et al. (2014) proposed that the definition of cyberbullying comprises four components. First, the behavior needs to be aggressive and intentional on the part of the perpetrator. Second, intentional aggressive behavior needs to be carried out repeatedly. Third, cyberbullying behavior occurs between a victim and perpetrator who are unequal in power. Finally, the behavior needs to occur through a form of electronic technology. Despite these four components resonating with characteristics of traditional bullying, researchers at times fail to address all these components when measuring cyberbullying, thus impacting how it is measured. Some of the literature suggests that there is an overlap between traditional bullying and cyberbullying, within the definitional sense and the experience involved. For example, research from the UK and Italy suggests that victims of traditional bullying are also victims of cyberbullying, as their aggressors use the online space to target their victims in different ways (Baldry et al., 2017; Wolke et al., 2017). Recent research from a large UK national sample found that cyberbullying presents very few new victims in comparison to traditional bullying, but rather offers a new medium for bullies to continue targeting victims in a different manner (Wolke et al., 2017). This study supports the notion that cyberbullying extends bullying beyond the school playground, highlighting a continuity on the roles of bullies and victims between the physical and digital world (e.g., in a sample of Italian students, the role continuity approach was also found; Baldry et al., 2017). This suggests the definition of cyberbullying should be considered through the lens of traditional bullying definitions, extending the traditional criteria into the online environment. On the other hand, other scholars in the USA and Ireland argue traditional bullying and cyberbullying should be considered as two distinct issues (Brown et al., 2014; Corcoran & Mc Guckin, 2014). For example, Brown et al. (2014) highlighted there is little evidence to suggest an overlap between the two forms of bullying, but rather place more importance on the contextual factors (e.g., age and gender) to develop our understanding of different cyberbullying behaviors. However, this has not found support in subsequent literature (Olweus & Limber, 2018). In addition, the fluidity of cyberbullying means aggressors can utilize different online mediums to target their victims, as well as bullying through different realms (e.g., public and private conversations) (Pabian et al., 2018).

3 Definitional issues of cyberbullying

Scholars have debated if the characteristics of traditional face-to-face bullying extend to cyberbullying. Some researchers argue that traditional bullying criteria should be applied to define and explain cyberbullying behaviors, differentiated by the use of digital technologies (Olweus, 2013; Smith, 2019). Definitional challenges can emerge using traditional bullying criteria to define cyberbullying, particularly the debate surrounding repetition and imbalance of power (Smith, 2015). While some may consider cyberbullying to be an extension of traditional bullying, the definitional features in cyberbullying are more ambiguous.

3.1 Intentionality

In traditional bullying, intentionality would involve aggressive behavior that is intended to inflict harm or hurt another individual. However, in cyberbullying, the notion of intent can be viewed as more ambiguous. For example, studies in England have found that in the online domain, young people often perceive the actions of the aggressor as fun or jokes, with no intent to cause harm (O'Brien & Moules, 2013; Tarapdar et al., 2013). Similarly, in the UK young people also see that acts of cyberbullying could be regarded as banter, implicating the difficulty of applying features of intentionality to the issue of cyberbullying (Betts & Spenser, 2017).

In the context of cyberbullying, the role of intention is somewhat ambiguous. For example, intention within cyberbullying may refer to the degree that the perpetrator is aware of what they are doing and the extent to which they understand that their actions are causing harm to the victim (Menesini & Nocentini, 2009). However, in cyberbullying, how the victim perceives the act is also important in characterizing cyberbullying because a victim may perceive an act to be cyberbullying, even though it may not have been intended to be one (Nocentini et al., 2010). In addition to this, Langos (2012) proposed that the repetition of an incident and intention to cause harm can be linked elements in cyberbullying. Langos suggested that when perpetrators repeat the behavior, this might illustrate an intention to harm, by implicating the perpetrators knowledge that they are causing harm to the victim. Therefore, it appears the interpretation of intent in the context of cyberbullying continues to present an issue in the definition of cyberbullying, with intent regarded as a fluid factor that can be influenced by contextual and external factors (Dennehy et al., 2020).

3.2 Repetition

In traditional bullying, the criteria of repetition can be clearly portrayed when bullying occurs consistently in the school playground. In other words, repetition is represented by the number of acts recorded and/or witnessed. However, in cyberbullying, repetition is more difficult to categorize. For example, repetition in cyberbullying could consist of a single one-off event that is viewed numerous times, a repetitive act by a single perpetrator, or the same aggressive act perpetrated by other individuals beyond the initial perpetrator (Slonje et al., 2013). While a single incident of cyber aggression would not meet the definitional aspect of cyberbullying, a single incident could be viewed and/or shared multiple times by other people online, thus meeting the repetition aspect of cyberbullying (Elledge et al., 2013; Slonje et al., 2013). Dooley et al. (2009) noted that even a single message or image sent to somebody online can lead to widespread humiliation due to the increased accessibility for other people to see it, and further disseminate the message/image. This suggests that while the concept of repetition is more ambiguous in cyberbullying, it could be that repetition is presented online in a different manner to that of traditional bullying; through the continual availability of victimization online (Kowalski et al., 2017).

In the online environment, cyberbullying may readily "snowball" out of the initial control of the bully (Slonje et al., 2013, p. 26). For instance, an offensive image or video clip of someone posted on a website only once may still reflect the essence of repetition. This is because many people may access the website where the image has been posted and go on to make further comments about the image or even circulate it to others (Smith,

2015). Therefore, a single aggressive act can result in continued and widespread ridicule and humiliation for the young person being targeted. As such, it has been suggested that rather than repetition being a distinct criterion of cyberbullying, it should instead be used as an indicator for the severity of the situation (Rodkin & Fischer, 2012). As noted by Dooley et al. (2009), when a single act of cyberbullying is viewed repeatedly by different audiences, this can present an increased impact on the victim via repeated exposure and humiliation. This suggests that for cyberbullying, even though the act itself does not need to be repeated, it is the repeated negative impact for the victim that meets the repetitive criteria. As such, Dooley et al. (2009) argued that the content of the cyberbullying act is more important. For example, if a cyberbullying act is perpetrated in a public domain, and is potentially accessible to anyone, this can have immediate and prolonged consequences for the victim. Therefore, while repetition may not be an explicitly distinct feature of cyberbullying (Hemphill & Heerde, 2014; Moreno et al., 2018), it is the repeated exposure to the victim and the associated impact that may offer a better explanation regarding how repetition is portrayed in the online environment.

3.3 Power imbalance

The third element associated with traditional bullying is the power imbalance between the victim and perpetrator, where the perpetrator has actual or perceived power compared to the victim. While this may have been portrayed through physical strength or social status in the school playground, power imbalance in the online domain is more difficult to identify. In the context of cyberbullying, such power dynamics need to be considered more carefully. While psychological power dynamics may still exist in the online domain, unique online features may offer a better explanation for how power imbalances work within cyberbullying. For example, young people engaged in bullying behaviors online may have increased power from greater technological skills and use these skills to target their victims in sophisticated ways (Corcoran & Mc Guckin, 2014). In addition, the anonymity feature of online communication means that bullies can target their victims while concealing their identity. The fact that bullies can conceal their identity when they target victims gives bullies a real or perceived sense of power where the victim feels defenseless (Moreno et al., 2018; Slonje et al., 2013).

However, the notion of power dynamics in cyberbullying can be more challenging to define. For example, victims of traditional bullying that would suffer victimization due to a physical or psychological power imbalance from the bully can use online mediums to target their bullies via anonymous means, thus reversing the power imbalance that may have been traditionally portrayed in the school environment (Wolke et al., 2017). The nature of power imbalance in cyberbullying is potentially different from that of face-to-face bullying. There is some debate in the literature regarding how power imbalances are portrayed in the context of cyberbullying. For example, a study conducted in Israel presents the view that power imbalances are absent in the online environment as the technology and digital devices remove features of power and status (Lapidot-Lefler & Dolev-Cohen, 2015). However, other researchers argue that while physical strength or status are removed in the online environment, unique features associated with technology such as anonymity and digital skills act as tools to create an imbalance of power between the perpetrator and victim (Langos, 2012; Slonje et al., 2013; Vandebosch & Van Cleemput, 2008). While a study in Australia has shown that young people regard power imbalance to be less applicable in the definitional aspect of cyberbullying (Dredge et al., 2014), another study has found that power imbalance was the most important criterion when defining cyberbullying from a sample of 11–17-year-olds across six European countries (Menesini et al., 2012). In addition, more recent research has highlighted that teachers do not perceive power imbalance as an important criterion when discussing how cyberbullying should be defined (Macaulay et al., 2018, 2020a).

On the other hand, the notion of power imbalance is also similar between the two forms of bullying. For example, in traditional bullying, perpetrators may exclude victims from friendship groups in order to gain a sense of power (Vandebosch & Van Cleemput, 2008). Similarly, in cyberbullying, perpetrators can also exclude victims from online groups and activities to gain this sense of power. However, others argue that digital technology acts as a barrier between the victim and perpetrator, and so the notion of power imbalance does not exist in the online domain (Lapidot-Lefler & Dolev-Cohen, 2015). Despite this view, others believe that the power imbalance can still exist in the online domain but is represented in a different manner to what would typically be portrayed in the physical world. For example, in traditional bullying, the criterion of imbalance of power can be presented through social status within a peer group or physical strength. In cyberbullying, these factors are minimized or even eliminated due to the anonymity aspect of online communication (Smith, 2015). However, it is the anonymity and public nature of cyberbullying that represents an imbalance of power (i.e., the victim can be unaware of the perpetrators' identity) (Smith, 2015; Thomas et al., 2015).

Together, this suggests a lack of consensus regarding the precise definition of cyberbullying, which creates additional challenges for research and practical applications in the real world due to variations in how cyberbullying is defined and measured. While there are some similarities between cyberbullying and traditional bullying, there are elements of cyberbullying that make it unique compared to traditional forms of bullying. For example, in cyberbullying the perpetrator can conceal their identity and remain anonymous more easily, and the nature of cyberbullying can vary accordingly to the publicity of the incident (i.e., how many people can see the incident varies).

4 Unique features of cyberbullying

In contrast to researchers who propose cyberbullying is an extension of traditional bullying (e.g., Olweus, 2013), other researchers argue that cyberbullying represents a distinct form of bullying that should be considered separately from its traditional counterpart (e.g., Pieschl et al., 2015). For example, as noted by Dooley et al. (2009), cyberbullying needs to be regarded as its own form of bullying so that the unique features can be fully explored and examined. In addition, as cyberbullying is occurring in a new space and is changing in line with the advancement of digital technology, it needs to be regarded as its own phenomenon (Huang & Chou, 2010). In support of this notion, Pieschl et al. (2015) argued that any of the defining features of traditional bullying may not be represented in the same manner when it comes to cyberbullying, as discussed earlier in this chapter. As such, cyberbullying needs to be defined and regarded as its own form of bullying to examine and clarify how the traditional characteristics extend in the online space, and unique features associated with cyberbullying.

4.1 Anonymity

One of the unique features associated with cyberbullying is that perpetrators can conceal their identity and 'hide behind the screen' (MacKay, 2012, p. 32). The opportunity for perpetrators to remain anonymous when targeting victims online could give a greater sense of power to the perpetrator, thus meeting the power imbalance criterion, but also means perpetrators feel less accountable for their actions.

One possible reason why a perpetrator might choose to remain anonymous is due to the perception of not being able to get caught for their actions (Olweus & Limber, 2018). Victims of cyberbullying may be defenseless as they are blind to who is targeting them, further amplifying the negative impact of cyberbullying (Vandebosch & Van Cleemput, 2008). This idea is also suggested in a study from Canada by Mishna et al. (2009, p. 1224) where perpetrators of cyberbullying: 'can be anyone, even someone next door'. Therefore, perpetrators of cyberbullying can gain a sense of power through anonymity as they can hide their identity (Raskauskas & Stoltz, 2007; Slonje et al., 2013). However, while anonymity is easier to achieve with cyberbullying compared to traditional bullying, scholars suggest it should remain as a unique feature rather than a defining attribute (Campbell & Bauman, 2018).

4.2 Publicity

In addition to anonymity, cyberbullying is also largely characterized by the increased dissemination and potential audience that witness these acts online. Compared to traditional bullying which would often occur within the school in front of a group of peers, cyberbullying can be witnessed by a potentially infinite audience (Smith, 2015). Research has suggested that cyberbullying may vary according to the publicity, often distinguished across public (i.e., visible to anyone), semi-public (i.e., visible to those in a group), and private (i.e., visible by the bully and victim only) (Dooley et al., 2009). In the context of victimization, such experiences are regarded to be more severe by young people when more people have witnessed the situation (Slonje & Smith, 2008). In addition, the victim may not know the true size of the audience that has witnessed their victimization, which may increase feelings of humiliation, anxiety, and reduce levels of self-esteem (Pieschl et al., 2015). For traditional bullying, the audience needs to be in the same physical space and environment, whereas for cyberbullying, the audience could be potentially infinite according to different public spaces, hence potentially increasing the negative impact for the victim (Tokunaga, 2010). In addition, this idea of publicity associated with cyberbullying has also been implicated in why young people engage in cyberbullying. For example, research in the USA has suggested that young people will change their behavior online when in the presence of a larger audience (Rafferty & Vander Ven, 2014). This online pressure and the struggles to maintain power in the online environment impacts on cyberbullying involvement. In addition, teachers in the UK perceive public instances of cyberbullying to be more severe for the victim and create additional challenges when managing cyberbullying (Macaulay et al., 2020b). This suggests the role of publicity is an important criterion to consider in the definition of cyberbullying.

Aligned to this notion of publicity is that as cyberbullying is not restrained to any geographical location or time, perpetrators can target victims at any time, with a potentially global audience associated with public spaces in the digital world (Thomas et al., 2015). This constant availability to target victims 24/7 means victims of cyberbullying are potentially unable to escape (Selkie et al., 2016). These unique features associated with cyberbullying suggest the complexity of these behaviors and the need to address how young people respond to cyberbullying, but also how those in the educational community identify and manage this issue.

5 The different forms of cyberbullying

In addition to exploring the definitional aspect of cyberbullying, it is also important to consider the different types of cyberbullying behaviors that young people are involved in. Cyberbullying can take many different forms, and one of the most recognized and used classifications was that proposed by Willard (2007). When examining the classification of different types of cyberbullying, Willard proposed eight different types of behavior: flaming, harassment, denigration, impersonation, outing and trickery, exclusion, cyberstalking, and cyberthreats. The term flaming involves posting insults and negative messages often on social networking sites directed at individuals but also small groups. On the other hand, harassment involves repeated intimidation to a targeted individual over a repeated period. The term denigration involves unfairly spreading comments about an individual with the goal of disrupting their social status. As such, victims of denigration are often unaware they are being targeted initially. Impersonation defines those who portray themselves online as their targeted victim, with the aim of portraying the victim in a negative light. Willard also proposed that outing and trickery are linked behaviors associated with cyberbullying. For example, outing involves the dissemination of personal information beyond the initial intended audience, and trickery involves disclosing sensitive information when they believe they are only for the recipient. In the context of cyberbullying, the term exclusion involves the perpetrator purposefully excluding the victim from online groups or activities. Finally, cyberstalking involves repeated acts on an individual which is more persistent than harassment, and cyberthreats comprise of direct and indirect threats online.

One of the main issues studying the forms of cyberbullying is that as technology advances, the forms of cyberbullying and the way these are perpetrated will continue to change and evolve. For example, earlier research in this area from England reported telephone calls and emails to be the most prevalent types of cyberbullying (Slonje & Smith, 2008; Smith et al., 2008). In addition, in a sample of 1211 young people from the Netherlands aged 11–13 years, abusive comments and rumor spreading were the most common types of cyberbullying, often perpetrated through emails and instant messenger (DeHue et al., 2008). Further to this, a study in England exploring the mediums used within cyberbullying over a five-year period found that the most preventable types used include emails, instant massager, and social network sites (Rivers & Noret, 2010). Recently, findings from 42 countries have shown that engagement in social media use is related to cyberbullying victimization and perpetration, suggesting accessibility to social media poses a risk for young people in relation to cyberbullying (Craig et al., 2020).

Despite several different types of cyberbullying behaviors that have been identified in this chapter, one recent study in Israel suggests that the most frequent behavioral acts of cyberbullying occur on social networking sites, with increased use of the internet and social networking sites associated with increased cyberbullying victimization (Aizenkot, 2020). In addition, types of cyberbullying behaviors can manifest as written-verbal behaviors (e.g., during a phone call, in one to one or group text message, or on social media) or as visual behaviors (e.g., such as posting or sharing media without consent, photographs, or videos; Nocentini et al., 2010; Palladino et al., 2015). This suggests that the application of different digital technologies has provided numerous opportunities for young people to target victims online across a plethora of media (Willard, 2011).

6 The prevalence of cyberbullying

The variability in cyberbullying conceptualizations and assessment measures has led to inconsistencies in reported prevalence (Kowalski et al., 2019; Volk et al., 2017).

Early research exploring the prevalence of cyberbullying in England provides some context in this regard. For example, Smith et al. (2008) conducted two studies using the same questionnaire to examine the prevalence of cyberbullying in young people aged 11–16-years-old. In study 1, the initial pilot study involving 92 students, 22.2% of the students reported experiencing cyberbullying, with 6.6% of the students experiencing it often, and 15.6% experiencing cyberbullying once or twice. Extending on this pilot study, study 2 involving 2533 young people aged 11–16-years-old found that a total of 17.2% (nearly one in five pupils) of the students had experienced cyberbullying, with 5.3% reported experiencing cyberbullying in the last week or month, 5.1% in the last term, and 3.1% in the last year (Smith et al., 2008). These findings are supported by longitudinal data of 2500 11–13-year-olds in England examining the occurrence of receiving aggressive and threatening texts or emails which found an increase in reported prevalence over a five-year period (Rivers & Noret, 2010). In 2002, 13% of students had reported receiving an aggressive or threatening text or email, rising to 16.4% in 2004, with the overall trend of cyberbullying involvement increasing (Rivers & Noret, 2010). In addition to this, in a sample of 325 students, it was reported that 11% had been cyberbullied (Ackers, 2012). Despite prior research suggesting the prevalence of cyberbullying to be gradually growing, recent research using a random sample of 120,115 adolescents aged 15 years, found that traditional bullying is actually more prevalent, with cyberbullying prevalence reported to be less than 1% (Przybylski &

Bowes, 2017). However, the literature has identified challenges measuring cyberbullying due to variations in understanding (Olweus & Limber, 2018), and so prevalence reports are prone to be misinterpreted.

These reported trends in the prevalence and gradual increase in cyberbullying involvement also reflect the accessibility to information and communication technologies. As more young people have access to online mediums and technologies, this is also likely to increase vulnerability to cyberbullying involvement. For example, compared to 2015, children aged 5–15 years are more prone to go online and own their own digital devices to access online applications (Ofcom, 2019). This can increase their vulnerability to online risks, including cyberbullying. In the UK, Ofcom (2019) reported that 93% of 8–11-year-olds have access to the internet for approximately 13 h a week, and the amount of young people having their own social media profiles is also continually increasing. In addition, the report also identified that access to the internet, time spent online each week, and the prospect of having a social media profile also increased by age (Ofcom, 2019). Again, this suggests that vulnerability to cyberbullying involvement is also potentially increasing as the accessibility to online platforms continues to grow.

While the trend of cyberbullying suggested the prevalence was gradually increasing, more recent reports suggest the issue may not be as prevalent as previously reported (Przybylski & Bowes, 2017). The variability in prevalence rates of cyberbullying are due, in large part, to the inconsistencies in how cyberbullying is defined, and the time parameter used to measure cyberbullying occurrence (Kowalski et al., 2019). In addition, one study reported in the literature found that young people in the UK choose not to report or disclose cyberbullying, attributed to fear of consequences when reporting cyberbullying (Betts & Spenser, 2017). This highlights the complex issues surrounding the prevalence of cyberbullying. Despite some of the complex issues on prevalence, cyberbullying is present across all ages of young people, particularly reported to be frequent in early to mid-adolescence (Slonje & Smith, 2008; Tokunaga, 2010).

7 The impact of cyberbullying

When it comes to the impact of cyberbullying, there is a controversial debate in the literature when considering how this may be similar or dissimilar to that of traditional bullying. For example, some researchers argue the unique features associated with cyberbullying amplify the impact and severity than that of traditional bullying (Campbell et al., 2013), whereas other researchers suggest the impact of cyberbullying may be overrated and so may be similar to that of traditional bullying. Considering the impact of cyberbullying, scholars have identified an array of emotional, social, and educational consequences associated with those young people who have been victimized online.

The emotional impact of being cyberbullied has been implicated in the literature. For example, involvement in cyberbullying has been linked to a deterioration in psychosocial adjustment, associated with depressive symptoms (Selkie et al., 2016). A meta-analysis across 131 studies found cyberbullying involvement can lead to depression, anxiety, and low self-esteem (Kowalski et al., 2014). While victims of cyberbullying may experience embarrassment and feelings of fear or anxiety (Raskauskas & Stoltz, 2007), in some cases, prolonged exposure to cyberbullying can lead to suicidal thoughts and attempts (Hinduja & Patchin, 2010, 2019). In addition to the emotional impact of cyberbullying, victims can also suffer socially. For example, victims of cyberbullying have reported increased social anxiety (Juvonen & Gross, 2008) and lower self-esteem (Brewer & Kerslake, 2015; Patchin & Hinduja, 2010). Furthermore, while young people may use digital technology to seek out social support or explore their identity (Leung, 2011), victims of traditional bullying are also reported to use digital technology to target their traditional face-to-face perpetrators (König et al., 2010). This suggests that cyberbullying involvement has an impact on increasing the prevalence of deviant and anti-social behavior (Kowalski et al., 2014).

In addition to the emotional and social impact of cyberbullying, victims also experience an impact on their education. For example, findings suggest that those involved in cyberbullying as a victim report a deterioration in academic achievement and attainment (Beale & Hall, 2007; Marsh et al., 2010). As such, this academic impact can lead to a deterioration in academic grades, and an increase in rule-breaking behavior (Ybarra & Mitchell, 2007). In particular, individuals that are victimized have reduced confidence in the teachers' ability to address the situation (Sourander et al., 2010). Research has shown that cyberbullying involvement can impact on children's school attainment, due to fear of cyberbullying experiences in the school setting (West, 2015), with perpetration experience associated with negative school attitudes (Pyżalski, 2012). These consequences illustrate the variability and impact cyberbullying experiences can have on the victim.

8 Conclusion

The universal presence of digital technologies and accessibility to the internet enables children and adolescents to benefit from fast and efficient communication, but also presents numerous online dangers, specifically cyberbullying. There is a clear overlap between traditional bullying and cyberbullying that leads some scholars to believe that cyberbullying is an extension of traditional face-to-face bullying (Olweus, 2013). However, the unique characteristics that online communication provides create a difference between traditional bullying and cyberbullying (Smith, 2015). Cyberbullying can be characterized online by the anonymity and publicity of the incident. However, while these unique features may offer a better explanation of cyberbullying from a definitional stance, scholars also suggest they should remain as unique features rather than specific definitional criteria due to the complexities of cyberbullying behavior (Campbell & Bauman, 2018). The prevalence of cyberbullying is ambiguous, based on inconsistent reports. With the impact of cyberbullying presenting serious consequences for those involved, further attention needs to address cyberbullying in the context of young people and consider the impact of cyberbullying across the lifespan.

References

Ackers, M. J. (2012). Cyberbullying: Through the eyes of children and young people. *Educational Psychology in Practice, 28*(2), 141–157. https://doi.org/10.1080/02667363.2012.665356.

Aizenkot, D. (2020). Social networking and online self-disclosure as predictors of cyberbullying victimization among children and youth. *Children and Youth Services Review, 119*. https://doi.org/10.1016/j.childyouth.2020.105695, 105695.

Baldry, A. C., & Farrington, D. P. (2004). Evaluation of an intervention program for the reduction of bullying and victimization in schools. *Aggressive Behavior: Official Journal of the International Society for Research on Aggression, 30*(1), 1–15. https://doi.org/10.1002/ab.20000.

Baldry, A. C., Farrington, D. P., & Sorrentino, A. (2017). School bullying and cyberbullying among boys and girls: Roles and overlap. *Journal of Aggression, Maltreatment & Trauma, 26*(9), 937–951. https://doi.org/10.1080/10926771.2017.1330793.

Bauman, S. (2014). *Cyberbullying: What counselors need to know*. John Wiley & Sons.

Bauman, S., & Bellmore, A. (2015). New directions in cyberbullying research. *Journal of School Violence, 14*(1), 1–10. https://doi.org/10.1080/15388220.2014.968281.

Beale, A. V., & Hall, K. R. (2007). Cyberbullying: What school administrators (and parents) can do. *The Clearing House: A Journal of Educational Strategies, Issues and Ideas, 81*(1), 8–12. https://doi.org/10.3200/TCHS.81.1.8-12.

Betts, L. R., & Spenser, K. A. (2017). "People think it's a harmless joke": Young people's understanding of the impact of technology, digital vulnerability and cyberbullying in the United Kingdom. *Journal of Children and Media, 11*(1), 20–35. https://doi.org/10.1080/17482798.2016.1233893.

Brewer, G., & Kerslake, J. (2015). Cyberbullying, self-esteem, empathy and loneliness. *Computers in Human Behavior, 48*, 255–260. https://doi.org/10.1016/j.chb.2015.01.073.

Brown, C. F., Demaray, M. K., & Secord, S. M. (2014). Cyber victimization in middle school and relations to social emotional outcomes. *Computers in Human Behavior, 35*, 12–21. https://doi.org/10.1016/j.chb.2014.02.014.

Campbell, M., & Bauman, S. (Eds.). (2018). *Reducing cyberbullying in schools: International evidence-based best practices* Academic Press.

Campbell, M. A., Slee, P. T., Spears, B., Butler, D., & Kift, S. (2013). Do cyberbullies suffer too? Cyberbullies' perceptions of the harm they cause to others and to their own mental health.

School Psychology International, 34(6), 613–629. https://doi.org/10.1177/0143034313479698.
Corcoran, L., & Mc Guckin, C. (2014). Addressing bullying problems in Irish schools and in cyberspace: A challenge for school management. *Educational Research, 56*(1), 48–64. https://doi.org/10.1080/00131881.2013.874150.
Craig, W., Boniel-Nissim, M., King, N., Walsh, S. D., Boer, M., Donnelly, P. D., et al. (2020). Social media use and cyber-bullying: A cross-national analysis of young people in 42 countries. *Journal of Adolescent Health, 66*(6), S100–S108. https://doi.org/10.1016/j.jadohealth.2020.03.006.
DeHue, F., Bolman, C., & Völlink, T. (2008). Cyberbullying: Youngsters' experiences and parental perception. *Cyberpsychology & Behavior, 11*(2), 217–223. https://doi.org/10.1089/cpb.2007.0008.
Dennehy, R., Meaney, S., Walsh, K. A., Sinnott, C., Cronin, M., & Arensman, E. (2020). Young people's conceptualizations of the nature of cyberbullying: A systematic review and synthesis of qualitative research. *Aggression and Violent Behavior, 51*. https://doi.org/10.1016/j.avb.2020.101379, 101379.
Dooley, J. J., Pyżalski, J., & Cross, D. (2009). Cyberbullying versus face-to-face bullying: A theoretical and conceptual review. *Zeitschrift für Psychologie/Journal of Psychology, 217*(4), 182–188. https://doi.org/10.1027/0044-3409.217.4.182.
Dredge, R., Gleeson, J., & De la Piedad Garcia, X. (2014). Cyberbullying in social networking sites: An adolescent victim's perspective. *Computers in Human Behavior, 36*, 13–20. https://doi.org/10.1016/j.chb.2014.03.026.
Elledge, L. C., Williford, A., Boulton, A. J., DePaolis, K. J., Little, T. D., & Salmivalli, C. (2013). Individual and contextual predictors of cyberbullying: The influence of children's provictim attitudes and teachers' ability to intervene. *Journal of Youth and Adolescence, 42*(5), 698–710. https://doi.org/10.1007/s10964-013-9920-x.
Finkelhor, D. (2014). Commentary: Cause for alarm? Youth and internet risk research–a commentary on Livingstone and Smith (2014). *Journal of Child Psychology and Psychiatry, 55*(6), 655–658.
Hemphill, S. A., & Heerde, J. A. (2014). Adolescent predictors of young adult cyberbullying perpetration and victimization among Australian youth. *Journal of Adolescent Health, 55*(4), 580–587. https://doi.org/10.1016/j.jadohealth.2014.04.014.
Hinduja, S., & Patchin, J. W. (2010). Bullying, cyberbullying, and suicide. *Archives of Suicide Research, 14*(3), 206–221. https://doi.org/10.1080/13811118.2010.494133.
Hinduja, S., & Patchin, J. W. (2019). Connecting adolescent suicide to the severity of bullying and cyberbullying. *Journal of School Violence, 18*(3), 333–346. https://doi.org/10.1080/15388220.2018.1492417.
Huang, Y. Y., & Chou, C. (2010). An analysis of multiple factors of cyberbullying among junior high school students in Taiwan. *Computers in Human Behavior, 26*(6), 1581–1590. https://doi.org/10.1016/j.chb.2010.06.005.
Juvonen, J., & Gross, E. F. (2008). Extending the school grounds?—Bullying experiences in cyberspace. *Journal of School Health, 78*(9), 496–505. https://doi.org/10.1111/j.1746-1561.2008.00335.x.
König, A., Gollwitzer, M., & Steffgen, G. (2010). Cyberbullying as an act of revenge? *Journal of Psychologists and Counsellors in Schools, 20*(2), 210–224.
Kowalski, R., Giumetti, G. W., & Limber, S. P. (2017). Bullying and cyberbullying among rural youth. In *Handbook of rural school mental health* (pp. 231–245). Cham: Springer.
Kowalski, R. M., Giumetti, G. W., Schroeder, A. N., & Lattanner, M. R. (2014). Bullying in the digital age: A critical review and meta-analysis of cyberbullying research among youth. *Psychological Bulletin, 140*(4), 1073. https://doi.org/10.1037/a0035618.
Kowalski, R. M., Limber, S. P., & McCord, A. (2019). A developmental approach to cyberbullying: Prevalence and protective factors. *Aggression and Violent Behavior, 45*, 20–32. https://doi.org/10.1016/j.avb.2018.02.009.

Langos, C. (2012). Cyberbullying: The challenge to define. *Cyberpsychology, Behavior and Social Networking, 15*(6), 285–289.

Lapidot-Lefler, N., & Dolev-Cohen, M. (2015). Comparing cyberbullying and school bullying among school students: Prevalence, gender, and grade level differences. *Social Psychology of Education, 18*(1), 1–16. https://doi.org/10.1007/s11218-014-9280-8.

Lenhart, A. (2015). *Teens, social media & technology overview 2015. Pew Research Center: Internet.* Science & Tech.

Leung, L. (2011). Loneliness, social support, and preference for online social interaction: The mediating effects of identity experimentation online among children and adolescents. *Chinese Journal of Communication, 4*(4), 381–399. https://doi.org/10.1080/17544750.2011.616285.

Macaulay, P. J., Betts, L. R., Stiller, J., & Kellezi, B. (2018). Perceptions and responses towards cyberbullying: A systematic review of teachers in the education system. *Aggression and Violent Behavior, 43*, 1–12. https://doi.org/10.1016/j.avb.2018.08.004.

Macaulay, P. J., Betts, L. R., Stiller, J., & Kellezi, B. (2020a). "It's so fluid, it's developing all the time": Pre-service teachers' perceptions and understanding of cyberbullying in the school environment. *Educational Studies, 46*(5), 590–606. https://doi.org/10.1080/03055698.2019.1620693.

Macaulay, P. J., Betts, L. R., Stiller, J., & Kellezi, B. (2020b). 'The more public it is, the more severe it is': Teachers' perceptions on the roles of publicity and severity in cyberbullying. *Research Papers in Education*, 1–28. https://doi.org/10.1080/02671522.2020.1767183.

MacKay, A. W. (2012). *Respectful and responsible relationships: There's no app for that (The report of the Nova Scotia task force on bullying and cyberbullying).* Nova Scotia Task Force on Bullying and Cyberbullying.

Marsh, L., McGee, R., Nada-Raja, S., & Williams, S. (2010). Brief report: Text bullying and traditional bullying among New Zealand secondary school students. *Journal of Adolescence, 33*(1), 237–240. https://doi.org/10.1016/j.adolescence.2009.06.001.

Menesini, E., & Nocentini, A. (2009). Cyberbullying definition and measurement: Some critical considerations. *Zeitschrift für Psychologie/Journal of Psychology, 217*(4), 230–232. https://doi.org/10.1027/0044-3409.217.4.230.

Menesini, E., Nocentini, A., Palladino, B. E., Frisén, A., Berne, S., Ortega-Ruiz, R., et al. (2012). Cyberbullying definition among adolescents: A comparison across six European countries. *Cyberpsychology, Behavior and Social Networking, 15*(9), 455–463.

Mishna, F., Saini, M., & Solomon, S. (2009). Ongoing and online: Children and youth's perceptions of cyber bullying. *Children and Youth Services Review, 31*(12), 1222–1228. https://doi.org/10.1016/j.childyouth.2009.05.004.

Moreno, M. A., Suthamjariya, N., & Selkie, E. (2018). Stakeholder perceptions of cyberbullying cases: Application of the uniform definition of bullying. *Journal of Adolescent Health, 62*(4), 444–449. https://doi.org/10.1016/j.jadohealth.2017.11.289.

Nocentini, A., Calmaestra, J., Schultze-Krumbholz, A., Scheithauer, H., Ortega, R., & Menesini, E. (2010). Cyberbullying: Labels, behaviours and definition in three European countries. *Journal of Psychologists and Counsellors in Schools, 20*(2), 129–142.

O'Brien, N., & Moules, T. (2013). Not sticks and stones but tweets and texts: Findings from a national cyberbullying project. *Pastoral Care in Education, 31*(1), 53–65. https://doi.org/10.1080/02643944.2012.747553.

Ofcom. (2019). *Children and parents: Media use and attitudes report.* London: Office of Communications. Retrieved from https://www.ofcom.org.uk/__data/assets/pdf_file/0023/190616/children-media-use-attitudes-2019-report.pdf.

Olweus, D. (1978). *Aggression in the schools: Bullies and whipping boys.* London: Hemisphere.

Olweus, D. (1993). *Bullying at school: What we know and what we can do.* Oxford: Blackwell Publishers.

Olweus, D. (2013). School bullying: Development and some important challenges. *Annual Review of Clinical Psychology, 9*, 751–780. https://doi.org/10.1146/annurev-clinpsy-050212-185516.

Olweus, D., & Limber, S. P. (2018). Some problems with cyberbullying research. *Current Opinion in Psychology, 19*, 139–143. https://doi.org/10.1016/j.copsyc.2017.04.012.

Pabian, S., Erreygers, S., Vandebosch, H., Van Royen, K., Dare, J., Costello, L., et al. (2018). "Arguments online, but in school we always act normal": The embeddedness of early adolescent negative peer interactions within the whole of their offline and online peer interactions. *Children and Youth Services Review, 86*, 1–13. https://doi.org/10.1016/j.childyouth.2018.01.007.

Palladino, B. E., Nocentini, A., & Menesini, E. (2015). Psychometric properties of the Florence cyberbullying-cybervictimization scales. *Cyberpsychology, Behavior and Social Networking, 18*(2), 112–119. https://doi.org/10.1089/cyber.2014.0366.

Patchin, J. W., & Hinduja, S. (2006). Bullies move beyond the schoolyard: A preliminary look at cyberbullying. *Youth Violence and Juvenile Justice, 4*(2), 148–169. https://doi.org/10.1177/1541204006286288.

Patchin, J. W., & Hinduja, S. (2010). Cyberbullying and self-esteem. *Journal of School Health, 80*(12), 614–621. https://doi.org/10.1111/j.1746-1561.2010.00548.x.

Pieschl, S., Kuhlmann, C., & Porsch, T. (2015). Beware of publicity! Perceived distress of negative cyber incidents and implications for defining cyberbullying. *Journal of School Violence, 14*(1), 111–132. https://doi.org/10.1080/15388220.2014.971363.

Przybylski, A. K., & Bowes, L. (2017). Cyberbullying and adolescent well-being in England: A population-based cross-sectional study. *The Lancet Child & Adolescent Health, 1*(1), 19–26. https://doi.org/10.1016/S2352-4642(17)30011-1.

Pyżalski, J. (2012). From cyberbullying to electronic aggression: Typology of the phenomenon. *Emotional and Behavioural Difficulties, 17*(3–4), 305–317. https://doi.org/10.1080/13632752.2012.704319.

Rafferty, R., & Vander Ven, T. (2014). "I hate everything about you": A qualitative examination of cyberbullying and on-line aggression in a college sample. *Deviant Behavior, 35*(5), 364–377. https://doi.org/10.1080/01639625.2013.849171.

Raskauskas, J., & Stoltz, A. D. (2007). Involvement in traditional and electronic bullying among adolescents. *Developmental Psychology, 43*(3), 564. https://doi.org/10.1037/0012-1649.43.3.564.

Rivers, I., & Noret, N. (2010). I h8 u': findings from a five-year study of text and email bullying. *British Educational Research Journal, 36*(4), 643–671. https://doi.org/10.1080/01411920903071918.

Rodkin, P. C., & Fischer, K. (2012). Cyberbullying from psychological and legal perspectives. *Missouri Law Review, 77*, 619.

Selkie, E. M., Fales, J. L., & Moreno, M. A. (2016). Cyberbullying prevalence among US middle and high school–aged adolescents: A systematic review and quality assessment. *Journal of Adolescent Health, 58*(2), 125–133.

Slonje, R. K., & Smith, P. (2008). Cyberbullying: Another main type of bullying?: Personality and social sciences. *Scandinavian Journal of Psychology, 49*(2), 147–154. https://doi.org/10.1111/j.1467-9450.2007.00611.x.

Slonje, R., Smith, P. K., & Frisén, A. (2013). The nature of cyberbullying, and strategies for prevention. *Computers in Human Behavior, 29*(1), 26–32. https://doi.org/10.1016/j.chb.2012.05.024.

Smith, P. K. (2015). The nature of cyberbullying and what we can do about it. *Journal of Research in Special Educational Needs, 15*(3), 176–184. https://doi.org/10.1111/1471-3802.12114.

Smith, P. K. (2016). School-based interventions to address bullying. *Eesti Haridusteaduste Ajakiri. Estonian Journal of Education, 4*(2), 142–164. https://doi.org/10.12697/eha.2016.4.2.06a.

Smith, P. K. (2019). Research on cyberbullying: Strengths and limitations. In H. Vandebosch, & L. Green (Eds.), *Narratives in research and interventions on cyberbullying among young people*. Cham: Springer.

Smith, P. K., Mahdavi, J., Carvalho, M., Fisher, S., Russell, S., & Tippett, N. (2008). Cyberbullying: Its nature and impact in secondary school pupils. *Journal of Child Psychology and Psychiatry, 49*(4), 376–385. https://doi.org/10.1111/j.1469-7610.2007.01846.x.

Sourander, A., Klomek, A. B., Ikonen, M., Lindroos, J., Luntamo, T., Koskelainen, M., et al. (2010). Psychosocial risk factors associated with cyberbullying among adolescents: A population-based study. *Archives of General Psychiatry, 67*(7), 720–728. https://doi.org/10.1001/archgenpsychiatry.2010.79.

Tarapdar, S., Kellett, M., & People, Y. (2013). Cyberbullying: Insights and age-comparison indicators from a youth-led study in England. *Child Indicators Research, 6*(3), 461–477. https://doi.org/10.1007/s12187-012-9177-z.

Thomas, H. J., Connor, J. P., & Scott, J. G. (2015). Integrating traditional bullying and cyberbullying: Challenges of definition and measurement in adolescents–a review. *Educational Psychology Review, 27*(1), 135–152. https://doi.org/10.1007/s10648-014-9261-7.

Tokunaga, R. S. (2010). Following you home from school: A critical review and synthesis of research on cyberbullying victimization. *Computers in Human Behavior, 26*(3), 277–287. https://doi.org/10.1016/j.chb.2009.11.014.

Vandebosch, H., & Van Cleemput, K. (2008). Defining cyberbullying: A qualitative research into the perceptions of youngsters. *Cyberpsychology & Behavior, 11*(4), 499–503. https://doi.org/10.1089/cpb.2007.0042.

Volk, A. A., Veenstra, R., & Espelage, D. L. (2017). So you want to study bullying? Recommendations to enhance the validity, transparency, and compatibility of bullying research. *Aggression and Violent Behavior, 36*, 34–43. https://doi.org/10.1016/j.avb.2017.07.003.

West, D. (2015). An investigation into the prevalence of cyberbullying among students aged 16–19 in post-compulsory education. *Research in Post-Compulsory Education, 20*(1), 96–112. https://doi.org/10.1080/13596748.2015.993879.

Willard, N. E. (2007). *Cyberbullying and cyberthreats: Responding to the challenge of online social aggression, threats, and distress*. Research Press.

Willard, N. (2011). School response to cyberbullying and sexting: The legal challenges. *Brigham Young University Education and Law Journal, 75*.

Wolke, D., Lee, K., & Guy, A. (2017). Cyberbullying: A storm in a teacup? *European Child & Adolescent Psychiatry, 26*(8), 899–908. https://doi.org/10.1007/s00787-017-0954-6.

Ybarra, M. L., & Mitchell, K. J. (2007). Prevalence and frequency of internet harassment instigation: Implications for adolescent health. *Journal of Adolescent Health, 41*(2), 189–195. https://doi.org/10.1016/j.jadohealth.2007.03.005.

CHAPTER

9

The impact of cyberbullying across the lifespan

Lucy R. Betts

Department of Psychology, Nottingham Trent University, Nottingham, United Kingdom

1 Introduction

While cyberspace can be regarded as a place of positive discourse and a positive nurturing environment where individuals can express themselves (Davis, Randall, Ambrose, & Orand, 2015) and online behavior can promote social responsibility and encourage caring and respect (Cassidy, Jackson, & Brown, 2009), researchers and educators have identified several risks associated with digital technology use. One such risk is cyberbullying (Sabella, Patchin, & Hinduja, 2013). Cyberbullying is recognized as one of the unintended consequences of technology use (Tokunaga, 2010). The impacts of involvement in face-to-face bullying have been widely documented (Aluede, Adeleke, Omoike, & Afen-Akpaida, 2008; Vanderbilt & Augustyn, 2010; Wolke & Lereya, 2015) but comparably less is known about the impacts of involvement in cyberbullying, especially across the lifespan.

There is a perception that society has moved away from the notion that involvement in bullying is simply part of growing up but rather there is a growing recognition of the potential longer-term impacts of experiencing cyberbullying (Anderson & Sturm, 2007). However, despite this change in society's perception about the impact of involvement in bullying, research with American high school teachers suggests that 25% of the 66 teachers asked thought that involvement in cyberbullying did not have lasting consequences and that it 'prepares students for life' (Stauffer, Heath, Coyne, & Ferrin, 2012). Despite the perceptions held by some of the teachers who participated in Stauffer et al.'s study, there is growing

Cybersecurity and Cognitive Science
https://doi.org/10.1016/B978-0-323-90570-1.00014-0

evidence that cyberbullying is a stressor that can have several impacts (Nixon, 2014). Therefore, this chapter will review some of these impacts.

The impacts of involvement in cyberbullying are believed to be social and emotional in nature and may vary according to the potential size of the audience that witnessed the cyberbullying behavior (Dooley, Pyżalski, & Cross, 2009). Further, the impacts of involvement in cyberbullying may also extend beyond the digital arena. For example, research has reported that the impact of involvement in cyberbullying spills over into the school environment (Betts, Houston, Steer, & Gardner, 2017; Betts, Spenser, & Gardner, 2017; Bhat, 2008; Spears, Slee, Owens, & Johnson, 2009).

It is also likely that the impacts of cyberbullying will vary according to the role that an individual adopts within an episode of cyberbullying (sometimes termed the participant role). Researchers have identified six different roles that individuals fulfill within face-to-face bullying incidents: victim, bully, reinforcer of the bully, defender of the bully, defender of the victim, and outsider (Salmivalli, Lagerspetz, Björkqvist, Österman, & Kaukiainen, 1996). Individuals can also fulfill more than one role at the same time for example bully/victim, in this case individuals engage in and also experience bullying. Researchers have also identified similar roles for cyberbullying (e.g., Bayraktar, Machackova, Dedkova, Cerna, & Ševčíková, 2015; Wachs, 2012) with individuals fulfilling victim, bully, and bully/victim roles. Individuals who fulfill the victim role are those who experience cyberbullying as the target of the behavior, individuals who fulfill the bully role are those who engage in cyberbullying behaviors directed towards others, and individuals who fulfill the bully/victim role simultaneously engage in cyberbullying behavior directed towards others and are the target for such behavior. In addition to these roles, some individuals may fulfill the bystander role and witness cyberbullying episodes. Individuals fulfilling the not involved role tend to experience very low levels of cyberbullying behavior as a victim and engage in very low levels of cyberbullying behavior as a bully (Betts, Gkimitzoudis, Spenser, & Baguley, 2017). Therefore, this chapter will explore the impacts of involvement in cyberbullying for those fulfilling the bully, victim, bully/victim, bystander, and not involved participant roles. Exploring the impacts of cyberbullying across these various roles is important because as Campbell, Slee, Spears, Butler, and Kift (2013) note, the impact of cyberbullying is not merely seen on those who are victims. Specifically, Campbell et al. highlight that those who engage in cyberbullying and fulfill the bully role also experience mental health difficulties and depressive symptoms associated with their involvement in cyberbullying.

In addition to discussing the impact of cyberbullying across the various cyberbullying roles, this chapter will also take a developmental perspective and consider the impacts of cyberbullying across the lifespan. Although researchers have reported that experiences of cyberbullying as a victim tend to peak at the age of 14 (Ortega, Elipe, Mora-Merchán, Calmaestra, & Vega, 2009), some studies have explored cyberbullying across the lifespan. For example, Ševčíková and Šmahel (2009) examined experiences of cyberbullying in 12- to 88-year-olds from the Czech Republic and found that those most likely to be a victim of cyberbullying were adolescents and young adults whereas those who were most likely to be a cyberbully were younger and older adolescents. However, more recent research examining cyberbullying across the lifespan using data collected from over 20,000 18- to 97-year-olds from New Zealand found that cyberbullying was experienced by participants of all ages but the prevalence of being a victim decreased with age (Wang, Yogeeswaran, Andrews, Hawi, & Sibley, 2019). Similarly, data collected in 2014 as part of the Canadian General Social Survey on Victimization documented a linear decrease in reports of cyberbullying with age (Kim, Boyle, & Georgiades, 2017). Although previous research has identified a negative relationship between cyberbullying and age (Kim et al., 2017; Wang et al., 2019), other studies have suggested that a non-linear relationship exists (Barlett & Chamberlin, 2017). Barlett and Chamberlin reported a quadratic relationship between age and cyberbullying such that cyberbullying increased from youth to emerging adulthood and then decreased after this time. Together these studies provide a rationale for considering the impact of cyberbullying into adulthood because adults continue to be involved in cyberbullying. Further Jenaro, Flores, and Frías (2018) argue that researchers should consider the impact of cyberbullying in adult samples because it is expected that the prevalence of cyberbullying will increase in adults as the current generation of adolescents age due to the prevalence of cyberbullying that is currently reported in this age group.

Alongside considering the impact of cyberbullying in older samples including adults, this chapter will also review the literature that considers the impact of cyberbullying during childhood. Although fewer studies have explored the impact of involvement in cyberbullying for children prior to adolescence, there is emerging evidence that this age group is spending increasing amounts of time using technology (Goode, Fomby, Mollborn, & Limburg, 2020) and that they are at greater risk of experiencing some forms of cyberbullying (Aizenkot & Kashy-Rosenbaum, 2019). For example, elementary school-aged children from Israel reported experiencing higher levels of cyberbullying as a victim through school-based WhatsApp groups compared to middle and high school-aged students (Aizenkot & Kashy-Rosenbaum, 2019). Therefore, this chapter will discuss the research literature that explores the impact of involvement in cyberbullying across the various roles associated with cyberbullying across the lifespan. Finally, the chapter will consider the challenges associated with trying to understand the impact of cyberbullying across the lifespan.

2 The impact of involvement in cyberbullying during the elementary years

A handful of studies have started to examine the impact of involvement in cyberbullying for children during the elementary school years. Research with children in England suggests that the prevalence of cyberbullying increases across elementary school, with older pupils reporting greater levels of experiencing cyberbullying as a victim in the form of being harassed by instant messenger compared to younger children (Monks, Robinson, & Worlidge, 2012). However, recent research with Chinese elementary school children who completed measures of cyberbullying five times over two years reported a reduction in involvement in cyberbullying as a victim and bully over the two-year period (Zhang, Heubner, & Tian, 2021). Despite the variation in the reported prevalence rates of involvement in cyberbullying during the elementary school years, there is evidence that experiencing cyberbullying during elementary school is associated with psychosocial adjustment. For example, American sixth-grade children who experienced higher levels of cyberbullying as a victim also reported experiencing higher levels of loneliness, lower rates of optimism about peer relationships, having fewer friends, and lower levels of social acceptance (Jackson & Cohen, 2012). Similarly, 7- to 11-year-olds from the UK also reported that those who experienced cyberbullying as a victim would most frequently experience feeling worried (58%), afraid (58%), upset (57%), stressed (48%), angry (40%), depressed (40%), and embarrassed (27%; Monks et al., 2012). Moreover, Monks et al. also highlighted that only 16% of children in their sample reported that they thought someone who experienced cyberbullying would not be affected by their experiences.

In addition to the cross-sectional evidence of the impact of experiencing cyberbullying during childhood, a longitudinal study conducted in America has reported similar associations between experiencing cyberbullying and psychosocial adjustment over time. Children who reported experiencing higher levels of cyberbullying as a victim initially reported that they experienced a greater number of depressive symptoms, lower self-esteem, and less school connectiveness 18 months later (DePaolis & Williford, 2019). However, there was no evidence that experiencing cyberbullying as a victim predicted the children's perceptions of their peers over time suggesting that being a cyberbullying victim during the elementary school years only impacts on some aspects of psychosocial adjustment. As well as exploring the impact of involvement in cyberbullying as a victim, there is some emerging evidence that involvement in the other roles in cyberbullying has an impact during the elementary school years. For example, considering those who fulfilled the witness role, witnessing elevated levels of cyberbullying was associated with higher levels of depressive symptoms and social anxiety in 8- to 12-year-olds from America (Doumas & Midgett,

2020). Together, these research studies suggest that there are impacts for elementary school children who are involved in cyberbullying although this evidence is limited for some of the roles associated with cyberbullying. The next section of the chapter considers the impact of involvement in cyberbullying during the adolescent years.

3 The impact of involvement in cyberbullying during adolescence

The impacts of involvement in cyberbullying during the adolescent years have received significant attention in the literature. Adolescents from England aged 11- to 15-years-old, recognized and described how the potential impacts of involvement in cyberbullying are wide ranging (Betts & Spenser, 2017). There is evidence that experiencing cyberbullying as a victim has impacts that extend across wellbeing and perceptions of school and learning for adolescents.

Focusing on wellbeing, research with Japanese seventh to ninth-grade students found that those who experienced higher levels of cyberbullying as a victim were also likely to experience poorer mental health (Imamura et al., 2009). Although there is concurrent evidence of the association between experiences of cyberbullying as a victim and mental wellbeing, longitudinal research with secondary school-aged students from the Netherlands did not find a comparable relationship over time for all students but rather the relationship between experiences of cyberbullying as a victim and mental wellbeing was moderated by gender (Bannink, Broeren, van de Looij-Jansen, de Waart, & Raat, 2014). Experiences of cyberbullying as a victim were not associated with mental health over a two-year period in males when controlling for baseline mental health. However, for females, experiencing cyberbullying as a victim was associated with poorer mental health two years later, after controlling for baseline mental health.

Focusing on specific aspects of mental health, research with American youth found that 12- to 17-year-olds who experience cyberbullying as a victim also experience increased levels of anxiety (Juvonen & Gross, 2008). Similar patterns have been found for social anxiety with 11-to-16-year-olds from America with those youth who experience higher levels of cyberbullying as a victim also experiencing higher levels of social anxiety but not depression (Dempsey, Sulkowski, Nichols, & Storch, 2009). However, more recent research with 10- to 13-year-olds from Sweden found that the relationship between experiencing cyberbullying as a victim and bully/victim and anxiety was mediated by perceived social support from families and teachers (Hellfeldt, López-Romoero, & Andershed, 2020). Specifically, greater levels of social support for those experiencing cyberbullying as a victim or bully/victim reduced the probability with which these individuals reported anxiety symptoms. Further, experiencing higher levels of social support also served to increase levels of subjective wellbeing among those youth who experience cyberbullying as a victim and bully/victim. Together, these studies suggest that experiences of cyberbullying are associated with greater levels of anxiety but that the level of social support an individual receives from their family and their teachers can serve as a protective buffer in this relationship.

Experiences of cyberbullying have the potential to elicit strong negative feelings and emotions (Spears et al., 2009). A study with 84 American 13- to 18-year-olds reported that 93% of the participants who disclosed experiencing cyberbullying as a victim said that the cyberbullying had made them feel sad, hopeless, or depressed (Raskauskas & Stoltz, 2007). There also appears to be variation in the emotions adolescents report feeling following experiencing cyberbullying as a victim according to the medium through which the cyberbullying occurred. Ortega et al. (2009) explored the emotional categories associated with experiences of cyberbullying as a victim with Spanish 12- to 17-year-olds. Ortega et al. reported that those adolescents who had experienced cyberbullying as a victim via a mobile telephone felt alone, defenseless, depressed, stressed, afraid, and embarrassed. However, those adolescents who reported experiencing cyberbullying as a victim via the internet stated they felt alone, defenseless, depressed, stressed, afraid, embarrassed, worried, and upset.

Recently research has explored the impact of cyberbullying involvement for emotional and behavioral outcomes in a single study with American 6th to 12th grade students (Kim, Colwell, Kata, Boyle, & Georgiades, 2018). Kim et al. reported that experiencing cyberbullying as a victim predicted behavioral and emotional outcomes; however, the nature of the outcomes varied according to gender. For females, the relationship was stronger between experiencing cyberbullying as a victim and emotional outcomes (assessed as depression and anxiety) whereas, for males, the relationship between experiencing cyberbullying as a victim and behavioral outcomes (assessed as conduct disorder and oppositional defiant disorder) was stronger. Similarly, Foody, McGuire, Kuldas, and O'Higgins Norman (2019) examined involvement in cyberbullying as a victim, bully, and bully/victim and wellbeing in a sample of 12- to 16-year-olds from Ireland and provided further evidence of variation of impact according to the cyberbullying role. Foody et al. found that individuals who fulfilled the cyber victim role reported higher levels of depression and emotional problems than individuals who fulfilled the cyberbully role. Further, compared to cyber victims and adolescents not involved in cyberbullying, individuals who fulfilled the cyberbully role engaged in significantly less prosocial behaviors. Also, individuals who fulfilled the cyberbully role also reported significantly more conduct problems compared to those individuals who were not involved in cyberbullying. Finally, those individuals who fulfilled the cyberbully/victim role reported lower levels of friendship quality than those not involved in cyberbullying suggesting that involvement in cyberbullying also impacts on adolescents' peer relationships as well as their emotional wellbeing.

There is also evidence that involvement in cyberbullying as a victim, bully, or bully/victim is associated with other internalizing behaviors beyond emotional wellbeing. For example, with a sample of 12- to 19-year-olds from the Netherlands with intellectual and developmental disabilities, Didden et al. (2009) reported that experiencing cyberbullying as a victim was associated with lower self-esteem and higher depressive symptoms. Further, those youth who reported engaging in higher levels of cyberbullying as a bully via a mobile telephone also reported experiencing lower levels of self-esteem. The relationship between experiencing cyberbullying as a victim, depressive symptoms, and self-worth may be mediated by attribution styles and vary according to gender based on research with 11- to 15-year-olds from the United Kingdom (Betts, Houston, et al., 2017). Specifically, for males and females experiencing higher levels of cyberbullying as a victim predicted reporting elevated negative attributions which, in turn, predicted elevated levels of depressive symptoms. Also, males and females experiencing higher levels of cyberbullying as a victim predicted reduced levels of positive attributions and elevated levels of positive attributions predicted lower levels of depressive symptoms. For males, negative attributions mediated the relationship between experiencing cyberbullying as a victim and self-worth such that experiencing higher levels of cyberbullying as a victim predicted more negative attributions which, in turn, predicted reduced levels of self-worth. For females, experiencing higher levels of cyberbullying as a victim predicted lower levels of positive attribution and holding more positive attributions predicted greater self-worth. Considering individuals who fulfill the bully/victim role in cyberbullying, Gradinger, Strohmeier, and Spiel (2009) reported that Australian 14- to 19-year-olds who were bully/victims were more likely to experience elevated levels of depressive symptoms and somatic symptoms. Together, these studies highlight the association between involvement in cyberbullying and various indicators of internalizing behaviors. The research also highlights the complexity of these associations through the presence of significant mediators and moderators.

Other impacts of involvement in cyberbullying for adolescents extend to externalizing behaviors such as aggression, especially for those who fulfills the bully role in cyberbullying. For example, Calvete, Orue, Estévez, Villardón, and Padilla (2010) reported that Spanish 12- to 17-year-olds who engaged in higher levels of cyberbullying as a bully also reported elevated levels of proactive aggression, justification of violence, and exposure to violence. Similarly, Carvalho, Branquinho, and Gasper de Matos

(2021) found that Portuguese 10- to 19-year-olds who were involved in cyberbullying as a bully had a higher involvement in fights compared to those individuals who experienced cyberbullying as a victim. There is also evidence that those individuals who fulfill the bully/victim role in cyberbullying experience elevated levels of externalizing behaviors. Australian 14- to 19-year-olds who were cyberbully/victims also reported engaging in higher levels of reactive aggression (i.e., an aggressive act in response to a threat) and instrumental aggression (i.e., an aggressive act motivated by a specific goal; Gradinger et al., 2009). Carvalho et al. (2021) also found that cyberbully/victims reported greater levels of alcohol consumption compared to individuals who were cyberbullying victims.

There are several high-profile cases that attribute being the victim of cyberbullying as one of the reasons why adolescents have committed suicide (Pendergrass & Wright, 2014). Research evidence suggests that there is an association between involvement in cyberbullying and self-harm and suicidal ideation. For example, 7th to 12th grade Canadian students who experienced cyberbullying as a victim had higher suicidal ideation, plans, and attempts compared to students who had not experienced cyberbullying (Sampasa-Kanyinga, Roumeliotis, & Xu, 2014). However, the relationships were fully mediated by depression such that elevated levels of cyberbullying experiences as a victim predicted elevated levels of depression, and levels of depression predicted suicidal ideation, plans, and attempts making the direct path between these outcomes and experiences of cyberbullying as a victim non-significant.

Researchers have also explored how experiencing poly-victimization (i.e., experiencing more than one type of victimization at the same time) influences the relationship between cyberbullying and suicide. One such study was conducted by Sincliar, Bauman, Poteat, Koenig, and Russell (2012) with American middle and high school youth and explored the cumulative effects of cyber harassment and bias-based harassment. For the purpose of their study, Sinclair et al. defined bias-based harassment as harassment relating to race, ethnicity, or sexual orientation. Sinclair et al. reported that those youth experiencing cyber harassment as a victim without experiencing bias-based harassment as a victim were 3.82 times more likely to have attempted suicide than those who had not experienced such harassment. As evidence of the cumulative effect of harassment types, when youth experienced both cyber-harassment and bias-based harassment as a victim, the likelihood of attempting suicide rose to 7.85 times.

A recent meta-analysis of research conducted with samples up to the age of 25 explored the relationship between involvement in cyberbullying, self-harm, suicidal behaviors, and suicide (John et al., 2018). John et al.'s analysis highlight variation in the likelihood of these behaviors according to the role an individual fulfills in cyberbullying. Data from across the 35 studies reviewed revealed that compared to those who were not victims of cyberbullying, victims of cyberbullying were 2.35 times more likely to self-harm, 2.10 times more likely to exhibit suicidal behaviors, 2.57 times more likely to attempt suicide, and 2.15 times more likely to have suicidal thoughts. Also, compared to those who were not cyberbullies, individuals who engaged in cyberbullying were 1.21 times more likely to exhibit suicidal behaviors and 1.23 times more likely to experience suicidal ideation. Despite the various studies and meta-analyses that have reported an association between involvement in cyberbullying and suicidal ideation, it is important to acknowledge that not all studies report significant associations between these variables. For example, Bannink et al. (2014) reported no association between cyberbullying involvement as a victim and suicidal ideation over 2 years in adolescents from the Netherlands. Bannink et al. account for this lack of association by arguing that their sample had had comparably limited time to experience cyberbullying as a victim.

Along with the wellbeing and psychosocial impacts of involvement in cyberbullying during adolescence, researchers have also explored the potential impact of cyberbullying on adolescents' experiences of school. For example, Raskauskas and Stoltz (2007) reported that those American adolescents who experienced cyberbullying as a victim said that their experiences of cyberbullying made them afraid to go to school. Further, 11- to 15-year-olds from the United Kingdom who reported experiencing higher levels of cyberbullying as a victim also had lower levels of school liking (Betts, Houston, et al., 2017) and this relationship was mediated by negative and positive attribution styles, with variations according to gender. For males and females, experiencing greater levels of cyberbullying as a victim predicted higher negative attribution styles, and more negative attribution styles predict lower levels of school liking. In females, the relationship between experiencing cyberbullying as a victim and school liking was also mediated by positive attribution style such that females who experienced higher levels of cyberbullying as a victim reported lower levels of positive attributions and having more positive attributions, in turn, predicted elevated levels of school liking. Research has also provided evidence that there is variation in how Turkish adolescents involved in cyberbullying perceive school, their teachers, and their peers at school according to the roles they fulfill in cyberbullying (Bayar & Uçanok, 2012). Adolescents who were not involved in cyberbullying perceived school and their teachers more positively than those adolescents who fulfilled the bully and bully/victim roles in cyberbullying. The adolescents who fulfilled the cyber victim role also reported that they perceived their teachers more positively than those who fulfilled the cyberbully role. Focusing on perceptions of peers at school, adolescents who were not involved in cyberbullying perceived peers at school more positively than those who fulfilled the cyberbully role, while not involved adolescents and those who fulfilled the cyberbully role perceived peers more positively than adolescents who fulfilled the cyber victim role.

Researchers have also discussed how the impact of involvement in cyberbullying extends beyond perceptions of school and relationships with teachers and peers to behaviors relating to the school. For example, American 10- to 15-year-olds who experienced harassment as a victim online also reported having two or more detentions or suspensions at school and skipping school during the last year (Ybarra, Espelage, & Mitchell, 2007). Further, those adolescents who reported experiencing internet harassment as a victim were 8 times more likely to report carrying a weapon to school in the last 30 days. Considering those who fulfill the cyberbully role, Carvalho et al. (2021) reported that Portuguese 10- to 19-year-olds who engaged in higher levels of cyberbullying reported liking school less compared to individuals who experienced cyberbullying as a victim. A complex pattern of results emerged according to gender and role in cyberbullying in the relationship between involvement in cyberbullying and perceptions of learning and school in 11- to 15-year-olds from England (Betts, Spenser, & Gardner, 2017). For young women, involvement in cyberbullying as a victim, bully, or bully/victim negatively predicted perceptions of learning and school and this relationship was mediated by perceived peer acceptance. Specifically, greater involvement in cyberbullying as a victim, bully, or bully/victim predicted lower levels of perceived peer acceptance and higher levels of perceived peer acceptance predicted more positive perceptions of learning and school. For young men, fulfilling the bully/victim role negatively predicted perceptions of learning and school. Focusing on academic attainment, Li (2007) reported that half of their sample of Canadian middle school students who were cyberbullying victims reported they had above average grades at school whereas less than 35% of those who fulfilled the cyberbully role reported that they had above average grades at school.

Together, the research discussed in this section of the chapter highlights the range of impacts that involvement in cyberbullying has for adolescents across the various participant roles associated with cyberbullying. There is evidence that these impacts extend beyond psychosocial adjustment indicators to include aspects of school adjustment. Further, the research suggests that the relationship between involvement in cyberbullying and the indicators of adjustment are complex and are often mediated by other variables. The next section of the chapter will consider the research that discusses the impacts of involvement in cyberbullying for emerging adults.

4 The impact of involvement in cyberbullying during emerging adulthood

There is growing evidence of the impact of involvement in cyberbullying during emerging adulthood and this evidence suggests similar impacts to those previously discussed in this chapter. For example, college students in England aged 16- to 19-years-old who experienced cyberbullying as a victim reported experiencing elevated levels of negative emotions (West, 2015). Specifically, participants who experienced cyberbullying as a victim also reported that they felt angry (64%), hurt (58%), sad (54%), depressed (44%), embarrassed (38%), anxious (37%), and afraid (26%). In addition to reporting experiencing negative emotions associated with experiencing cyberbullying as a victim, the impacts also extended into other measures of adjustment with West's participants reporting that experiencing cyberbullying as a victim impacted on their experiences at college including their learning/academic performance (42%), attendance at college (30%), and not wanting to go back to college (29%). Participants also reported that their experiences of cyberbullying as a victim had impacts on their ability to develop relationships online (33%) and in the 'real' world (42%), and their diet/eating patterns (42%) and sleeping (48%). Using interpretative phenomenological analysis of American college students' experiences of cyberbullying, Rivituso (2014) reported the following themes: (a) repeated instances of cyberbullying leading to feelings of vulnerability and fear; (b) distrust of technology and mistrust of people; (c) the value of friends in college and their impact on victim self-esteem; (d) self-control in response to lack of control over cyberbullying instances; and (e) feelings of stress, depression, and embarrassment and frustration leading to self-blame.

Research with university students has highlighted the impacts for those who experience cyberbullying as a victim. For example, university students from America who were the victims of cyberbullying said that the impact of their experiences extended to a tendency to not interact with new friends, not trusting others, and suicidal ideation (Crosslin & Crosslin, 2014). More recent research with university students from Myanmar highlighted how the impact of experiencing cyberbullying extended into other areas of psychosocial adjustment (Khine et al., 2020). For example, students who experienced higher levels of cyberbullying as a victim also reported experiencing greater difficulties in concentrating and understanding lectures, and substance abuse. Research also highlights that there are potential gender differences in the impact of experiencing cyberbullying for Spanish university students (Felipe-Castaño et al., 2019). Male university students who reported experiencing cyberbullying as a victim had higher levels of anxiety, hostility, and psychoticism than females who experienced cyberbullying. Another individual difference that has been reported to influence the impact of experiencing cyberbullying for emerging adults is self-esteem (Balakrishan & Fernandez, 2018). Balakrishan and Fernandez reported that emerging adults from Malaysia who had higher levels of self-esteem and who were victims of cyberbullying were more likely to feel angry following experiencing cyberbullying and were more likely to disclose their experiences compared to those with lower levels of self-esteem. Aligned to the variation in disclosure of cyberbullying experiences, there is also variation in how emerging adults cope with cyberbullying. Alipan, Skues, and Theiler (2021) reported that although Australian emerging adults discussed a range of coping styles utilized as a mechanism to deal with cyberbullying, there was an acknowledgment that coping styles can have an impact on wellbeing. For example, while avoidance coping styles such as alcohol use, drug use, or self-harm may reduce the distress associated with cyberbullying these coping styles also have the potential to exacerbate emotional distress over time.

As with other developmental stages, there is evidence that those emerging adults who fulfill the different roles within cyberbullying experience different impacts based on their experiences. For American female students aged 18- to 25, any involvement in cyberbullying had increased odds of experiencing depression, while involvement in cyberbullying as a bully was associated with greater odds of depression and problem alcohol use, and involvement as a bully/victim increased the odds for depression compared to those not involved (Selkie, Kota, Chan, & Moreno, 2015). Although researchers have primarily focused on the experiences of college and university students when considering the impact of involvement in cyberbullying, a recent study by Jenaro et al. (2018) examined the impact of involvement in cyberbullying with Spanish adults aged 18 to 40 with intellectual disabilities. Similar to other studies, Jenaro, Flores, Vega et al. reported that adults who experienced higher levels of cyberbullying as a victim reported that these experiences had a negative impact on them. Together these studies highlight that involvement in cyberbullying during emerging adulthood has a variety of impacts across relationships, engagement with learning, and psychosocial adjustment, and these impacts appear to be similar to those impacts reported during other developmental stages. The next section of the chapter will consider the impact of involvement in cyberbullying during adulthood.

5 The impact of involvement in cyberbullying during adulthood

Although a recent systematic review undertaken by Jenaro, Flores, and Frías (2018) highlights that the impacts of involvement in cyberbullying during adulthood may be similar to those experienced by younger samples, the research examining the impacts of cyberbullying during adulthood has tended to focus on the impact of the involvement in workplace cyberbullying across a range of professions and experience levels. Research with trainee doctors in the United Kingdom found that those trainee doctors who reported experiencing higher levels of cyberbullying as a victim also reported having lower job satisfaction and higher levels of mental strain (Farley, Coyne, Sprigg, Axtell, & Subramanian, 2015). However, the nature of these relationships was influenced by the participants' attributions of blame, with negative emotion mediating the relationship between experiences of cyberbullying as a victim and elevated levels of mental strain. Despite the research of Farley et al. reporting an association between experiencing workplace cyberbullying as a victim and job satisfaction, other studies have not reported similar associations. For example, for Australian white-collar employees, there was no significant association between experiencing workplace cyberbullying as a victim and job satisfaction (Snyman & Loh, 2015). However, there was a significat relationship between experiencing workplace cyberbullying as a victim and stress that was mediated by optimism. Specifically, experiencing workplace cyberbullying as a victim negatively predicted optimism and optimism negatively predicted stress. Experiencing workplace cyberbullying as a victim also resulted in elevated levels of perceived stress and perceived stress, in turn, predicted job dissatisfaction in Australian white-collar employees (Loh & Snyman, 2020). Together, these studies highlight that involvement in workplace cyberbullying as a victim can impact on aspects of job performance and wellbeing.

Researchers have also explored the impact of experiencing cyberbullying during adulthood for relationships and broader aspects of wellbeing. For example, Lam, Ferlatte, and Salway (2019) with a sample of gay and bisexual Canadian males reported that experiencing cyberbullying as a victim was associated with wellbeing. Specifically, experiencing cyberbullying as a victim was also associated with experiences of antigay verbal violence, intimate partner violence, experiencing antigay physical violence, suicidality, and having no social support. Lam and colleagues' research provides further evidence for the conclusions drawn by Jenaro, Flores, and Frías (2018) that the impact of cyberbullying during adulthood is similar to the impacts experienced by younger participants but that the impacts of involvement in cyberbullying are influenced by personal and environmental factors. Further, following their review of the literature Jenaro, Flores, and Frías suggest that emotional intelligence and social support are the most influential factors that underpin the potential impact of involvement in cyberbullying for adults.

In summary, although there is variation in the reported prevalence rates across the lifespan, research does suggest that involvement in cyberbullying has similar impacts across the lifespan, although variation in the impacts is seen according to the participant role an individual fulfills in cyberbullying.

6 Challenges associated with understanding the impact of cyberbullying

Having considered the research evidence across the lifespan that discusses the impact of involvement in cyberbullying, the final section of the chapter will reflect on some of the challenges associated with understanding the impact of cyberbullying. Broadly these challenges can be considered to be conceptual and methodological.

Considering the conceptual challenges associated with understanding the impact of cyberbullying, there is of course the challenges associated with how cyberbullying has been defined and conceptualized in the literature which is discussed at length in the chapter by Macaulay, Betts, Stiller, and Kellezi (2022) but also challenges associated with the evolving nature of technology and the preferences for platforms. An example of the change in popularity of platforms was documented by Robards (2012) who highlighted the transition Australian youth made from MySpace to Facebook. While the popularity of platforms evolves, it is important to acknowledge preferences that reflect broader societal divisions. For example, Boyd (2011) argues that the American youth's preference for Facebook over MySpace potentially reflects societal divisions according to race and social inequality. Therefore, researchers should ensure that when exploring the impact of involvement in cyberbullying, they assess participants' experiences relevant to the platforms that they use.

Focusing on the methodological considerations of exploring the impact of involvement in cyberbullying these can be conceptualized in terms of the samples and also the study design. In terms of the samples, researchers need to ensure that the methods that they use to assess cyberbullying are age appropriate and reflect participants' ability to engage with the research, something that is a particular challenge when working with younger children (Ey, Taddeo, & Spears, 2015). Also, many of the studies that have explored cyberbullying have tended to do so with self-selecting samples and, as Chisholm (2014) notes, this represents a particular challenge. Specifically, because participants can self-select to take part in the research this may mean that only a particular group of individuals may be motivated to take part in the research, meaning that researchers may fail to capture the full experience of those involved with cyberbullying. Moreover, there is a tendency for researchers to explore cyberbullying in adults while at university and Jenaro, Flores, and Frías (2018) argue that more research needs to be undertaken with adult participants outside of the university context.

In terms of study design, the majority of studies that have been conducted to date have tended to be cross-sectional in nature with comparably fewer studies adopting a longitudinal design. Further, due to research into cyberbullying only beginning in the early 2000s (Bauman & Bellmore, 2015), fewer studies have used a longitudinal design that extends over a significant number of years. Consequently, this has led some authors to argue that the long-term effects of involvement in cyberbullying are not well understood because of the lack of longitudinal studies in this area (Jenaro, Flores, & Frías, 2018; Kowalski, Limber, & McCord, 2019). Similarly, Kwan et al. (2020) argue that the lack of longitudinal research in the area means that the extent to which mental health outcomes are the result of cyberbullying or the precursors of cyberbullying remains unclear. In addition to a call for longitudinal designs to explore the impact of cyberbullying, Nixon (2014) argues that it is important to consider the variables that potentially act as mediators or moderators in the relationship between cyberbullying and outcome measures, as not everyone who experiences cyberbullying is impacted in the same way.

A particular challenge associated with understanding the impacts of cyberbullying is that there is some evidence that the different forms of cyberbullying are associated with different impacts (Nixon, 2014). Therefore, researchers need to ensure that cyberbullying is assessed in such a way that these different forms are captured. Also, research to date has tended to focus on the impact of cyberbullying for those who experience cyberbullying as a victim. However, individuals can fulfill many roles in cyberbullying and the impact of fulfilling some of these roles is less well understood. For example, the impact of witnessing cyberbullying for those in a bystander role. Finally, Kowalski et al. (2019) argue that to better understand the impact of cyberbullying across the lifespan, research that makes direct comparisons between different age groups in the same study is needed.

7 Conclusion

In summary, this chapter has provided an overview of the research that has been conducted across the lifespan to explore the impact involvement in cyberbullying has on individuals. Although the prevalence rates of cyberbullying vary with age, regardless of age involvement in cyberbullying has an impact on psychosocial adjustment. The research also highlights that the impacts of involvement in cyberbullying extend beyond psychosocial adjustment to educational and workplace settings. There is also evidence that regardless of the role individuals fulfill within cyberbullying, the relationship between involvement in cyberbullying and adjustment is complex with factors such as social support serving as potential mediators and gender served as a moderator. The chapter has also considered some of the challenges researchers need to be mindful of when considering the impact of involvement in cyberbullying across the lifespan.

References

Aizenkot, D., & Kashy-Rosenbaum, G. (2019). Cyberbullying victimization in WhatsApp classmate groups among Israeli elementary, middle, and high school students. *Journal of Interpersonal Violence*. https://doi.org/10.1177/0886260519842860.

Alipan, A., Skues, J. L., & Theiler, S. (2021). "They will find another way to hurt you": Emerging adults' perceptions of coping with cyberbullying. *Emerging Adulthood, 9*, 22–34. https://doi.org/10.1177/2167696818816896.

Aluede, O., Adeleke, F., Omoike, D., & Afen-Akpaida, J. (2008). A review of the extent, nature, characteristics and effects of bullying behaviour in schools. *Journal of Instructional Psychology, 35*(2), 151.

Anderson, T., & Sturm, B. (2007). Cyberbullying: From playground to computer. *Young Adult Library Services, 5*, 24–27. winter 2007.

Balakrishan, V., & Fernandez, T. (2018). Self-esteem, empathy and their impacts on cyberbullying among young adults. *Telematics & Informatics, 35*, 2028–2037. https://doi.org/10.1016/j.tele.2018.07.006.

Bannink, R., Broeren, S., van de Looij-Jansen, P. M., de Waart, F. G., & Raat, H. (2014). Cyber and traditional bullying victimization as a risk factor for mental health problems and suicidal ideation in adolescents. *PLoS One, 9*(4), e94026.

Barlett, C. P., & Chamberlin, K. (2017). Examining cyberbullying across the lifespan. *Computers in Human Behavior, 71*, 444–449. https://doi.org/10.1016/j.chb.2017.02.009.

Bauman, S., & Bellmore, A. (2015). New directions in cyberbullying research. *Journal of School Violence, 14*, 1–10. https://doi.org/10.1080/15388220.2014.968281.

Bayar, Y., & Uçanok, Z. (2012). School social climate and generalized peer perception in traditional and cyberbullying status. *Educational Sciences: Theory & Practice, 12*, 2353–2358.

Bayraktar, F., Machackova, H., Dedkova, L., Cerna, A., & Ševčíková, A. (2015). Cyberbullying: The discriminant factors among cyberbullies, cybervictims, and cyber-victims in a Czech adolescent sample. *Journal of Interpersonal Violence, 30*, 3192–3216. https://doi.org/10.1177/0886260514555006.

Betts, L. R., Gkimitzoudis, T., Spenser, K. A., & Baguley, T. (2017). Examining the roles young people fulfil in five types of cyber bullying. *Journal of Social and Personal Relationships, 34*, 1080–1098. https://doi.org/10.1177/0265407516668585.

Betts, L. R., Houston, J. E., Steer, O. L., & Gardner, S. E. (2017). Adolescents' experiences of victimization: The role of attribution style and generalized trust. *Journal of School Violence, 16*, 25–48. https://doi.org/10.1080/15388220.2015.1100117.

Betts, L. R., & Spenser, K. A. (2017). "People think it's a harmless joke": Young people's understanding of the impact of technology, digital vulnerability and cyberbullying in the United Kingdom. *Journal of Children and Media, 11*, 20–35. https://doi.org/10.1080/17482798.2016.1233893.

Betts, L. R., Spenser, K. A., & Gardner, S. E. (2017). Adolescents' involvement in cyber bullying and perceptions of school: The importance of perceived peer acceptance for female adolescents. *Sex Roles, 77*, 471–481. https://doi.org/10.1007/s11199-017-0742-2.

Bhat, C. S. (2008). Cyber bullying: Overview and strategies for school counsellors, guidance officers, and all school personnel. *Australian Journal of Guidance & Counselling, 18*, 53–66.

Boyd, D. (2011). White flight in networked publics? How race and class shaped American teen engagement with MySpace and Facebook. In L. Nakamura, & P. A. Chow-White (Eds.), *Race after the internet* (pp. 203–222). Routledge.

Calvete, E., Orue, I., Estévez, A., Villardón, L., & Padilla, P. (2010). Cyberbullying in adolescents: Modalities and aggressors' profile. *Computers in Human Behavior, 26*, 1128–1135. https://doi.org/10.1016/j.chb.2010.03.017.

Campbell, M. A., Slee, P. T., Spears, B., Butler, D., & Kift, S. (2013). Do cyberbullies suffer too? Cyberbullies' perceptions of the harm they cause to others and to their own mental health. *School Psychology International, 34*, 613–629. https://doi.org/10.1177/0143034313479698.

Carvalho, M., Branquinho, C., & Gasper de Matos, M. (2021). Cyberbullying and bullying: Impact on psychological symptoms and well-being. *Child Indicators Research, 14,* 435–452. https://doi.org/10.1007/s12187-020-09756-2.

Cassidy, W., Jackson, M., & Brown, K. N. (2009). Sticks and stones can break my bones, but how can pixels hurt me?: Students' experiences with cyber-bullying. *School Psychology International, 30,* 383–402. https://doi.org/10.1177/0143034309106948.

Chisholm, J. F. (2014). Review of the status of cyberbullying and cyberbullying prevention. *Journal of Information Systems Education, 25,* 77–87.

Crosslin, K., & Crosslin, M. (2014). Cyberbullying at a Texas university—A mixed method approach to examining online aggression. *Texas Public Health Journal, 66,* 26–31.

Davis, K., Randall, D. P., Ambrose, A., & Orand, M. (2015). 'I was bullied too': Stories of bullying and coping in an online community. *Information, Communication & Society, 18,* 357–375. https://doi.org/10.1080/1369118X.2014.952657.

Dempsey, A. G., Sulkowski, M. L., Nichols, R., & Storch, E. A. (2009). Differences between peer victimization in cyber and physical settings and associated psychosocial adjustment in early adolescence. *Psychology in the Schools, 46,* 962–972. https://doi.org/10.1002/pits.20437.

DePaolis, K., & Williford, A. (2019). Pathways from cyberbullying victimization to negative health outcomes among elementary school students: A longitudinal investigation. *Journal of Child and Family Studies, 28,* 2390–2403. https://doi.org/10.1007/s10826-018-1104-6.

Didden, R., Scholte, R. H. J., Korzilius, H., de Moor, J. M. H., Vermeulen, A., O'Reilly, M., et al. (2009). Cyberbullying among students with intellectual and developmental disability in special education settings. *Developmental Neurorehabilitation, 12,* 146–151. https://doi.org/10.1080/17518420902971356.

Dooley, J., Pyżalski, J., & Cross, D. (2009). Cyberbullying versus face-to-face bullying: A theoretical and conceptual review. *Journal of Psychology, 217,* 182–188. https://doi.org/10.1027/0044-3409.217.4.182.

Doumas, D. M., & Midgett, A. (2020). The association between witnessing cyberbullying and depressive symptoms and social anxiety among elementary school students. *Psychology in the Schools, 58,* 622–637.

Ey, L.-A., Taddeo, C., & Spears, B. (2015). Cyberbullying and primary-school aged children: The psychological literature and the challenge for sociology. *Societies, 5,* 492–514. https://doi.org/10.3390/soc5020492.

Farley, S., Coyne, I., Sprigg, C., Axtell, C., & Subramanian, G. (2015). Exploring the impact of workplace cyberbullying on trainee doctors. *Medical Education, 49*(4), 436–443. https://doi.org/10.1111/medu.12666.

Felipe-Castaño, E., León-del-Barco, B., Polo-del-Rio, I., Mendo-Lázaro, S., Gómez-Carroza, T., & Fajardo-Bullón, F. (2019). Differential analysis of psychopathological impact of cyberbullying in university students. *Frontiers in Psychology, 10,* 1620. https://doi.org/10.3389/fpsyg.2019.01620.

Foody, M., McGuire, L., Kuldas, S., & O'Higgins Norman, J. (2019). Friendship quality and gender differences in association with cyberbullying involvement and psychological well-being. *Frontiers in Psychology, 10,* 1723. https://doi.org/10.3389/fpsyg.2019.01723.

Goode, J. A., Fomby, P., Mollborn, S., & Limburg, A. (2020). Children's technology time in two US cohorts. *Child Indicators Research, 13,* 1107–1132. https://doi.org/10.1027/0044-3409.217.4.205.

Gradinger, P., Strohmeier, D., & Spiel, C. (2009). Traditional bullying and cyberbullying: Identification of risk groups for adjustment problems. *Journal of Psychology, 217,* 205–213. https://doi.org/10.1027/0044-3409.217.4.182.

Hellfeldt, K., López-Romoero, L., & Andershed, H. (2020). Cyberbullying and psychological well-being in young adolescence: The potential protective mediation effects of social support from family, friends, and teachers. *International Journal of Environmental Research and Public Health, 17*, 1–16. https://doi.org/10.3390/ijerph17010045.

Imamura, A., Nishida, A., Nakazawa, N., Shimodera, S., Tanaka, G., Kinoshita, H., et al. (2009). Effects of cellular phone email use on the mental health of junior high school students in Japan. *Psychiatry and Clinical Neurosciences, 63*, 703. https://doi.org/10.1111/j.1440-1819.2009.02007.x.

Jackson, C. L., & Cohen, R. (2012). Childhood victimization: Modeling the relation between classroom victimization, cyber victimization, and psychosocial functioning. *Psychology of Popular Media Culture, 1*, 254–269. https://doi.org/10.1037/a0029482.

Jenaro, C., Flores, N., & Frías, C. P. (2018). Systematic review of empirical studies on cyberbullying in adults: What we know and what we should investigate. *Aggression and Violent Behavior, 38*, 113–122. https://doi.org/10.1016/j.avb.2017.12.003.

Jenaro, C., Flores, N., Vega, V., Cruz, M., Pérez, M. C., & Torres, V. A. (2018). Cyberbullying among adults with intellectual disabilities: Some preliminary data. *Research in Developmental Disabilities, 72*, 265–274. https://doi.org/10.1016/j.ridd.2017.12.006.

John, A., Glendenning, A. C., Marchant, A., Montgomery, P., Stewart, A., Wood, S., et al. (2018). Self-harm, suicidal behaviours, and cyberbullying in children and young people: Systematic review. *Journal of Medical Internet Research, 20*. https://doi.org/10.2196/jmir.9044, e129.

Juvonen, J., & Gross, E. F. (2008). Extending the school grounds?—Bullying experiences in cyberspace. *Journal of School Health, 78*, 496–505.

Khine, A. T., Saw, Y. M., Htut, Z. Y., Khaing, C. T., Soe, H. Z., Swe, K. K., et al. (2020). Assessing risk factors and impact of cyberbullying victimization among university student in Myanmar: A cross-sectional study. *PLoS One, 15*. https://doi.org/10.1371/journal.pone.0227051, e0227051.

Kim, S., Boyle, M. H., & Georgiades, K. (2017). Cyberbullying victimization and its association with health across the life course: A Canadian population study. *Canadian Journal of Public Health, 108*, e468–e474. https://doi.org/10.17269/CJPH.108.6175.

Kim, S., Colwell, S. R., Kata, A., Boyle, M. H., & Georgiades, K. (2018). Cyberbullying victimization and adolescent mental health: Evidence of differential effects by sex and mental health problem type. *Journal of Youth and Adolescence, 47*, 661–672. https://doi.org/10.1007/s10964-017-0678-4.

Kowalski, R. M., Limber, S. P., & McCord, A. (2019). A developmental approach to cyberbullying: Prevalence and protective factors. *Aggression and Violent Behavior, 45*, 20–32. https://doi.org/10.1016/j.avb.2018.02.009.

Kwan, I., Dickson, K., Richardson, M., MacDowall, W., Burchett, H., Stansfield, C., et al. (2020). Cyberbullying and children and young people's mental health: A systematic map of systematic reviews. *Cyberpsychology, Behavior and Social Networking, 23*, 72–82. https://doi.org/10.1089/cyber.2019.0370.

Lam, S., Ferlatte, O., & Salway, T. (2019). Cyberbullying and health: A preliminary investigation of the experiences of Canadian gay and bisexual men. *Journal of Gay & Lesbian Social Services, 31*, 3320352. https://doi.org/10.1080/10538720.2019.1596860.

Li, Q. (2007). New bottle but old wine: A research of cyberbullying in schools. *Computers in Human Behavior, 23*, 1777–1791. https://doi.org/10.1016/j.chb.2005.10.005.

Loh, J., & Snyman, R. (2020). The tangled web: Consequences of workplace cyberbullying in adult male and female employees. *Gender in Management: An International Journal, 35*, 567–584. https://doi.org/10.1108/GM-12-2019-0242.

Macaulay, P. J. R., Betts, L. R., Stiller, J., & Kellezi, B. (2022). An introduction to cyberbullying. In A. Moustafa (Ed.), *Cybersecurity, psychology, and cognitive science* Elsevier.

Monks, C. P., Robinson, S., & Worlidge, P. (2012). The emergence of cyberbullying: A survey of primary school pupils' perceptions and experiences. *School Psychology International, 33*, 477–491. https://doi.org/10.1177/0143034312445242.

Nixon, C. L. (2014). Current perspectives: The impact of cyberbullying on adolescent health. *Adolescent Health, Medicine and Therapeutics, 5*, 143–158. https://doi.org/10.2147/AHMT.S36456.

Ortega, R., Elipe, P., Mora-Merchán, J. A., Calmaestra, J., & Vega, E. (2009). The emotional impact on victims of traditional bullying and cyberbullying: A study of Spanish adolescents. *Journal of Psychology, 217*, 197–204. https://doi.org/10.1027/0044-3409.217.4.197.

Pendergrass, W. P., & Wright, M. (2014). Cyberbullied to death: An analysis of victims taken from recent events. *Issues in Information Systems, 15*, 132–140.

Raskauskas, J., & Stoltz, A. D. (2007). Involvement in traditional and electronic bullying among adolescents. *Developmental Psychology, 43*, 564–575. https://doi.org/10.1037/0012-1649.43.3.564.

Rivituso, J. (2014). Cyberbullying victimization among college students: An interpretative phenomenological analysis. *Journal of Information Systems Education, 25*, 71–75.

Robards, B. (2012). Leaving MySpace, joining Facebook: 'Growing up' on social network sites. *Continuum: Journal of Media & Cultural Studies, 26*, 385–398. https://doi.org/10.1080/10304312.2012.665836.

Sabella, R. A., Patchin, J. W., & Hinduja, S. (2013). Cyberbullying myths and realities. *Computers in Human Behavior, 29*(6), 2703–2711. https://doi.org/10.1016/j.chb.2013.06.040.

Salmivalli, C., Lagerspetz, K., Björkqvist, K., Österman, K., & Kaukiainen, A. (1996). Bullying as a group process: Participant roles and their relations to social status within the group. *Aggressive Behavior, 22*, 1–15. https://doi.org/10.1002/(SICI)1098-2337(1996)22:1<1::AID AB1>3.0.CO;2-T.

Sampasa-Kanyinga, H., Roumeliotis, P., & Xu, H. (2014). Associations between cyberbullying and school bullying victimization and suicidal ideation, plans and attempts among Canadian schoolchildren. *PLoS One, 9*. https://doi.org/10.1371/journal.pone.0102145, e102145.

Selkie, E., Kota, R., Chan, Y.-F., & Moreno, M. (2015). Cyberbullying, depression, and problem alcohol use in female college students: A multisite study. *Cyberpsychology, Behavior and Social Networking, 18*, 7–86. https://doi.org/10.1089/cyber.2014.0371.

Ševčíková, A., & Šmahel, D. (2009). Online harassment and cyberbullying in the Czech Republic: Comparison across age groups. *Journal of Psychology, 217*, 227–229. https://doi.org/10.1027/0044-3409.217.4.227.

Sincliar, K. O., Bauman, S., Poteat, V. P., Koenig, B., & Russell, S. T. (2012). Cyber and bias-based harassment: Associations with academic, substance use, and mental health problems. *Journal of Adolescent Health, 50*, 521–523. https://doi.org/10.1016/j.jadohealth.2011.09.009.

Snyman, R., & Loh, J.(. M. I.). (2015). Cyberbullying at work: The mediating role of optimism between cyberbullying and job outcomes. *Computers in Human Behavior, 53*, 161–168. https://doi.org/10.1016/j.chb.2015.06.050.

Spears, B., Slee, P., Owens, L., & Johnson, B. (2009). Behind the scenes and screens: Insights into the human dimension of covert and cyberbullying. *Journal of Psychology, 217*, 189–196. https://doi.org/10.1027/0044-3409.217.4.189.

Stauffer, S., Heath, M. A., Coyne, S. M., & Ferrin, S. (2012). High school teachers' perceptions of cyberbullying prevention and intervention strategies. *Psychology in the Schools, 49*(4), 352–367. https://doi.org/10.1002/pits.21603.

Tokunaga, R. S. (2010). Following you home from school: A critical review and synthesis of research on cyberbullying victimization. *Computers in Human Behavior, 26*, 277–287. https://doi.org/10.1016/j.chb.2009.11.014.

Vanderbilt, D., & Augustyn, M. (2010). The effects of bullying. *Paediatrics and Child Health, 20*(7), 315–320. https://doi.org/10.1016/j.paed.2010.03.008.

Wachs, S. (2012). Moral disengagement and emotional and social difficulties in bullying and cyberbullying: Differences by participant role. *Emotional and Behavioural Difficulties, 17*, 347–360. https://doi.org/10.1080/13632752.2012.704318.

Wang, M.-J., Yogeeswaran, K., Andrews, N. P., Hawi, D. R., & Sibley, C. G. (2019). How common is cyberbullying among adults? Exploring gender, ethnic, and age differences in the prevalence of cyberbullying. *Cyberpsychology, Behavior and Social Networking, 22*, 736–741. https://doi.org/10.1089/cyber.2019.0146.

West, D. (2015). An investigation into the prevalence of cyberbullying among students aged 16-19 in post-compulsory education. *Research in Post-Compulsory Education, 20*, 96–112. https://doi.org/10.1080/13596748.2015.993879.

Wolke, D., & Lereya, S. T. (2015). Long-term effects of bullying. *Archives of Disease in Childhood, 100*(9), 879–885. https://doi.org/10.1136/archdischild-2015-309491.

Ybarra, M. L., Espelage, D. L., & Mitchell, K. J. (2007). The co-occurrence of internet harassment and unwanted sexual solicitation victimization and perpetration: Associations with psychosocial indicators. *Journal of Adolescent Health, 41*, S31–S41. https://doi.org/10.1016/j.jadohealth.2007.09.010.

Zhang, D., Heubner, E. S., & Tian, L. (2021). Neuroticism and cyberbullying among elementary school students: A latent growth curve modelling approach. *Personality and Individual Differences, 171*, 110472. https://doi.org/10.1016/j.paid.2020.110472.

CHAPTER

10

Cyber situational awareness issues and challenges

Ulrik Franke[a,b], *Annika Andreasson*[a], *Henrik Artman*[a,c], *Joel Brynielsson*[a,c], *Stefan Varga*[a,d], *and Niklas Vilhelm*[a,e]

[a]KTH Royal Institute of Technology, Stockholm, Sweden, [b]RISE Research Institutes of Sweden, Kista, Sweden, [c]FOI Swedish Defence Research Agency, Stockholm, Sweden, [d]Swedish Armed Forces Headquarters, Stockholm, Sweden, [e]Norwegian National Security Authority, Sandvika, Norway

1 Introduction

Modern society is full of information technology, underpinning the operations of corporations, governments, and nongovernmental organizations, as well as the personal and social lives of billions of individuals. While there are obvious benefits to these digital tools, their widespread use also means that they must be increasingly dependable. As a result, enterprises of all sorts need to make informed decisions that uphold the dependability of digital services. If the information stored in digital systems cannot be trusted, or if the systems themselves cannot be reached when needed, the benefits of digitalization will not materialize.

1.1 Cyber risk management

Clearly, there are numerous risks to networks and the information in them (see Cebula, Popeck, & Young, 2014 for a taxonomy). One class of risk is systems and technology failures, internal or external, that occur without any adversarial intent. Another class of risk is the threat from adversaries; threat actors who can reason rationally and adapt to different defensive strategies. It is important to consider both classes of risk, and in

Cybersecurity and Cognitive Science
https://doi.org/10.1016/B978-0-323-90570-1.00015-2

particular not to forget about nonadversarial risks just because there is much talk about deliberate attacks. Accidents can be equally damaging.

However, while it is easy to make the theoretical distinction between these two classes, it is not equally clear in practice, where adversaries have every reason to conceal themselves and their actions as mere accidents. Indeed, there is some evidence to suggest that defenders do not always care about this distinction (Varga, Brynielsson, & Franke, 2018, 2021), and that decision makers assess the probabilities of the two classes of threats as very similar (Franke & Wernberg, 2020).

Properly managing these risks typically involves many different actions. Some are technical in nature, and involve things like setting up and automating updates, backups, encryption, access privileges, and network monitoring. However, it is important to realize that this is not enough. Organizational measures such as training of employees (McCrohan, Engel, & Harvey, 2010) and exercises in cyber incident management (Maennel, Ottis, & Maennel, 2017) are also important. Human error is often the root cause of incidents, and technical measures rarely suffice if a legitimate user can be tricked into assisting in an attack (Krombholz, Hobel, Huber, & Weippl, 2015). Furthermore, senior management needs to work actively with identifying how operations depend upon technology, and document this in a strategy that is regularly updated (Dunbar, 2012; Dutta & McCrohan, 2002). It is important to realize that such risk management work critically depends upon understanding one's own organization and operations—it is not something that can be outsourced.

1.2 Cyber situational awareness

The brief outline above suggests that prudent cyber risk management encompasses a multitude of decisions. To make the best decisions, however, decision makers need to have an accurate situational awareness (SA)—colloquially, to *know what is going on*. Such SA with respect to information technology and networks—cyber situational awareness (CSA)—is the topic of this chapter.

Somewhat more formally, our point of departure is the definition of SA that stems from Endsley (1988) and her investigation of how aircraft pilots understand the situation. Using her definition, SA is "the perception of the elements in the environment within a volume of time and space, the comprehension of their meaning, and the projection of their status in the near future" (Endsley, 1988, p. 792). Based on this definition, Endsley in later work defines three levels of SA: (i) perception, (ii) comprehension, and (iii) projection (Endsley, 1995). In this chapter, we adhere to the Endsley-inspired definition of CSA given in our previous work: "a subset of situational awareness, i.e., cyber situational awareness is the part of situational awareness which concerns the 'cyber' environment" (Franke & Brynielsson, 2014, p. 20).

Much of the existing CSA literature is very technologically oriented, that is, focused on awareness of the state of the technical artifacts that constitute the network. In the literature review we published in 2014, we found comparatively much work on CSA for industrial control systems, and for general algorithms and information fusion in intrusion detection systems (IDSs). By contrast, there was considerably less research done in other areas such as information exchange, the risks of cyber deception, and of cyber battle damage assessments in military operations. Evancich et al. (2014) suggested that aspects such as operational awareness and threat awareness are also relevant for CSA.

1.3 Common operational picture and situational awareness

Since the two are often discussed in parallel and sometimes confused, it is useful to make the distinction between a common operational picture (COP) and SA. The former is an artifact—such as a shared big screen, a whiteboard, or a map with symbols—which aims to provide stakeholders with a "common picture" of what is going on (the concept of a COP has military origins; Hager, 1997). The latter is a mental state—awareness of what is going on. Of course, the COP is designed, more or less deliberately, to facilitate SA (CSA, in our case), but even though the two are closely related, they are conceptually different. This COP/CSA distinction is reminiscent of how Gutzwiller, Hunt, and Lange (2016) argued that there is a need to differentiate (the products of) data fusion (typically a COP) from actual CSA, where human cognition is a central component.

The simplest way in which the COP facilitates SA is that an individual just looks at the COP to gain SA. Here, the COP serves as the "information warehouse" (Copeland, 2008) where relevant information is stored and computed. However, particularly when several individuals are involved, the COP can also be perceived as a process (Wolbers & Boersma, 2013), where the meaning and interpretation of information is constantly (re)negotiated as stakeholders interact. From this perspective, the COP is a "trading zone" (Wolbers & Boersma, 2013), and all the interactions within this zone facilitate SA (CSA, in our case).

1.4 Outline

In the rest of this chapter, we further develop and elaborate the themes introduced above. A recurring theme is the necessity to include sociocognitive and organizational factors, in addition to technology, to achieve the most relevant and useful kind of CSA, and further that an important challenge to be addressed concerns adversarial behavior.

First, however, in Section 2, some of the technological building blocks necessary for CSA are discussed. Once these are in place, the scene is set for a look at socio-cognitive aspects in Section 3. Based on the previous two sections, we proceed to organizational issues in Section 4. Section 5 then investigates the prospects for reasoning about adversarial actions. As mentioned earlier, not all cyber incidents are adversarial (accidents and honest mistakes abound), but when there is an adversary, this adds an important dimension to CSA. In Section 6, some directions for future research are proposed, before Section 7 concludes this chapter.

2 The technological perspective

It is obvious that CSA is dependent on information from technical sensors. In this section, we review some of the technologies necessary to adequately understand what is going on in the cyber domain. How can IDSs, techniques for visualization of large datasets, and fusion of information from different sources contribute to CSA?

2.1 Nonadversarial cyber incidents, design, and resilience

The strategies to build secure and dependable IT systems have evolved over the years. In the infancy of computing, hardware failure was behind many outages and incidents. However, with improved manufacturing techniques and quality control in computer components, as well as reliability designs such as redundant components with failover, hardware failure no longer accounts for a significant share of downtime (Genadis, 1996).

Instead, for the last three or four decades, the major culprits of nonadversarial incidents have been software errors and mistakes in IT administration (Gray, 1985, 1990). For example, an investigation of failures in web applications found some 80% to be caused by software or human error. Similarly, Gartner found that people and process failures were behind approximately 80% of "mission-critical" downtime (Malik & Scott, 2010).

Avoiding nonadversarial cyber incidents, thus, has evolved from a matter of building more reliable hardware to having quality checks on the software being used and finding ways to make sure that the people managing it adhere to appropriate processes. This is not the place to provide full introductions to the fields of software testing (see, e.g., Bertolino, 2007; Whittaker, 2000) or IT service management (see, e.g., Galup, Dattero, Quan, & Conger, 2009; Iden & Eikebrokk, 2013), but we do note that adequate CSA must include a basic understanding of these issues—not least to be able to distinguish nonadversarial events caused by mistakes and process failures from adversarial attacks.

2.2 Adversarial cyber incidents and defense against them

As for nonadversarial incidents, the strategies to defend against attacks have evolved over time. Li, Huang, Wang, and Li (2019, p. 2) identified four paradigms. The first paradigm occurred before the 1960s. Here security was a design feature; it was assumed that a solid hardware and software architecture would prevent attacks. In the 1970s and 1980s, security assistance systems, such as, for example, IDSs, were added to the existing systems to provide protection from attacks. Later in the 1990s, more sophisticated approaches emerged. A common theme for these, arguably, was that they aimed to be proactive. Attacks were modeled using techniques such as state graphs and attack trees (Mauw & Oostdijk, 2005; Schneier, 1999), allowing preparations for defense in advance. The latest and present paradigm, according to Li et al. (2019), commenced after the turn of the millennium. This paradigm prescribes that a wider perspective than before has to be taken into account to improve defensive efforts. Li et al. (2019) argued that today it is necessary to try and predict future trends with techniques such as complex network theory (Wan, Cao, Chen, & Huang, 2017; Xu, 2014), game theory (Attiah, Chatterjee, & Zou, 2018), etc. These new requirements have emerged not because the probabilities of chance events have changed in a significant way, but rather because of the continuous improvement of adversarial (cyber) attack methods.

2.3 The expanding scope of cyber situational awareness

As illustrated earlier, strategies both for (i) resilience against nonadversarial incidents and (ii) security against adversarial attacks are ever-expanding in scope. It seems clear that the cyber defense community has abandoned the position that security could be solved by design only, once and for all. Instead, the scope of adequate CSA now also includes knowledge of attackers and the *modus* of attacks, as well as factors beyond the technical network itself.

For example, with increasing use of third-party cloud services, maintaining awareness of *internet connectivity* is of vital interest (Gunawi et al., 2016). Are there planned outages? Are there incidents in other places that may result in lower quality of service as the load rebalances (Omer, Nilchiani, & Mostashari, 2009)? Have necessary precautions been taken, and is there a way to insure against the residual risk that cannot be managed by technical means (Franke, 2018)?

Another example concerns the *insider threat*. To know whether an unauthorized employee is moving around at the physical premises should be of interest for cyber defenders, because such an individual could have gained physical access to the IT system and pose a (cyber) threat. Similarly, if a suspicious hardware device is seen in an office room, that is also of interest. These two examples are mentioned by Vielberth, Menges, and Pernul (2019), and illustrate the point that CSA is not only about "cyber" events—things going on in the physical world can be just as important. In the longer term, vetting and recruitment processes of cybersecurity personnel to ensure that competent and trustworthy personnel with the required psychological profiles is hired, are also beneficial for cybersecurity (Chen et al., 2014).

2.4 Intrusion detection systems

As mentioned earlier, the IDS is one the most important technical tools used for CSA with respect to adversarial attacks. An IDS collects data from networks or files on devices, and then determines if this data contains any indicators of compromise. Broadly, IDSs can be classified into two categories: (i) signature-based systems and (ii) anomaly-based systems (Khraisat, Gondal, Vamplew, & Kamruzzaman, 2019). The first category uses signatures of already known threats to detect them when encountered again. The second category instead looks for anomalies that may indicate that something is wrong in the network as a result of an attack. While different in principle, most commercial systems are a combination of both. Furthermore, while it might naïvely be assumed that signature-based systems are unable to detect previously unencountered threats, this is not the case. On the contrary, it has been shown that a rule-based IDS can detect attacks not described by its set of rules, though its detection rate is lower than for known attacks (Holm, 2014).

To successfully detect an incident the IDS depends on capturing the correct data and interpreting it correctly. Current approaches to detecting intrusions can be summarized as the use of (i) known pattern templates, (ii) threatening behavior templates, (iii) traffic analysis, (iv) statistical anomaly detection, and (v) state-based detection (Bass, 2000). Decisions about where data should be collected and at what rate are of critical importance. Adding more collection points in the infrastructure and more frequent data captures increase resource requirements. However, in real-world implementations trade-offs have to be made with respect to data collection (Werlinger, Hawkey, Muldner, Jaferian, & Beznosov, 2008).

A significant challenge to the use of an IDS is its detection characteristics: the rate of false positives (threats that are reported, but not real) and false negatives (threats that are real, but not reported). Too many reports might overwhelm system administrators, but too few might not provide adequate information. The sensitivity to which indicators should trigger a report is a balance between these two diverging metrics (Sommestad & Franke, 2015). Improving these rates is, therefore, an active research area (Goeschel, 2016; Spathoulas & Katsikas, 2010). However, it is also noteworthy that even though an IDS is a technical tool, it matters who uses it: competent system administrators are important for IDS usage to be effective (Sommestad & Hunstad, 2013). This serves as a cautionary reminder about the importance of including human factors when designing and using technical solutions for CSA.

To generate the best possible assessment of whether attacks are ongoing, an IDS has to combine many different kinds of data. However, combining data and information from heterogeneous origins and producing reports useful to human decision makers is challenging (Bass, 2000). A collection of low-level responses from security mechanisms and other systems is not easily interpreted in isolation and typically needs to be put into a larger context and be analyzed to provide reliable information about the system. Still, inferences about the systems can be assumed to be correct to a greater degree of certainty if supported by more evidence. Conversely, errors are more reliably spotted if information on the same issue is gathered from multiple sources.

2.5 Intrusion detection systems and explainable artificial intelligence

With recent advances in artificial intelligence and in particular deep neural networks (DNNs), it is not surprising that these techniques have been applied to IDSs (Kang & Kang, 2016; Kim, Shin, Jo, & Kim, 2017; Roy, Mallik, Gulati, Obaidat, & Krishna, 2017). Since manual analysis of indicators of compromise tends to be very labor-intensive, the possibility to automate such analysis using AI methods is very attractive.

However, DNNs typically use a huge parametric space with hundreds of layers and millions of parameters and are thus for practical purposes considered "black boxes" which are very difficult to interpret for humans (Arrieta et al., 2020). This lack of transparency has led to an intense discussion—in the scientific community and in society at large—about the risk that black box AI systems will make biased or erroneous decisions in the absence of meaningful human supervision. For example, many AI systems have been found to exhibit biases disadvantaging, for example, poorer people and those from minorities (Nature, 2016).

For DNN-based IDSs to become more useful and trusted to make critical decisions autonomously, the explainability issues mentioned earlier need to be addressed. Within the field of explainable AI (XAI), methods are developed to make AI (and in particular DNN) systems less opaque. Review articles include Guidotti et al. (2018) and Du, Liu, and Hu (2019). Though many questions remain unsolved, advances within XAI are widely acknowledged as a critical feature for the practical deployment of AI models (Arrieta et al., 2020). Such research should also include studies of explainability from the perspective of human-computer interaction (Abdul, Vermeulen, Wang, Lim, & Kankanhalli, 2018).

2.6 Network-centric and domain-centric cyber situational awareness

It seems intuitively plausible that a computer network that is set up with state-of-the-art cyber defense software fully loaded with relevant and timely IDS signatures and the likes, will fare better than a system that lacks up-to-date threat information. Similarly, it also seems likely that a network defended by individuals who possess a deep understanding of the techniques, tactics and procedures of potential adversaries, along with their underlying motivations and rationales (Brynielsson, Franke, Tariq, & Varga, 2016), are able to perform a better job in defending their network than people who do not know anything about these things. Unfortunately, to the best of our knowledge, neither of these two propositions have been rigorously examined through research efforts. We note that the former question, which comprises the "on-the-network" fight (Borum, Felker, Kern, Dennesen, & Feyes, 2015), draws more academic interest than the latter more tactical and strategic dimensions of cyber intelligence (Borum et al., 2015).

Cyber threat intelligence (CTI) is often discussed with the tacit assumption that the "use case" is cyber defense from the context of cyber analysts. However, cyber defense tasks can be performed at multiple managerial levels: ranging from the operational all the way up to the strategic level. Therefore, CTI should be thought of in terms of multiple levels as well (Mattern, Felker, Borum, & Bamford, 2014). Cyber analysts need to understand the human side of the threat, and not only the technical manifestations of it (Mattern et al., 2014). This means that analysts need to understand underlying motivations and rationales for people who are behind attacks on a strategic level, but also the role of humans as offensive cyber operators on the tactical and operational levels. In line with this thinking, Zheng and Lewis (2015) assert that there are two basic categories of cyber threat information (to share), first: technical threat indicators (e.g., IP addresses, specific strings of data, file hashes, exploit toolkits or payloads, adversary tactics, techniques and procedures), and second: contextual threat intelligence (e.g., exploit targets, exfiltrated content, incident details, and specific courses of action). The technical indicators provide direct descriptive information related to the actual breach, while the contextual indicators provide more indirect information such as related effects and other pieces of explanatory information. In line with this distinction, we propose that it is useful to distinguish between *network-centric* and *domain-centric* CSA, as outlined in Table 1.

The network-centric category mainly relates to awareness of technological (network) level data and information. Data and information at this level include low-level indicators of compromise, such as known malware hash values and command and control servers, IP addresses, and so on

TABLE 1 Network-centric and domain-centric CSA.

	Network	Domain
Focus	Technical	Other (organization, mission)
Scope	Narrow (computers and network)	Wide
Automation	Yes	No
Analysis	Fusion	Aggregation, semantic meaning
Human intervention required	No (yes)	Yes
Timescale	Short term	Long term

(Zheng & Lewis, 2015). Furthermore, Brown, Gommers, and Serrano (2015) point out that related "network"-level intelligence management processes probably can be fully automated, as discussed in the previous section. Network-centric awareness may lead to immediate corrective actions, such as the patching of vulnerabilities or other "stop the bleeding"-kind of activities that will improve the level of cybersecurity in the short term (Ahmad, Desouza, Maynard, Naseer, & Baskerville, 2020).

By contrast, the domain-centric category relates to awareness of issues in the organizational and threat dimensions, including information or intelligence about more complex matters such as, for example, descriptions of threat actors and their motivations, modes of attack, and other characteristics. Brown et al. (2015) also mention information about the use of different malware families *over time*. A deeper understanding of this kind of information can result in more profound organizational changes, such as amendments of current security strategies, processes, and workflows, that will lead to improvements of the overall cybersecurity posture in the long term.

To further clarify this distinction, consider how different types of specific information can be assigned to one or the other of the two categories. We do this by using STIX, the *Structured Threat Information eXpression* language and serialization format, since it is the de facto standard for exchange of cyber threat intelligence (Sauerwein, Sillaber, Mussmann, & Breu, 2017). In STIX, we find a taxonomy of relevant information elements for cyber defense in the so-called STIX domain objects (SDOs).

Hence, we propose that network-centric CSA should correspond to the following example SDOs: *indicator*, *observed data*, *tool*, and *vulnerability*. These categories are formed by what we may call atomic information elements. Here we mean atomic in the sense of foundational and minuscule. Atomic information (elements) should leave no or little room for interpretation—it is either right or wrong. Further, our chosen atom metaphor is appropriate because it also alludes to the possibility of combining multiple atoms to infer a high-level understanding of the phenomena under investigation.

The domain-centric CSA can be built upon the analysis of atomic information elements and higher-order pieces of information or intelligence. Corresponding STIX SDO examples are: *campaign*, *course of action*, and *threat actor*. To make sense of this type of information, richer descriptions are required. At the moment, these SDOs need to be processed and acted on by humans who can understand, and discuss, their meaning. This sets the scene for the discussion about socio-cognitive aspects in the next section.

3 The socio-cognitive perspective

From the previous section, we are now acquainted with some of the cyber information available to decision makers. But how do human decision makers process and combine this information? How do we reach CSA when working in teams? What pitfalls need to be avoided?

Endsley's definition of SA and, by inheritance, our definition of CSA as given in Section 1 is individualistic. There is a world, the individual can pay attention to or fail to pay attention to it, and (C)SA is in some sense about mirroring the world within the individual mind. However, much work with complex problems and systems, with a great emphasis on security and reliability that requires significant and fast action, is conducted in teams (Artman, 2000; Artman & Wærn, 1999). This *teamwork* introduces an additional and important perspective. While some aspects of teamwork are addressed in Section 4, we start this exposition of the socio-cognitive perspective with some considerations about team SA (Demir, McNeese, & Cooke, 2017), before considering the implications of the cyber environment for (i) perception, (ii) comprehension, and (iii) projection into the future.

3.1 Team cyber situational awareness

In most teams working to manage cyber incidents, responsibilities are distributed for different, albeit overlapping, tasks which must somehow be coordinated if problems are to be solved. Coordination in such cases is usually a continuous process where different team members tell each other what they are paying attention to or looking for, and how they understand the situation. In these communications between different members, a shared model of understanding of the course of events, and possibly also explanatory models of why they arose, is also created (Artman, Brynielsson, Johansson, & Trnka, 2011).

The physical world is governed by the laws of physics. Based on examining physical objects and their relationship, one can create fairly reliable models and causal relationships. Social systems are not equally amenable to description by laws, because they are based on intangible relationships such as trust, confidence, experience, social hierarchies, knowledge, and the communication of these between people. In our context, this highlights the difficulty of understanding and communicating about a specific attack and the intentions behind it (Artman et al., 2011). Since you can never know for sure another person's intentions with specific actions, you have to make assumptions. Such assumptions may well be based on experience and knowledge, but are nevertheless based on detached prejudices and hypotheses.

Such a shared model—created for the moment and in order to understand a complex process—risks becoming so influential that it in itself becomes an indisputable reality for the group. This may lead the group to look more for evidence confirming what is (thought to be) known than for contradicting evidence. For individuals, this is known as *confirmation bias* (Nickerson, 1998; Tversky & Kahneman, 1974), whereas the corresponding group phenomenon is sometimes called *groupthink*. This concept has famously been used to explain decisions made during the Bay of Pigs invasion, the Vietnam war, and the Watergate scandal (Kramer, 1998; Raven, 1998).

3.2 Perception

Starting from Endsley's definition, what does it mean to translate it to the cyber environment, as we do in this chapter? Endsley's SA concept is developed for aviation. In aviation (and more generally, in most physical environments), perception—of views, sounds, vibrations, etc.—cannot be avoided. In the cockpit, the pilot can see what is happening in the physical space either through the cockpit window or through instruments, which mediate elements in the environment beyond visual range. Auditory aspects are also important to be able to assess the state of the aircraft. Other sensory input such as G-forces also play a significant role.

This differs from the cyber environment, where perception only takes place once appropriate sensors—firewalls, IDSs, vigilant users—are in place. Log file analysis systems that warn of unusual and critical events can, perhaps, be likened to a radar system, but the systems must be designed to actively search for unusual elements or events. Radar systems search for physical materials that can be distinguished from other physical materials. In the cyber world, however, all elements and all events are of the same immaterial nature. What remains are intangible processes that must differ from the normal state of the system one is trying to protect. Anyone who tries to attack and break into the digital system knows this, of course, and can continuously delete any digital tracks to avoid detection. This cleaning of digital tracks also differs greatly from how a physical element can be masked. Furthermore, an attacker can introduce false tracks to confuse. The attacker can impersonate a less competent actor or, conversely; a competent person who is trying to attack a particular part of the system may act so as to hide that he/she is "actually" trying to intrude on another part of the system, which may also be less defended. Thus, while SA in the physical environment is *perceptually driven*, CSA is rather *search-driven*.

In aviation, these elements typically appear some time before there is an actual crisis—and the pilot will have at least some time to act and react—while elements and events in cyberspace might appear only at the moment of imminent attack. At least, this is the case for network-centric CSA as discussed in Section 2; it may be less so for domain-centric CSA. In the cyber world, the constraints of time and space are less binding. For example, a person can be on the other side of the globe and attack a system without any significant time constraints.

3.3 Comprehension

The next level in Endsley's model is about understanding what the different elements or events mean. The discussion above has already touched on meanings and meaning because in cyberattacks there is always an acting person and this person has intentions with his/her actions. However, the intention may vary; some attack a system or organization for criminal reasons, others do it to learn and test their skills, and some do it to warn the organization about loopholes in the system. There are certainly many other purposes for a cyberattack as well (Tariq, Brynielsson, & Artman, 2012). As we have already mentioned, adversarial attacks mean that as a defender of the system or organization, one must have an understanding of who the opponent is, what intentions they may have, and how many and well-organized they are. Events and elements that can be traced in log files or during ongoing attacks contain weak signals to be used for the purpose of determining the intentions and goals of the opponent. As we have already pointed out, this means that CSA must to a large extent rely on creating plausible hypotheses and constantly examining the course of events.

3.4 Projection into the future

Such hypotheses in turn govern how one understands future events and can act to prevent these from happening. A team working to counter future potential events is well advised not only to understand the intentions of the opponent, but also its skills and capacity to carry out attacks. One must therefore to a large extent preempt the opponent's future activities and block the possibilities for further intrusion. In the physical world, there is usually a certain time difference between starting an activity until you can complete it. In the cyber world, this time difference is minimized, which makes it difficult to take the time to reflect on a potential course of events—there is, hence, a risk of "extinguishing ongoing fires" rather than preventing the fire from spreading.

3.5 The challenge of a complex nonphysical environment

When trying to understand complex dynamic systems where there are many interrelated dependencies—often in chains of events and where it is not obvious how elements, subsystems, and individual systems relate to each other—there is a risk that we create simplified mental models that work well in the short run, but have unknown and unintended adverse effects in the long run. Among natural systems, the climate is an obvious example, where effects occur in long causal chains which are difficult to fully grasp. Human use of raw materials releases carbon dioxide into the atmosphere, which in turn increases the temperature, which in turn causes animals and plants to migrate, etc. Among social systems, the economy is a good example, where the possibility or impossibility of central planning was a hotly contested topic in the early 20th century (and revisited time and again since), and many arguments focus precisely on the impossibility for any single individual to gather all the data and properly understand all the long causal chains involved in economic transactions (Cottrell & Cockshott, 1993; Hayek, 1982; Yeager, 1994). In such cases, there is a risk that decision makers and analysts create such a simplified worldview that even though a single good is closely monitored and achieved, other elements are ignored, to the detriment of other important desiderata. This phenomenon is known as *encapsulation* (Dörner & Schaub, 1994).

Discussing CSA and the intangible cyber environment, the risk of encapsulation must be considered: do analysts and decision makers have oversimplified conceptions of what is going on, what they themselves try to achieve, and what the attackers are doing? Such oversimplifications can also be entrenched by groupthink.

4 The organizational perspective

Different stakeholders need different forms of CSA. For example, an operator in a security operations center (SOC) has a different time perspective than the CEO. But sometimes the operator might need to escalate a question quickly to the CEO. How should an organization be designed to facilitate the kind of information flows and mandates that enable making correct and timely decisions?

4.1 Operational level

At first glance, CSA usually seems to be discussed in relation to analysts in a SOC. It is often in the SOC that adversarial cyberattacks are detected, responded to, and recovered from by the SOC staff. But SOC analysts are not the only members of an organization with a need for CSA. Understanding the complex inner workings of the SOC is, indeed, a prioritized undertaking, but the cybersecurity efforts of an organization should not stop there. One way of addressing CSA in organizations is to define the CSA requirements of different professional security roles, as proposed by, among others, Gutzwiller, Dykstra, and Payne (2020). One idea suggested is to use the *Workforce Framework for Cybersecurity (NICE Framework)*—developed by the US National Institute of Standards and Technology (NIST)—as a starting point (Petersen, Santos, Smith, Wetzel, & Witte, 2020). The framework is an effort to establish what different professional cybersecurity roles entail, without being dependent on where the role is situated. Each role has skills and tasks associated with it. This would result in generic CSA requirements that organizations could use as a starting point for establishing what information is needed for the work performed by different cybersecurity operators within the organization.

4.2 Tactical and strategic level

However, in an organization there are other decision makers in need of CSA than the immediate cybersecurity workforce. When an organization is under attack, information about the present situation often needs to reach higher up in the decision chain, as the event could pose a potential threat to ongoing operations and even the enterprise as a whole. Goals, timelines, and CSA requirements vary with the roles in the organizational hierarchy. The distinction between network-centric and domain-centric CSA as illustrated in Table 1 can be seen as an example of this. As pointed out by McKenna, Staheli, and Meyer (2015), organization members like the SOC analyst, SOC manager, CIO, and CEO, all need CSA to a varying degree. While under attack, the core issues and goals of the SOC analyst are quite different from the core issues and goals of the CEO. While one focuses on what is going on in the network, the other focuses on what that might mean in the wider context of future operations. In this linear organization example, information flows up in the hierarchy and decisions flow down.

4.3 Crisis management

It should not be forgotten, though, that there are roles not directly involved in the "vertical" line organization or that are part of the NICE framework roles, that nevertheless have a need for CSA. As an example, in the face of larger cyber events, special crisis management organizations are often activated, involving members from several different parts of an organization. While cybersecurity employees are involved in the immediate incident response, crisis management teams are simultaneously working on issues such as business continuity, stakeholder communications, and media relations. These teams, to a varying degree, also need CSA to be able to make the best decisions in going forward. Organizations need to ensure that each team has the CSA required to fulfill their goals, which is often complicated by the conflicting need to restrict regular users' access to sensitive information due to security concerns (Tariq, Brynielsson, & Artman, 2014). Aligned with the goals, the operators should have the mandate to make the decisions required to fulfill those goals.

4.4 Methods for research and organizational design

The method proposed by Endsley and Jones (2011) to capture SA requirements, is a goal-directed task analysis (GDTA). In a GDTA, the goals, decisions, and information requirements of different operators are identified. From there, an organization could design information sharing infrastructures to aid the CSA of different teams and operators with a need to know "what is going on" in the cyber domain. This could indeed be a way forward for organizations to determine where there are shared information requirements, and establish routines and ways of working where information reaches all the decision makers who need it. This reduces the risk that decision makers needing the same information collect it from sources of different quality, unbeknownst of each other.

5 Reasoning about adversarial behavior

Not all cyber incidents are adversarial, but the line can be difficult to draw—partly because an adversary might want to disguise attacks as mere accidental outages. Previous research suggests that in some circumstances decision makers do not take adversarial behavior into account to the extent that would be prudent (Varga et al., 2018, 2021). How can CSA be improved by reasoning about adversarial behavior to better identify things like diversions?

5.1 Threat modeling

There are numerous cyber threat models that aim to inform about cyber threats. Adam Shostack (2014) lists a few in his book *Threat modeling: Designing for security*. A few examples include the Threat Agent Risk Assessment (TARA) framework from the Intel corporation (Rosenquist, 2009) that lists 21 different threat agents; the OCTAVE Allegro information security risk assessment methodology that emphasizes the use of threat scenarios (Caralli, Stevens, Young, & Wilson, 2007); and the Military Activities and Cyber Effects (MACE) taxonomy that lists eight different adversary types (Bernier, 2013). The adversary types are described in terms of skill levels, maliciousness, motivation, and methods used. There are plenty of other models available.

These models generally draw a picture of various attackers based on certain characteristics. The idea is that conclusions about the threat against one's own organization should be drawn and put to use. Precisely *how* such inferences should be made and transformed into action, however, is far from trivial. This aspect does not appear to be extensively discussed in the literature. The term "threat actors" is mentioned a total of six times in the book *Cyber threat intelligence* (Dehghantanha, Conti, & Dargahi, 2018), but its relevance and meaning are not elaborated on beyond the fact that threat actors exist. Nor is the specifics of threat actors a subject for treatment in the book *Information security practices: Emerging threats and perspectives* (Traoré, Awad, & Woungang, 2017). Furthermore, in an extensive analysis of information security data sources by Sauerwein, Pekaric, Felderer, and Breu (2019), threat actor information was not mentioned at all. In conclusion, there appears to be a strong emphasis on information about vulnerabilities and other threats rather than on details on threat actor information in the contemporary literature about cyber threat information and intelligence. It seems that there is a sense of dutiful obligation to mention the presence of threat actors, but then to leave it at that.

5.2 Usefulness of threat actor information

Knowledge about potential adversaries' goals and motives, for example, their target selection priorities, as well as details about how they conduct their cyber operations, will surely help cyber defenders to prioritize their defensive efforts. A corporation, for example, with a range of products or services that almost certainly is not of interest for a nation state, might with some certainty assume that they will not have to dimension their cybersecurity to withstand that particular threat actor. Furthermore, detailed threat actor knowledge might help defenders to predict adversarial actions in sustained cyberattack campaigns.

To verify and indicate whether practitioners actually value and use threat actor information, we interviewed 10 experts about whether they perceived a need for threat actor information or not in a team cyber defense exercise (see the "Appendix" section). The experts *did not* unanimously ask for information about threat actors. Two respondents pointed out that atomic information (e.g., IP addresses) was considered to be enough. One respondent *explicitly declined* to receive information about threat actor identities. The remaining seven respondents expressed a wish to receive cyber threat actor information, although sometimes in vague terms. They generally asked for adversary goals and motives, as well as for details about their *modus operandi*. Moreover, two respondents raised the point that threat actor information is probably more useful at the strategic managerial level (for the purpose of determining threat sources, e.g., attribution), than at the technical level (for hands-on threat detection and mitigation). This is in line with the observation in Section 4 that different stakeholders need different information depending on their roles.

5.3 Cyber threat intelligence

Inquiries into how cyber threat information can and should be selected, processed and shared have emerged as a research field called cyber threat intelligence, CTI (Mavroeidis & Bromander, 2017; Sauerwein et al., 2017; Shin & Lowry, 2020; Tounsi & Rais, 2018; Wagner, Mahbub, Palomar, & Abdallah, 2019). CTI at large, hence, aims to design (cyber) information processing capabilities that help decision makers to leverage information in order to enable sensible cyber defense-related decisions to ultimately remedy the fundamentally disadvantageous position of the cyber defender (Mohaisen, Al-Ibrahim, Kamhoua, Kwiat, & Njilla, 2017). Burger, Goodman, Kampanakis, and Zhu (2014) propose a taxonomy for cyber threat information that encompasses the full scope of intelligence information exchange between stakeholders. The taxonomy has five layers that contain information elements ranging from straightforward to more complex. The layers are *transport*, *session*, *indicators*, *intelligence*, and *5W1H*. At one end of the spectrum, the lowermost transport-level involves information about the movement of bytes, for example, data streams that represent the cyber threat intelligence between enterprises. At the other end, we find the uppermost 5W1H-layer. Information here, which is fed from the underlying layers, aims to answer questions such as *who*, *what*, *when*, *where*, *why*, and *how*? Hence, it seeks to answer the overarching question of attribution: who or what organization is responsible for the threat?

5.4 Game-theoretic approaches

Based on what has been presented earlier in the chapter, this section serves to provide a basis for what would be required for an organization to be able to gain full understanding of the cyber threat, with a special emphasis on adversarial behavior, and how such a model can be used as the underlying model to be used for obtaining and sustaining CSA with respect to adversaries.

In Sections 2.1 and 2.2, a distinction was made between nonadversarial and adversarial cyber incidents, where nonadversarial incidents refer to different types of IT-related system errors while adversarial attacks concern some form of willful-thinking opponent being behind the attack. However, as discussed in Section 2.1, it is not easy to distinguish between these two cases because a willful-thinking opponent will typically be interested in hiding his/her intentions by, for example, masking them in terms of ordinary system outages or other shortcomings that can be related to the organization's ordinary information technology architecture. There is thus a need to be able to maintain and reason about the situation with respect to two conceivable, but widely differing, models where one model is based on technical problems and the other on a rational-thinking opponent.

Section 2.3 further raised the issue of an attack's "modus," which exemplifies that adversarial attackers can follow widely differing modus operandi: is, for example, a single bank attack "solely" about stealing money, or is it part of a more far-reaching operation in which a highly capable state actor performs the attack as part of a larger attack with the aim to destabilize the country's payment system? This rhetorical question points out that it is not enough to just distinguish between nonadversarial and adversarial cyber threats, but that the question is more difficult: responding to a state-sponsored cyber operation is very different from dealing with a "script kiddie." Cyber operations undertaken by foreign powers are typically characterized by their focus, longevity, (close to) unlimited resources, and professional organization with specialists covering each field (target identification, infrastructure, operators, support, etc.).

Thus far, in our quest for modeling the cyber threat, an increasingly complex three-level modeling scale can be discerned along the lines of: (i) systems and technology failures, (ii) threats from reasoning, but financially restricted, adversaries, and (iii) threats from state actors. However, other dimensions also need to be addressed. Section 2.3 also discusses the insider threat, which forms the basis for a completely different type of rational-thinking opponent. Such an adversary can cause great harm even without significant IT skills, and the insider probably also has a completely different set of core values compared to the external adversaries discussed previously. Hence, this calls for a different model than the three-level model anticipated above.

The depicted situation is characterized by the fact that it is game-theoretical in that it largely deals with "reasoning about reasoning." Furthermore, there is uncertainty as to which model, or "game" according to game theory jargon, actually applies, for example, the "insider model" and the "technology failure model" are very different, and we would like to incorporate the fact that we are uncertain about which model really applies. Moreover, even if an enduring state-sponsored attack is unlikely, we would still like to model that possibility and update the likelihood as new intelligence arrives.

In game theory, adhering to established notation (Camerer, 2003; Myerson, 1991; Schelling, 1960; von Neumann & Morgenstern, 1944), the model is called a *game*, consisting of "the *strategies* each of several *players* have, with precise rules for the order in which players choose strategies, the information they have when they choose, and how they rate the desirability, or *utility*, of resulting outcomes" (Brynielsson, 2006, p. 47). In our context, the players consist of (i) our own organization and (ii) the adversary that we would like to reason about given the intelligence at hand; our own strategies consist of the different means to defend ourselves that we have at our disposal; the adversarial strategies consist of the possible attack vectors that our organization exposes; and the utilities denote the payoffs that the players experience given certain outcomes, that is, payoff values need to be defined for each combination of outcomes, and therefore the payoff values in a sense define the whole game model.

Now, as highlighted above, in our context the game models for different attackers are likely to be very different in terms of game rules, strategies, and payoffs, that is, the strategies available to an insider will be totally different from the strategies available to a state-sponsored cyber operation, and the utilities, that is, how these different attackers value different outcomes, will also be very different. Assuming a Bayesian prior probability distribution representing what players believe about other players, the previously described game model can be extended into a so-called *Bayesian game*, as described by Harsanyi (1967–1968). Using such a game—conceptually a probability distribution over the relevant game models—it is possible to model the envisioned situation where the involved actors not only reason about the opponent's reasoning, but also models their reasoning about its own and the other actors' reasoning about who the opponent actually is in terms of the rules of the game and the opponent's capabilities, beliefs, desires, and intentions, thereby making it possible to model the full complexity of the cyber threat.

So far in the discussion, we have defined the model top-down, but the bottom-up perspective is just as important. As already noted in passing, the utility/payoff values define the whole game and are equally important to consider. These values must be related to the information assets and information technology that make up the organization. This gives rise to the threat intelligence needed to feed the model. As mentioned in Section 2.6, there are a number of different types of IT sensors that give rise to both technical threat indicators and contextual threat intelligence. Based on our previous work, this intelligence ought to be used as input to feed the envisioned Bayesian game with utility values in a bottom-up fashion (Brynielsson & Arnborg, 2006).

6 Research directions

While much is known about CSA, there are also many open questions. In particular, previous research has a technological focus, while the cognitive and organizational areas have received comparatively less attention. In the following, we list a few directions where we believe valuable research contributions ought to be made.

First, it would be valuable to reassess the applicability of current SA models for the cyber environment. As discussed in Section 3, current models were developed and adapted to solve problems within the physical realm, but the cyber domain is different. Current models require distinctly defined "situations." Such situations are hard to delimit in the cyber environment. Furthermore, current models have poor applicability for network-centric CSA: the timescale, with, for example, "computational speeds," cannot always be timely comprehended, much less acted upon, by humans. Models may not be optimal for domain-centric CSA either. Here, the timescale may instead be too long, and the "situations" become hard to delimit because of this. It might be that ordinary risk management practices that aim to reduce or mitigate risks for operations can be adapted for this purpose.

Second, it would be interesting to study the team aspect of CSA further. As argued in Section 3, analyzing how CSA is obtained in a command and control center means that one must to a great extent understand not only the digital but also the social interaction. Studying distributed CSA, where different team members with different skills and different tasks collaborate towards a common goal, requires that an event can be followed from the start, monitoring how it is perceived, and how understanding is jointly created. There are few studies that show how these processes work and how information is transformed over time.

Third, there is a need to develop CSA measurement techniques and rigorous experimental designs. For CSA research to become an evidence-based research area based on how people actually act, a number of practices and principles need to be developed. Descriptive studies are required to be able to understand how the social processes are formed based on existing information and how they affect how one assumes a position for counterattack. Because such contexts are highly sensitive and secretive, access may be difficult to obtain for researchers, but such studies would still have great value. Furthermore, experimental systems where teams work against each other under realistic but controlled circumstances are required, as suggested previously (Brynielsson et al., 2016).

Fourth, the existing literature lacks good empirical investigations of whether good CSA actually improves the overall cybersecurity stance or not. While it is very plausible that better decisions are made if informed by better CSA, such a simplistic wording hides details that might matter, for example, how much does long-term, domain-centric, CSA contribute to short-term incident management? And if time is short, as is often the case, are there some indicators or pieces of evidence that are more important—give a higher return on the invested time and mental resources—than others? Empirically based studies of such questions would have great practical value.

Fifth, and finally, the game-theoretic model described in Section 5.4 provides a promising framework for studying the multiple uncertainties that cyber situations entail. It would be valuable to develop it further; first theoretically and then empirically to test its applicability on real problems.

7 Conclusions

This chapter has argued that CSA is best understood by combining three complementary points of view: the technological, the socio-cognitive, and the organizational perspectives. Much of the existing literature focuses mostly on the first of these, perhaps assuming that issues related to cognition and organization are best handled through technical solutions. However, even if this would be true, such technical solutions cannot be built in the first place if the socio-cognitive and organizational perspectives are ignored when requirements are set and prioritized. The reasoning set forth in Sections 2–4 gives ample evidence for why an organization aiming to achieve adequate CSA needs to consider all three perspectives from the very start of the life cycle of any technical support systems.

Cyber incidents can be nonadversarial mistakes and errors, or adversarial attacks. Oftentimes, articles in the literature fail to acknowledge this duality, immediately leaping for one or the other strand. However, both are important, and neither should a priori be ignored—especially since a capable adversary might well hide behind the semblance of inconspicuous errors. Throughout the chapter, we have argued for the importance of maintaining both perspectives.

That said, the presence of an adversary poses particular challenges. "Thinking about thinking" is the title of the first chapter in Heuer's

(1999) seminal work on intelligence analysis—a mandatory must-read throughout the intelligence community. Reasoning about and maintaining an understanding of a thinking adversary have always been central to intelligence, and within the cyber domain, this is more important than ever: in today's online cyber world, it is harder than ever to form a reasonable hypothesis of what is going on, that is, to maintain a high level of CSA. It is also evident from our previous work (Varga et al., 2018, 2021) that this capability is both needed and sought for. To design such a model similar to what has previously been proposed for the intelligence domain (Brynielsson, Horndahl, Kaati, Mårtenson, & Svenson, 2009) and make it part of the cyber analyst's toolbox, is an important research undertaking on our current research agenda.

Acknowledgments

We would like to thank Gazmend Huskaj, Swedish Defence University, for the joint effort in conducting interviews with Locked Shields 2019 cyber defense exercise participants.

Appendix A Interview methodology

To determine the need for knowledge about adversaries for cyber defense analysts, we interviewed 10 Swedish expert participants from the cyber defense exercise Locked Shields 2019. The interviews were held between 1 and 2 months after the exercise. The exact question asked was: *Do you have any use for information about the threat actors themselves?*[a] The concept "threat actor," was neither defined specifically, nor discussed prior to the question. Two interviews were carried out while having two respondents from the same organization in the same room simultaneously. The interview protocol contained several other questions that were not related to the research presented in this chapter.

The annual cyber defense exercise Locked Shields is the world's largest unclassified defensive exercise. It is hosted by the NATO Cooperative Cyber Defence Centre of Excellence. The training audience consists of several so-called blue teams that are tasked to defend critical infrastructure. The exercise is also a competition, and the blue teams are awarded points depending on their performance. The Swedish blue team in 2019 was composed of around 60 cybersecurity experts. They represented both governmental agencies (70%) and commercial companies (30%). The team was highly motivated, and finished in third place out of 24 international competing teams.

[a] Authors' translation from Swedish: *Har du nytta av information om hotaktörer i sig?*

The 10 interviewees all either had leadership positions, or were designated subject matter experts according to the following:

- One was the team leader and one was the deputy team leader of the Swedish blue team.
- One was the chief of staff.
- Four people were group leaders for various functional groups (e.g., Windows, Linux, etc.), which each consisted of a number of personnel.
- One was a staff member responsible for intelligence issues.
- One was a staff member who performed threat-hunting and hash-cracking (via a GPU cluster). This individual was also responsible for real-world logistical support and general technical support.
- One was a technical expert on the domain name system (DNS) and the Apple Mac operating system computing environment.

A.1 Interview results

A synopsis of the answers from the 10 interviewees to the question *Do you have any use for information about the threat actors themselves?* follows:

Respondent 1: Yes, it was useful to hear from the red team after the exercise. It would be good to get the [adversary's] overarching goals and objectives, to understand what they are going to focus on. It would be good to understand their technical capabilities. If they use publicly known code, it would be useful to know which ones. It would even be useful if they use hitherto unknown code. As much threat actor information as possible would have been good, like: "This is how they code their trojans!"; "This is how they are setting up their command and control servers!"; and "This is how they hide themselves!" But, there would probably be issues of secrecy in a real scenario. I don't know how, and if, an intelligence service would share such information, and whether they plan to do so or not…

Respondent 2: It was a game, so that was an artificial situation. Maybe in real life… I would like to know what they are after, and what methods they use. In this case we were given a set of threat actors, and we knew roughly what we could expect. Then we identified additional threat actors. Some might have a use for it [threat actor information]. I didn't.

Respondent 3: I did have use for cyber threat actor information, especially TTPs [tactics, techniques, and procedures]. What are the attackers after? If you know the doctrine of the attackers, e.g., if they employ *maskirovka* [deception], then you can interpret some of the observed attacks as potential smoke screens to protect even more nefarious attacks. We could then prioritize our assets to counter this type of threat. Another thing is to pay attention to what is *not* done at a specific time; these kind of observations might also reveal something about the motivations of the attackers.

Respondent 4: I did not have any use for that kind of information. We only focused on attackers' IPs, and thought that it would be an easy thing for them to change attack platforms, while remaining hostile. But, it would have been good to have an appreciation of the threat actor's "modus," in order to adapt our behavior.

Respondent 5: We did not get too much information about threat actors or their capabilities [in the exercise]. We had to make some assumptions. We still gathered that an operating principle for the adversary was to distract and have us concentrate our resources on nonessential tasks. The same question at my real job: Yes, information about capabilities, tactics and tools would be of interest. It would be good to know whether an attacker can or cannot perform certain things. It would be good to know their capabilities as well as their level of knowledge, e.g., if and how they can acquire privileges [in accounts] to execute certain things. How they attack us; how they perform lateral movement; where they get a foothold, and understand their perspectives. Where will they find the weakest part of our defense? We could use a bit of reverse thinking here: What would we go after if we were to be the attacker? [In the exercise] we managed to successfully predict what accounts the attackers would attack. So, if you have a good understanding of what the attackers want, it is easier to protect yourself.

Respondent 6: You had to see what was happening to be able to predict what was going to happen next. You had to know what the attacker was after. We would need [intelligence] analyses, and not only "raw data." It is not helpful to only be informed about events; we also need to understand their consequences.

Respondent 7: We had performed an analysis based on earlier exercises. We assumed that the same threat actor was in play, but with partly new "weapons." But what were they after? Threat actor information was not the most important thing for us [at the technical level]; it is more useful for the strategic level. On the technical level, we did not miss so much about the actors.

Respondent 8: We were not good at that! We did not get good info on the threat actors and their capabilities. We made assumptions. Some attacks were done to tie up our resources. It would be good to have info about goals and capabilities of threat actors so we could prioritize our own resources. We also found out that some users were cooperating with the red team. The actual identity of the threat actors is not important.

Respondent 9: The threat actors were not distinct. What will they do (given their capabilities)? Can we counter that? Maybe not that

important at the "technical" level? Attribution is important on the political level, though.
Respondent 10: No, that type of information is not interesting in this scenario. At my normal work, where I know my own system, it would be of use so that I could focus my attention to look out for certain things, and to be proactive. But I would assert that threat actor information is of no use for an ordinary company. I think a very mature organization is required in order to handle such threat information in a good way.

A.2 Interview validity and reliability

The potential pool of interviewees was the total number of Swedish participants in the exercise ($N = 63$). The selection criteria for interviewees were that they had to: (i) have some senior position in the exercise, (ii) have adequate expertise and experience in the field, and (iii) be available for interviews. All interviewees were drawn from the Stockholm region geographical area. Further, all respondents were working with practical network defense measures, that is, they were not working with managerial and leadership issues other than to a small extent.

We did not clarify the meaning of the term "threat actor" in the interviews. The received responses clearly showed how respondents interpreted its meaning in different ways. Rather, several respondents indicated that they immediately assumed that we were asking about (adversarial) tactics, techniques, and procedures, TTPs.

Guest, Bunce, and Johnson (2006) claimed that there virtually did not exist any guidelines for determining nonprobabilistic sample sizes. They then investigated the property of data saturation, that is, how various themes accumulate in a series of interviews in homogeneous groups. They found that saturation occurred within the first 12 interviews, that is, that the number of emergent themes were saturated and that the information value added in the consecutive interviews was negligible. Furthermore, they identified that elements of emerging "metathemes" could be discerned as early as in the sixth interview.

References

Abdul, A., Vermeulen, J., Wang, D., Lim, B. Y., & Kankanhalli, M. (2018). Trends and trajectories for explainable, accountable and intelligible systems: An HCI research agenda. In *Proceedings of the 2018 CHI conference on human factors in computing systems (CHI 2018)* (pp. 1–18). New York, NY: ACM. https://doi.org/10.1145/3173574.3174156.

Ahmad, A., Desouza, K. C., Maynard, S. B., Naseer, H., & Baskerville, R. L. (2020). How integration of cyber security management and incident response enables organizational learning. *Journal of the Association for Information Science and Technology*, *71*(8), 939–953. https://doi.org/10.1002/asi.24311.

Arrieta, A. B., Díaz-Rodríguez, N., Del Ser, J., Bennetot, A., Tabik, S., Barbado, A., ... Herrera, F. (2020). Explainable artificial intelligence (XAI): Concepts, taxonomies, opportunities and challenges toward responsible AI. *Information Fusion, 58*, 82–115. https://doi.org/10.1016/j.inffus.2019.12.012.

Artman, H. (2000). Team situation assessment and information distribution. *Ergonomics, 43*(8), 1111–1128. https://doi.org/10.1080/00140130050084905.

Artman, H., Brynielsson, J., Johansson, B. J. E., & Trnka, J. (2011). Dialogical emergency management and strategic awareness in emergency communication. In *Proceedings of the eighth international conference on information systems for crisis response and management (ISCRAM 2011)* (pp. 1–9). http://urn.kb.se/resolve?urn=urn:nbn:se:kth:diva-61800.

Artman, H., & Wærn, Y. (1999). Distributed cognition in an emergency co-ordination center. *Cognition, Technology & Work, 1*(4), 237–246. https://doi.org/10.1007/s101110050020.

Attiah, A., Chatterjee, M., & Zou, C. C. (2018). A game theoretic approach to model cyber attack and defense strategies. In *Proceedings of the 2018 IEEE international conference on communications (ICC 2018)* (pp. 1–7). Piscataway, NJ: IEEE. https://doi.org/10.1109/ICC.2018.8422719.

Bass, T. (2000). Intrusion detection systems and multisensor data fusion. *Communications of the ACM, 43*(4), 99–105. https://doi.org/10.1145/332051.332079.

Bernier, M. (2013). *Military activities and cyber effects (MACE) taxonomy*. Technical Memorandum DRDC CORA TM 2013-226. Ottawa, Canada: Centre for Operational Research and Analysis, Defence Research and Development Canada. https://cradpdf.drdc-rddc.gc.ca/PDFS/unc218/p803340_A1b.pdf.

Bertolino, A. (2007). Software testing research: Achievements, challenges, dreams. In *Proceedings of future of software engineering (FoSE 2007)* (pp. 85–103). Piscataway, NJ: IEEE. https://doi.org/10.1109/FOSE.2007.25.

Borum, R., Felker, J., Kern, S., Dennesen, K., & Feyes, T. (2015). Strategic cyber intelligence. *Information and Computer Security, 23*(3), 317–332. https://doi.org/10.1108/ICS-09-2014-0064.

Brown, S., Gommers, J., & Serrano, O. (2015). From cyber security information sharing to threat management. In *Proceedings of the 2015 ACM workshop on information sharing and collaborative security (WISCS 2015)* (pp. 43–49). New York, NY: ACM. https://doi.org/10.1145/2808128.2808133.

Brynielsson, J. (2006). *A gaming perspective on command and control* (Doctoral dissertation). Stockholm, Sweden: Royal Institute of Technology. http://urn.kb.se/resolve?urn=urn:nbn:se:kth:diva-4029.

Brynielsson, J., & Arnborg, S. (2006). An information fusion game component. *Journal of Advances in Information Fusion, 1*(2), 108–121. http://urn.kb.se/resolve?urn=urn:nbn:se:kth:diva-24204.

Brynielsson, J., Franke, U., Tariq, M. A., & Varga, S. (2016). Using cyber defense exercises to obtain additional data for attacker profiling. In *Proceedings of the 2016 IEEE conference on intelligence and security informatics (ISI 2016)* (pp. 37–42). Piscataway, NJ: IEEE. https://doi.org/10.1109/ISI.2016.7745440.

Brynielsson, J., Franke, U., & Varga, S. (2016). Cyber situational awareness testing. In B. Akhgar, & B. Brewster (Eds.), *Combatting cybercrime and cyberterrorism: Challenges, trends and priorities* (pp. 209–233). Cham, Switzerland: Springer. https://doi.org/10.1007/978-3-319-38930-1_12.

Brynielsson, J., Horndahl, A., Kaati, L., Mårtenson, C., & Svenson, P. (2009). Development of computerized support tools for intelligence work. In *Proceedings of the 14th international command and control research and technology symposium (14th ICCRTS)*. Washington, DC: US Department of Defense CCRP. http://urn.kb.se/resolve?urn=urn:nbn:se:kth:diva-89594.

Burger, E. W., Goodman, M. D., Kampanakis, P., & Zhu, K. A. (2014). Taxonomy model for cyber threat intelligence information exchange technologies. In *Proceedings of the 2014 ACM workshop on information sharing and collaborative security (WISCS 2014)* (pp. 51–60). New York, NY: ACM. https://doi.org/10.1145/2663876.2663883.

Camerer, C. F. (2003). *Behavioral game theory: Experiments in strategic interaction*. Princeton, NJ: Princeton University Press.

Caralli, R. A., Stevens, J. F., Young, L. R., & Wilson, W. R. (2007). *Introducing OCTAVE Allegro: Improving the information security risk assessment process*. Technical Report CMU/SEI-2007-TR-012. Pittsburgh, PA: Software Engineering Institute, Carnegie Mellon University. https://doi.org/10.1184/R1/6574790.v1.

Cebula, J. J., Popeck, M. E., & Young, L. R. (2014). *A taxonomy of operational cyber security risks version 2*. Technical Note CMU/SEI-2014-TN-006. Pittsburgh, PA: Software Engineering Institute, Carnegie Mellon University. https://doi.org/10.1184/R1/6571784.v1.

Chen, T. R., Shore, D. B., Zaccaro, S. J., Dalal, R. S., Tetrick, L. E., & Gorab, A. K. (2014). An organizational psychology perspective to examining computer security incident response teams. *IEEE Security & Privacy, 12*(5), 61–67. https://doi.org/10.1109/MSP.2014.85.

Copeland, J. (2008). *Emergency response: Unity of effort through a common operational picture*. Strategy Research Project. Carlisle, PA: US Army War College. https://apps.dtic.mil/sti/citations/ADA479583.

Cottrell, A., & Cockshott, W. P. (1993). Calculation, complexity and planning: The socialist calculation debate once again. *Review of Political Economy, 5*(1), 73–112. https://doi.org/10.1080/09538259300000005.

Dehghantanha, A., Conti, M., & Dargahi, T. (Eds.). (2018). *Cyber threat intelligence*. Cham, Switzerland: Springer. https://doi.org/10.1007/978-3-319-73951-9.

Demir, M., McNeese, N. J., & Cooke, N. J. (2017). Team situation awareness within the context of human-autonomy teaming. *Cognitive Systems Research, 46*, 3–12. https://doi.org/10.1016/j.cogsys.2016.11.003.

Dörner, D., & Schaub, H. (1994). Errors in planning and decision-making and the nature of human information processing. *Applied Psychology, 43*(4), 433–453. https://doi.org/10.1111/j.1464-0597.1994.tb00839.x.

Du, M., Liu, N., & Hu, X. (2019). Techniques for interpretable machine learning. *Communications of the ACM, 63*(1), 68–77. https://doi.org/10.1145/3359786.

Dunbar, T. (2012). The first steps to managing cyber-risk. *Risk Management, 59*(8). http://www.rmmagazine.com/articles/article/2012/10/05/-The-First-Steps-to-Managing-Cyber-Risk-.

Dutta, A., & McCrohan, K. (2002). Management's role in information security in a cyber economy. *California Management Review, 45*(1), 67–87. https://doi.org/10.2307/41166154.

Endsley, M. R. (1988). Situation awareness global assessment technique (SAGAT). In *Proceedings of the IEEE 1988 national aerospace and electronics conference (NAECON 1988): Vol. 3* (pp. 789–795). Piscataway, NJ: IEEE. https://doi.org/10.1109/NAECON.1988.195097.

Endsley, M. R. (1995). Toward a theory of situation awareness in dynamic systems. *Human Factors, 37*(1), 32–64. https://doi.org/10.1518/001872095779049543.

Endsley, M. R., & Jones, D. G. (2011). *Designing for situation awareness: An approach to user-centered design*. Boca Raton, FL: CRC Press. https://doi.org/10.1201/b11371.

Evancich, N., Lu, Z., Li, J., Cheng, Y., Tuttle, J., & Xie, P. (2014). Network-wide awareness. In A. Kott, C. Wang, & R. F. Erbacher (Eds.), *Cyber defense and situational awareness* (pp. 63–91). Cham, Switzerland: Springer. https://doi.org/10.1007/978-3-319-11391-3_5.

Franke, U. (2018). Cyber insurance against electronic payment service outages. In *Proceedings of the 14th international workshop on security and trust management (STM 2018)* (pp. 73–84). Cham, Switzerland: Springer. https://doi.org/10.1007/978-3-030-01141-3_5.

Franke, U., & Brynielsson, J. (2014). Cyber situational awareness: A systematic review of the literature. *Computers & Security, 46*, 18–31. https://doi.org/10.1016/j.cose.2014.06.008.

Franke, U., & Wernberg, J. (2020). A survey of cyber security in the Swedish manufacturing industry. In *Proceedings of the 2020 international conference on cyber situational awareness, data analytics and assessment (CyberSA 2020)* (pp. 1–8). Piscataway, NJ: IEEE. https://doi.org/10.1109/CyberSA49311.2020.9139673.

Galup, S. D., Dattero, R., Quan, J. J., & Conger, S. (2009). An overview of IT service management. *Communications of the ACM, 52*(5), 124–127. https://doi.org/10.1145/1506409.1506439.

Genadis, T. C. (1996). A cost optimization model for determining optimal burn-in times at the module/system level of an electronic product. *International Journal of Quality & Reliability Management, 13*(9), 61–74. https://doi.org/10.1108/02656719610150623.

Goeschel, K. (2016). Reducing false positives in intrusion detection systems using data-mining techniques utilizing support vector machines, decision trees, and naive Bayes for off-line analysis. In *IEEE SoutheastCon 2016* (pp. 1–6). Piscataway, NJ: IEEE. https://doi.org/10.1109/SECON.2016.7506774.

Gray, J. (1985). *Why do computers stop and what can be done about it?* Technical Report 85.7. Cupertino, CA: Tandem Computers, Inc.

Gray, J. (1990). A census of Tandem system availability between 1985 and 1990. *IEEE Transactions on Reliability, 39*(4), 409–418. https://doi.org/10.1109/24.58719.

Guest, G., Bunce, A., & Johnson, L. (2006). How many interviews are enough? An experiment with data saturation and variability. *Field Methods, 18*(1), 59–82. https://doi.org/10.1177/1525822X05279903.

Guidotti, R., Monreale, A., Ruggieri, S., Turini, F., Giannotti, F., & Pedreschi, D. (2018). A survey of methods for explaining black box models. *ACM Computing Surveys, 51*(5), 1–42. https://doi.org/10.1145/3236009.

Gunawi, H. S., Hao, M., Suminto, R. O., Laksono, A., Satria, A. D., Adityatama, J., & Eliazar, K. J. (2016). Why does the cloud stop computing? Lessons from hundreds of service outages. In *Proceedings of the seventh ACM symposium on cloud computing (SoCC 2016)* (pp. 1–16). New York, NY: ACM. https://doi.org/10.1145/2987550.2987583.

Gutzwiller, R. S., Dykstra, J., & Payne, B. (2020). Gaps and opportunities in situational awareness for cybersecurity. *Digital Threats: Research and Practice, 1*(3), 1–6. https://doi.org/10.1145/3384471.

Gutzwiller, R. S., Hunt, S. M., & Lange, D. S. (2016). A task analysis toward characterizing cyber-cognitive situation awareness (CCSA) in cyber defense analysts. In *Proceedings of the 2016 IEEE international multi-disciplinary conference on cognitive methods in situation awareness and decision support (CogSIMA 2016)* (pp. 14–20). Piscataway, NJ: IEEE. https://doi.org/10.1109/COGSIMA.2016.7497780.

Hager, R. S. (1997). *Current and future efforts to vary the level of detail for the common operational picture* (Master's thesis). Monterey, CA: Naval Postgraduate School. https://apps.dtic.mil/sti/citations/ADA341674.

Harsanyi, J. C. (1967–1968). Games with incomplete information played by "Bayesian" players. *Management Science, 14*(3, 5, 7), 159–182, 320–334, 486–502. https://doi.org/10.1287/mnsc.14.3.159, https://doi.org/10.1287/mnsc.14.5.320, https://doi.org/10.1287/mnsc.14.7.486.

Hayek, F. A. (1982). Two pages of fiction: The impossibility of socialist calculation. *Economic Affairs, 2*(3), 135–142. https://doi.org/10.1111/j.1468-0270.1982.tb01416.x.

Heuer, R. J., Jr. (1999). *Psychology of intelligence analysis.* Washington, DC: Center for the Study of Intelligence, Central Intelligence Agency.

Holm, H. (2014). Signature based intrusion detection for zero-day attacks: (Not) a closed chapter? In *Proceedings of the 2014 47th annual Hawaii international conference on system sciences (HICSS 2014)* (pp. 4895–4904). Piscataway, NJ: IEEE. https://doi.org/10.1109/HICSS.2014.600.

Iden, J., & Eikebrokk, T. R. (2013). Implementing IT service management: A systematic literature review. *International Journal of Information Management, 33*(3), 512–523. https://doi.org/10.1016/j.ijinfomgt.2013.01.004.

Kang, M.-J., & Kang, J.-W. (2016). Intrusion detection system using deep neural network for in-vehicle network security. *PLoS ONE, 11*(6), 1–17. https://doi.org/10.1371/journal.pone.0155781.

Khraisat, A., Gondal, I., Vamplew, P., & Kamruzzaman, J. (2019). Survey of intrusion detection systems: Techniques, datasets and challenges. *Cybersecurity, 2*(1), 1–22. https://doi.org/10.1186/s42400-019-0038-7.

Kim, J., Shin, N., Jo, S. Y., & Kim, S. H. (2017). Method of intrusion detection using deep neural network. In *Proceedings of the 2017 IEEE international conference on big data and smart computing (BigComp 2017)* (pp. 313–316). Piscataway, NJ: IEEE. https://doi.org/10.1109/BIGCOMP.2017.7881684.

Kramer, R. M. (1998). Revisiting the Bay of Pigs and Vietnam decisions 25 years later: How well has the groupthink hypothesis stood the test of time? *Organizational Behavior and Human Decision Processes, 73*(2–3), 236–271. https://doi.org/10.1006/obhd.1998.2762.

Krombholz, K., Hobel, H., Huber, M., & Weippl, E. (2015). Advanced social engineering attacks. *Journal of Information Security and Applications, 22*, 113–122. https://doi.org/10.1016/j.jisa.2014.09.005.

Li, Y., Huang, G.-q., Wang, C.-z., & Li, Y.-c. (2019). Analysis framework of network security situational awareness and comparison of implementation methods. *EURASIP Journal on Wireless Communications and Networking, 2019*(205), 1–32. https://doi.org/10.1186/s13638-019-1506-1.

Maennel, K., Ottis, R., & Maennel, O. (2017). Improving and measuring learning effectiveness at cyber defense exercises. In *Proceedings of the 22nd Nordic conference on secure IT systems (NordSec 2017)* (pp. 123–138). Cham, Switzerland: Springer. https://doi.org/10.1007/978-3-319-70290-2_8.

Malik, B., & Scott, D. (2010). *How to calculate the cost of continuously available IT services.* Gartner Research Paper G00209007. Stamford, CT: Gartner, Inc.

Mattern, T., Felker, J., Borum, R., & Bamford, G. (2014). Operational levels of cyber intelligence. *International Journal of Intelligence and CounterIntelligence, 27*(4), 702–719. https://doi.org/10.1080/08850607.2014.924811.

Mauw, S., & Oostdijk, M. (2005). Foundations of attack trees. In *Proceedings of the eighth international conference on information security and cryptology (ICISC 2005)* (pp. 186–198). Berlin/Heidelberg, Germany: Springer. https://doi.org/10.1007/11734727_17.

Mavroeidis, V., & Bromander, S. (2017). Cyber threat intelligence model: An evaluation of taxonomies, sharing standards, and ontologies within cyber threat intelligence. In *Proceedings of the 2017 European intelligence and security informatics conference (EISIC 2017)* (pp. 91–98). Piscataway, NJ: IEEE. https://doi.org/10.1109/EISIC.2017.20.

McCrohan, K. F., Engel, K., & Harvey, J. W. (2010). Influence of awareness and training on cyber security. *Journal of Internet Commerce, 9*(1), 23–41. https://doi.org/10.1080/15332861.2010.487415.

McKenna, S., Staheli, D., & Meyer, M. (2015). Unlocking user-centered design methods for building cyber security visualizations. In *Proceedings of the 2015 IEEE symposium on visualization for cyber security (VizSec 2015)* (pp. 1–8). Piscataway, NJ: IEEE. https://doi.org/10.1109/VIZSEC.2015.7312771.

Mohaisen, A., Al-Ibrahim, O., Kamhoua, C., Kwiat, K., & Njilla, L. (2017). Rethinking information sharing for threat intelligence. In *Proceedings of the fifth ACM/IEEE workshop on hot topics in web systems and technologies (HotWeb 2017)* (pp. 1–7). New York, NY: ACM. https://doi.org/10.1145/3132465.3132468.

Myerson, R. B. (1991). *Game theory: Analysis of conflict.* Cambridge, MA: Harvard University Press.

Nature. (2016). More accountability for big-data algorithms. *Nature, 537*(7621), 449. https://doi.org/10.1038/537449a.

Nickerson, R. S. (1998). Confirmation bias: A ubiquitous phenomenon in many guises. *Review of General Psychology, 2*(2), 175–220. https://doi.org/10.1037/1089-2680.2.2.175.

Omer, M., Nilchiani, R., & Mostashari, A. (2009). Measuring the resilience of the global internet infrastructure system. In *Proceedings of the 2009 third annual IEEE international systems conference (SysCon 2009)* (pp. 156–162). Piscataway, NJ: IEEE. https://doi.org/10.1109/SYSTEMS.2009.4815790.

Petersen, R., Santos, D., Smith, M. C., Wetzel, K. A., & Witte, G. (2020). *Workforce framework for cybersecurity (NICE framework)*. NIST Special Publication 800-181 Revision 1. National Institute of Standards and Technology, US Department of Commerce. https://doi.org/10.6028/NIST.SP.800-181r1.

Raven, B. H. (1998). Groupthink, Bay of Pigs, and Watergate reconsidered. *Organizational Behavior and Human Decision Processes, 73*(2–3), 352–361. https://doi.org/10.1006/obhd.1998.2766.

Rosenquist, M. (2009). *Prioritizing information security risks with threat agent risk assessment*. IT@Intel White Paper. Intel Corporation.

Roy, S. S., Mallik, A., Gulati, R., Obaidat, M. S., & Krishna, P. V. (2017). A deep learning based artificial neural network approach for intrusion detection. In *Proceedings of the third international conference on mathematics and computing (ICMC 2017)* (pp. 44–53). Singapore: Springer. https://doi.org/10.1007/978-981-10-4642-1_5.

Sauerwein, C., Pekaric, I., Felderer, M., & Breu, R. (2019). An analysis and classification of public information security data sources used in research and practice. *Computers & Security, 82*, 140–155. https://doi.org/10.1016/j.cose.2018.12.011.

Sauerwein, C., Sillaber, C., Mussmann, A., & Breu, R. (2017). Threat intelligence sharing platforms: An exploratory study of software vendors and research perspectives. In *Proceedings of the 13th international conference on wirtschaftsinformatik (WI 2017)* (pp. 837–851). Atlanta, GA: Association for Information Systems (AIS).

Schelling, T. C. (1960). *The strategy of conflict*. Cambridge, MA: Harvard University Press.

Schneier, B. (1999). Attack trees. *Dr. Dobb's Journal, 24*(12), 21–29.

Shin, B., & Lowry, P. B. (2020). A review and theoretical explanation of the 'Cyberthreat-Intelligence (CTI) capability' that needs to be fostered in information security practitioners and how this can be accomplished. *Computers & Security, 92*, 1–16. https://doi.org/10.1016/j.cose.2020.101761.

Shostack, A. (2014). *Threat modeling: Designing for security*. Indianapolis, IN: Wiley.

Sommestad, T., & Franke, U. (2015). A test of intrusion alert filtering based on network information. *Security and Communication Networks, 8*, 2291–2301. https://doi.org/10.1002/sec.1173.

Sommestad, T., & Hunstad, A. (2013). Intrusion detection and the role of the system administrator. *Information Management & Computer Security, 21*(1), 30–40. https://doi.org/10.1108/09685221311314400.

Spathoulas, G. P., & Katsikas, S. K. (2010). Reducing false positives in intrusion detection systems. *Computers & Security, 29*(1), 35–44. https://doi.org/10.1016/j.cose.2009.07.008.

Tariq, M. A., Brynielsson, J., & Artman, H. (2012). Framing the attacker in organized cybercrime. In *Proceedings of the 2012 European intelligence and security informatics conference (EISIC 2012)* (pp. 30–37). Piscataway, NJ: IEEE. https://doi.org/10.1109/EISIC.2012.48.

Tariq, M. A., Brynielsson, J., & Artman, H. (2014). The security awareness paradox: A case study. In *Proceedings of the 2014 IEEE/ACM international conference on advances in social networks analysis and mining (ASONAM 2014)* (pp. 704–711). Piscataway, NJ: IEEE. https://doi.org/10.1109/ASONAM.2014.6921663.

Tounsi, W., & Rais, H. (2018). A survey on technical threat intelligence in the age of sophisticated cyber attacks. *Computers & Security, 72*, 212–233. https://doi.org/10.1016/j.cose.2017.09.001.

Traoré, I., Awad, A., & Woungang, I. (Eds.). (2017). *Information security practices: Emerging threats and perspectives*. Cham, Switzerland: Springer. https://doi.org/10.1007/978-3-319-48947-6.

Tversky, A., & Kahneman, D. (1974). Judgment under uncertainty: Heuristics and biases. *Science, 185*(4157), 1124–1131. https://doi.org/10.1126/science.185.4157.1124.

Varga, S., Brynielsson, J., & Franke, U. (2018). Information requirements for national level cyber situational awareness. In *Proceedings of the 2018 IEEE/ACM international conference on advances in social networks analysis and mining (ASONAM 2018)* (pp. 774–781). Piscataway, NJ: IEEE. https://doi.org/10.1109/ASONAM.2018.8508410.

Varga, S., Brynielsson, J., & Franke, U. (2021). Cyber-threat perception and risk management in the Swedish financial sector. *Computers & Security, 105*, 1–18. https://doi.org/10.1016/j.cose.2021.102239.

Vielberth, M., Menges, F., & Pernul, G. (2019). Human-as-a-security-sensor for harvesting threat intelligence. *Cybersecurity, 2*(1), 1–15. https://doi.org/10.1186/s42400-019-0040-0.

von Neumann, J., & Morgenstern, O. (1944). *Theory of games and economic behavior*. Princeton, NJ: Princeton University Press.

Wagner, T. D., Mahbub, K., Palomar, E., & Abdallah, A. E. (2019). Cyber threat intelligence sharing: Survey and research directions. *Computers & Security, 87*, 1–13. https://doi.org/10.1016/j.cose.2019.101589.

Wan, Y., Cao, J., Chen, G., & Huang, W. (2017). Distributed observer-based cyber-security control of complex dynamical networks. *IEEE Transactions on Circuits and Systems I: Regular Papers, 64*(11), 2966–2975. https://doi.org/10.1109/TCSI.2017.2708113.

Werlinger, R., Hawkey, K., Muldner, K., Jaferian, P., & Beznosov, K. (2008). The challenges of using an intrusion detection system: Is it worth the effort? In *Proceedings of the fourth symposium on usable privacy and security (SOUPS 2008)* (pp. 107–118). New York, NY: ACM. https://doi.org/10.1145/1408664.1408679.

Whittaker, J. A. (2000). What is software testing? And why is it so hard? *IEEE Software, 17*(1), 70–79. https://doi.org/10.1109/52.819971.

Wolbers, J., & Boersma, K. (2013). The common operational picture as collective sensemaking. *Journal of Contingencies and Crisis Management, 21*(4), 186–199. https://doi.org/10.1111/1468-5973.12027.

Xu, S. (2014). Cybersecurity dynamics. In *Proceedings of the 2014 symposium and bootcamp on the science of security (HoTSoS 2014)* (pp. 1–2). New York, NY: ACM. https://doi.org/10.1145/2600176.2600190.

Yeager, L. B. (1994). Mises and Hayek on calculation and knowledge. *The Review of Austrian Economics, 7*(2), 93–109. https://doi.org/10.1007/BF01101944.

Zheng, D. E., & Lewis, J. A. (2015). *Cyber threat information sharing: Recommendations for congress and the administration*. A Report of the CSIS Strategic Technologies Program. Washington, DC: Center for Strategic & International Studies. https://www.csis.org/analysis/cyber-threat-information-sharing.

CHAPTER

11

Development and application of the Information Security Core Human Error Causes (IS-CHEC) technique

Mark Evans[a], *Ying He*[b], *Leandros Maglaras*[a], *and Helge Janicke*[c]

[a]School of Computer Science and Informatics, De Montfort University, Leicester, United Kingdom, [b]School of Computer Science, University of Nottingham, Nottingham, United Kingdom, [c]Cyber Security Cooperative Research Centre, Joondalup, WA, Australia

1 Introduction

1.1 Chapter 1

1.1.1 Background

There is a belief in the literature that the human being is the weakest link in information security assurance (Cook, Maglaras, Smith, & Janicke, 2018; Maglaras et al., 2019; Shabut, Lwin, & Hossain, 2016) and as a result, there is interest in this topic (Dhakal, 2018; Twisdale, 2018), particularly due to the associated risk of data breaches (Ebenezer, 2017; Elisa, Yang, Chao, & Cao, 2018; Osei-Amanfi, 2018). Reason (1990) identified three categories of human errors, these being:

- a slip, which refers to a correct action but a faulty execution;
- a lapse, which is a lapse in memory or distraction causing the failed execution of a task;
- a mistake, which is defined as the correct execution of an incorrect intention.

https://doi.org/10.1016/B978-0-323-90570-1.00010-3

Unintentional human errors affect all sectors and environments that rely upon humans, including industry (Ani, He, & Tiwari, 2019; Burrell, 2019; Cook et al., 2018; Cope, Siewe, Chen, Maglaras, & Janicke, 2017; Dhakal, 2018; Drozd & Kapulin, 2017; Kauspadiene, 2019; Kauspadiene, Ramanauskaite, Cenys, Janulevicius, & Rastenis, 2018; Nobles, 2018; Safa et al., 2019; Twisdale, 2018), governments and the public sector (Drivas, Maglaras, Janicke, & Ioannidis, 2020; Elisa et al., 2018), critical infrastructure (Maglaras et al., 2019; Maglaras et al., 2018), health (Al-Barnawi, He, Maglaras, & Janicke, 2019; Cilliers, 2017; Ebenezer, 2017; Twisdale, 2018), social networks and systems (Ferrag, Maglaras, & Ahmim, 2017), autonomous vehicles (Linkov, Zámečník, Havlíčková, & Pai, 2019), mobile devices (Ferrag, Maglaras, Derhab, & Janicke, 2018), finance (Lee, 2018), marketing (Trim & Lee, 2019) business process outsourcing (Singh & Jalota, 2018) and peacekeeping (Robinson, Jones, Janicke, & Maglaras, 2019). As a result, it was found that the human factor and behavior and associated risk is an important component of effective information security assurance (Al-Barnawi et al., 2019; Alissa et al., 2018; Ani et al., 2019; Eling & Wirfs, 2019; Ferrag et al., 2017; Huian, Mironiuc, & Chiriac, 2018; Kauspadiene, 2019; Kauspadiene et al., 2018; Maglaras, Janicke, Jiang, & Crampton, 2020; Rindasu, 2018; Singh & Jalota, 2018; Snyman, Kruger, & Kearney, 2018). It is also clear that for the foreseeable future effective information security assurance cannot be provided by technology alone, for it also requires human involvement and consideration (Atoum & Otoom, 2017; Robinson et al., 2019; Safa et al., 2019; Shabut et al., 2016). Moreover, human vulnerabilities cannot be established through scanning as can be done for technological ones (Pham & Dang, 2018). Due to human fallibility, humans are targeted in order to undertake successful information security attacks and enable malware to propagate (Ferrag et al., 2018; Maglaras, Kim, et al., 2018; Rindasu, 2018).

A specific aspect of the human factor that is prominent within the literature (Cook et al., 2018; Drozd & Kapulin, 2017) is unintentional human error (Alissa et al., 2018; Elisa et al., 2018; Nobles, 2018). It is held that these unintentional human errors can lead to information security incidents and breaches (Burrell, 2019; Coffey, Smith, Maglaras, & Janicke, 2018; Drivas et al., 2020; Lamba, Singh, Singh, Dutta, & Muni, 2017; Maglaras et al., 2019; Nobles, 2018; Shabut et al., 2016; Twisdale, 2018), which can have a significant negative impact (Cilliers, 2017; Cook et al., 2018). As a result, initiatives that have been successfully established in the safety field, such as human reliability analysis and statistical quality control (Alissa et al., 2018; Atoum & Otoom, 2017; Twisdale, 2018), have been suggested as mitigation as well as more research in this area being proposed (Jackson, 2018; Lamba et al., 2017; Twisdale, 2018). Also, the source of new techniques and methods (Cilliers, 2017) have been proposed. It has also been recognized that basic compliance activities without confirmatory assurance are not sufficient (Cope et al., 2017) and that with changes to culture (Huian et al., 2018; Rindasu, 2018; Trim & Lee, 2019), effective training (Dhakal, 2018; Tarazan & Bostan, 2016) and employee monitoring (Linkov et al., 2019), people can both increase the security of controls (Cook et al., 2018; Huian et al., 2018) and defend against attacks (Shabut et al., 2016).

1.2 Motivation

The initial motivation driving the research was to understand holistic themes, proportions, and causes behind current information security weaknesses and information security incidents within participating organizations and compare these with published data. Once the initial understanding had been ascertained and the human factor identified, we looked outside of the field of information security to identify potential solutions that could be applicable and beneficial to the information security field. The initial research identified the human factor, specifically unintentional human error, rather than intentional action or technological, procedural, or physical failings as the direction for the research. Accordingly, human reliability analysis (HRA), as an established method applied globally in the safety field and high-reliability sectors, was drawn upon, tested, adapted, and applied. We set out to rigorously demonstrate and explore the perception that the human is the weakest link in information security assurance. The aim was also to understand when human error is the proximate cause and identify associated underlying systemic root causes and themes. This objective was achieved through reactive and proactive interaction with information security professionals and participating organizations as well as their employees. This included analysis and comparison with related published data and established mechanisms and practices that could be applied by organizations to deal effectively with this commonly misunderstood concern, weakness or problem. The study produced novel findings that enhanced current global information security knowledge whilst, through intervention, benefiting participating organizations and the associated general public and wider society that is dependent upon their services.

2 IS-CHEC technique

2.1 Chapter 2

The IS-CHEC technique is an adaptation of HEART HRA and has been developed to be applied within the information security field in both a retrospective manner related to incident management and a proactive one in terms of probabilistic risk assessment for the first time. This is due to the fact that there is currently no approach specifically aimed at managing the human error problem within the field of information security. The IS-CHEC technique comprises several core components, as shown in Fig. 1, with their respective relationships, which have been developed and enhanced over the research period. An expectation within both the private and public sector healthcare organizations during the real-time element of the research was that they established the primary root cause of an incident as quickly as possible. Specifically, both organizations had an expectation that the primary root cause of the incident was established within five working days through the use of the IS-CHEC technique to manage incidents in real time. This expedited the identification of incidents that related to human error and therefore required the application of the IS-CHEC technique. To achieve this five-day expectation, the 5-whys root cause analysis technique was undertaken and documented within the IS-CHEC incident process (Evans, He, Yevseyeva, & Janicke, 2018). As stated within the information security incident management

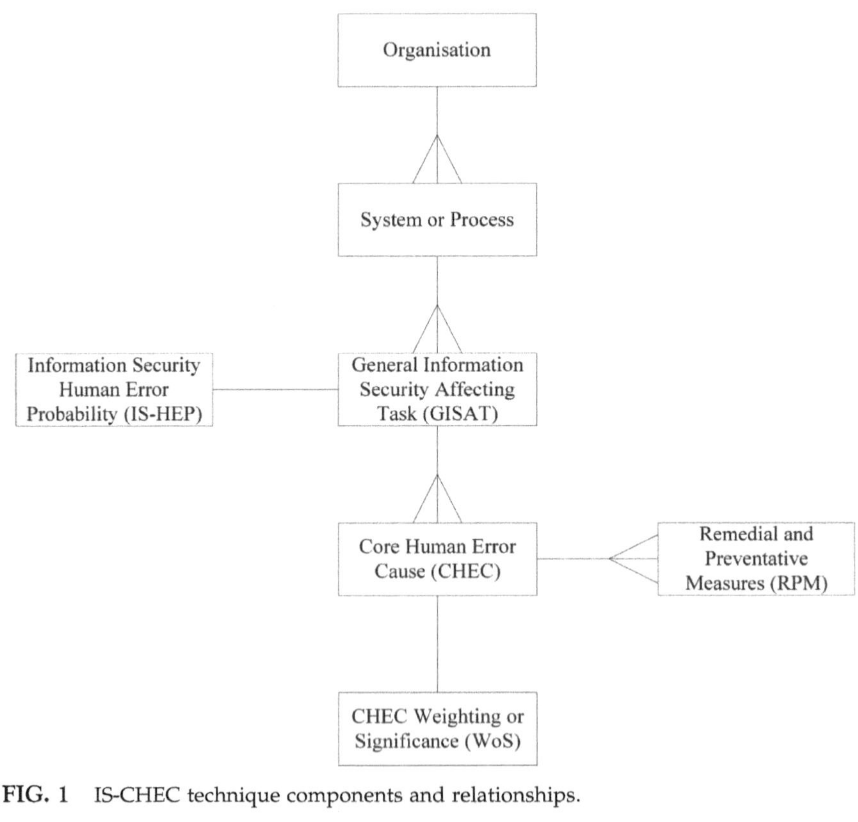

FIG. 1 IS-CHEC technique components and relationships.

literature (Grispos, 2016), the 5-whys technique can be adapted to a variety of domains, it does not require a considerable amount of training and is considered to be less stressful for participants.

2.2 System or process

Every organization is made up of many systems and processes to deliver their intended output. A system is defined as a set of elements or parts that is coherently organized and interconnected in a pattern or structure that produces a characteristic set of behaviors, often classified as its "function" or "purpose" (Meadows, 2008). A process is defined as a horizontal sequence of activities that transforms an input (need) into an output (result) to meet the needs of customers or stakeholders (Palmberg, 2009). These systems or processes, for example, could be responding to enquiries by customers that comprise numerous specific tasks to deliver the intended output. The IS-CHEC technique and this research have a focus on the capturing of systems and processes that have been affected by incidents, whilst also proactively identifying systems or processes that are susceptible to human error that could yield a security incident. This approach enabled the participating organizations to focus their attention on the systems and processes that attract the greatest risk exposure.

There was a difference applied between the participating public and private sector organizations that are worthy of note. The private sector organization had documented a list of all of its processes that were embedded within the incident management system as part of the research. This enabled consistent selection of processes affected by information security incidents, whilst the use of a drop-down list within the system enabled accurate reporting and reduced effort in report creation. The public sector organization did not have this level of maturity, which resulted in different terms being applied for systems or processes associated with incidents. The personnel involved in the study were advised at the outset of the study what constituted a system or process. This made reporting very difficult and limited the effectiveness of the technique in terms of focusing on incidents and preventing them from reoccurring.

2.3 General information security affecting task (GISAT)

Each system or process comprises many individual tasks, such as updating a system or record or sending an email. The GISATs were developed after the initial feasibility study. It was identified that the HEART Generic Task Types (GTT) (Williams, 1992) were difficult to understand by those personnel who were neither an information security professional nor a human factors professional. The HEART GTTs required interpretation in terms of the difficulty and complexity of a task. Consequently, it was decided to create a list of actual specific tasks that were being performed as identified initially from the research feasibility study (Evans, He, Maglaras, Yevseyeva, & Janicke, 2019; Evans, Maglaras, He, & Janicke, 2019) and the map these to the HEART GTTs (Table 1). The associated nominal human reliability scores for the mapped GTTs were adopted for the GISATs, as shown in Table 2. As each organization and working context is different it was necessary throughout the research to identify alternate GTTs, where possible, as shown in Table 2. The mapping and associated nominal unreliability scores were assessed on a monthly basis as part of ongoing governance and IS-CHEC report reviews.

TABLE 1 General Information Security Affecting Tasks (GISAT) and their mapping to the HEART technique (Evans et al., 2019b;Williams, 1992 ; Williams & Bell, 2016).

GISAT	HEART GTT mapping	Alternate GTT mapping
GISAT1—Sending an email	G	D
GISAT2—Entering, updating or deleting data within a system, file or document	D	E, G
GISAT3—Posting an item or information	E	D, G
GISAT4—Configuring a system	B	A, C, F
GISAT5—Administering a system	D	F, G
GISAT6—Scanning a document	D	E, G
GISAT7—Printing a document	D	E, F
GISAT8—Providing information verbally	G	D
GISAT9—Delivering information or equipment	G	D
GISAT10—Filing or sorting information	G	E, D
GISAT11—Reading or checking an email, file, document or item	G	A, C, D, E
GISAT12—Safeguarding information or equipment	G	D, E
GISAT13—Destroying information or equipment	D	A, E, G
GISAT14—Accessing a location or environment	G	
GISAT15—Faxing information	D	E, G
GISAT16—Sharing or handing over information or equipment in person	G	

TABLE 2 General Information Security Affecting Tasks (GISAT) and their associated nominal human unreliability (Williams, 1992; Williams & Bell, 2016).

HEART generic task type (GTT)	Nominal human unreliability	Nominal human unreliability (lower bound)	Nominal human unreliability (upper bound)
A—Totally unfamiliar, performed at speed with no real idea of likely consequences	0.41 41/100	0.2 1/5	0.85 17/20
B—Shift or restore the system to a new or original state on a single attempt without supervision or procedures	0.24 6/25	0.06 3/50	1.0 1
C—Complex task requiring a high level of comprehension and skill	0.17 17/100	0.05 1/20	0.6 3/5
D—Fairly simple task performed rapidly or given scant attention	0.06 3/50	0.02 1/50	0.19 19/100
E—Routine, highly practiced, a rapid task involving a relatively low level of skill	0.02 1/50	0.005 1/200	0.09 9/100
F—Restore or shift a system to original or new state following procedures, with some checking	0.001 1/1000	0.00002 1/50000	0.04 1/25
G—Completely familiar, well-designed, highly practiced, routine task occurring several times per hour, performed to highest possible standards by a highly motivated, highly trained and experienced person, totally aware of implications of failure, with time to correct the potential error, but without the benefit of significant job aids	0.002 1/500	0.0002 1/5000	0.01 1/100
H—Respond correctly to system command even when there is an augmented or automated supervisory system providing an accurate interpretation of system state	0.00002 1/50000	0.000006 3/500000	0.0009 9/10000
M—Miscellaneous task for which no description can be found	0.03 3/100	0.008 1/125	0.11 11/100

2.4 Core human error causes (CHEC)

The list of Core Human Error Causes (CHEC) can be seen in Table 3. They were initially developed from the HEART User Manual (Williams, 1992) Error Producing Conditions (EPC) and were adapted with the respective participating organizations over the course of the research, whilst also taking into account subsequent published materials (Williams & Bell, 2015). An applied principle was that every information security incident that was determined to have a root cause of human error must have had between 1 and 3 CHECs (Evans, He, Luo, et al., 2019b) recorded and an associated weighting or significance applied, as required by the HEART technique. For the CHECs, the HEART EPC strength value was adopted wherever possible to enable the HEART calculations to be undertaken successfully (Evans, Maglaras et al., 2019).

2.5 CHEC weighting or significance (WoS)

In accordance with the HEART technique (Williams, 1992), each identified error-producing condition (EPC) must be weighted in terms of its associated proportion of affect (APOA). This has been adapted to enable greater understanding by technique users. Following on from previous feasibility research and published works, Table 4 presents a graphical rating scale and associated ranges and rules.

2.6 Information security human error probability (IS-HEP)

During the participatory action research phase of the study (Evans, He, Luo, et al., 2019b), both the public and private sector organizations benefitted from clear reporting on whether the volumes of human error-related information security incidents were within, above, or below tolerance. In order to achieve an Information Security Human Error Probability (IS-HEP), the HEART technique, as an established mechanism in other fields, was used (Evans, Maglaras, He, & Janicke, 2016; Evans, Maglaras, He, & Janicke, 2019). A fictional example is presented below for information.

- GISAT mapping to HEART GTT (Associated Nominal Human Unreliability, e.g., 0.002(GTT G))
- Up to 3 CHECs selected (Associated Total HEART Strength, e.g., 11, 6, 3 (CHEC 2,9 & 39) & WoS, e.g., 0.3, 0.2, 0.1
- Establish assessed affect for each CHEC—(HEART Strength-1)*WoS +1, e.g., $(11-1)*0.3+1=4$, $(6-1)*0.2+1=2$, $(3-1)*0.1+1=1.2$
- Establish IS-HEP—Multiply each assessed affect and GTT nominal human unreliability, e.g., 0.002*4*2*1.2
- IS-HEP = 0.0192

TABLE 3 Core human error causes (CHEC) (Evans, He, Luo, et al., 2019b).

CHEC
CHEC1—Unfamiliarity with a task or situation which is potentially important but which only occurs infrequently or which is novel
CHEC2—A shortage of time available for error detection and correction
CHEC3—Too many alerts, notifications, messages, or inputs leading to important information not being seen or heard and acted upon
CHEC4—Too easy to switch off, disable, or incorrectly modify alerts, notifications, messages, or inputs leading to important information being missed or not acted upon
CHEC5—No means of communicating information in a form which can be understood and used
CHEC6—A mismatch between an operator's model of the world and that imagined by a designer
CHEC7—No obvious means of reversing an unintended action
CHEC8—Person performing the task is monitoring numerous incoming information channels at the same time, such as numerous computer monitors
CHEC9—Person performing the task required to learn a new technique, process, procedure, or way of working which differs in attitude or way of thinking to the previous one
CHEC10—The need to transfer specific knowledge from task to task without loss
CHEC11—The person performing the task does not fully understand the policy, standards, process, or procedures they are required to adhere to
CHEC12—The person performing the task does not understand the actual risk exposure
CHEC13—The system information communicated is inaccurate, unclear, or inappropriate
CHEC14—The system information which confirms that an action has been successfully completed, is delayed, takes too long, or does not happen
CHEC15—Inexperience of the person performing the task
CHEC16—Inaccurate or incomplete information communicated by procedures, or from a person to a person
CHEC17—Little or no independent checking or testing of output
CHEC18—A conflict between immediate and long-term objectives
CHEC19—Not enough information to allow completeness or accuracy checks to be undertaken
CHEC20—A mismatch between the educational achievement level of an individual and the requirements of the task
CHEC21—An incentive to use other more dangerous procedures

Continued

TABLE 3 Core human error causes (CHEC) (Evans, He, Luo, et al., 2019b)—cont'd

CHEC22—Little opportunity, such as rest breaks, to exercise mind and body outside the immediate confines of a job
CHEC23—Unreliable instrumentation used to communicate information leading to lack of trust and person performing the task ignoring information
CHEC24—A need for decision making which is beyond the capabilities or experience of the person performing the task
CHEC25—Unclear allocation of role and responsibility
CHEC26—No obvious way to keep track of progress during an activity
CHEC27—Task requirement exceeds the physical capabilities of the person performing the task
CHEC28—Person performing the task is unaware of its significance and their contribution to corporate objectives
CHEC29—High-level emotional stress
CHEC30—Evidence of ill-health amongst operatives, especially fever
CHEC31—Low workforce morale
CHEC32—Information displayed and how this is applied within procedures or working practices is not fully understood
CHEC33—A poor or hostile environment (below 75% of health or life-threatening severity)
CHEC34—Prolonged inactivity or highly repetitious low mental workload tasks
CHEC35—Disruption of normal work-sleep cycles
CHEC36—Pressure from someone else to increase the speed or pace at which a task is performed, beyond an individual's preferred pace and capability
CHEC37—Additional team members over and above those necessary to perform the task normally and satisfactorily
CHEC38—Age of personnel performing perceptual tasks requiring the ability to interpret or become aware of something through the senses (sight, hearing, taste, smell, or touch)
CHEC39—Little or no self-checking or testing of output
CHEC40—Lack of significant job aids
CHEC41—Distraction/Task Interruption
CHEC42—Time-of-Day

TABLE 4 CHEC Weighting or Significance (WoS) Graphical Rating Scale and associated ranges and rules (Evans, He, Maglaras, et al., 2019).

WoS graphical rating scale		**WoS ranges and rules**
1.0	Entirely the cause of the error	*Most significant CHEC WoS*
0.9		Range: 0–1.0
0.8		
0.7	Significant cause of the error	
0.6		*Second most significant CHEC WoS*
0.5	Moderate cause of the error	
0.4		Range: 1.0 minus Primary CHEC WoS
0.3		Rule: Cannot exceed Primary CHEC WoS
0.2	Insignificant cause of the error	*Least significant CHEC WoS*
0.1		Range: 1.0 minus (Primary CHEC WoS+Secondary CHEC WoS) Rule: Cannot exceed Secondary CHEC WoS
0	Not a cause of the error	Rule: Cannot exceed Secondary CHEC WoS

In addition to forecasting incident occurrences based on those recorded within the proactive use of the IS-CHEC phase of the research (Evans et al., 2019a), the IS-HEP calculation, established from a survey questionnaire deployed across operations teams within the private sector participating organization, was used to provide the likelihood element of the standard risk calculation alongside the impact. This was to establish whether the human error-related information security risk was above the organizational risk tolerance.

Over the course of the study, several adaptations were made to the IS-CHEC technique, which resulted in the predicted human error probability calculations reflecting actual incident frequencies. These adaptations involved creating specific GISATs to enable the accurate selection of tasks that were being performed at the time that the incident occurred, mapping of the GISATs to the 9 HEART Generic Task Types (GTT), the creation of alternate GISAT to GTT mappings to enable the technique to be agile in reflecting different environments and contexts as well as the creation of the WoS graphical rating scale and associated rules to apply the conservative approach required by the HEART technique. All of the above were presented to the private and public healthcare organizations via the IS-CHEC report at monthly governance meetings for review and validation.

2.7 Remedial and preventative measures

In order for the IS-CHEC to be applied and add beneficial value to the participating organizations, a set of remedial and preventative measures (RPM) was established over the course of the research (Evans, He, Luo, et al., 2019b). In order to promote and enable the application of effective RPMs aimed at both reducing the number of incident occurrences and enabling a consistent selection of controls resulting from incidents, all RPMs had an associated measure of effectiveness (Evans, He, Luo, et al., 2019b), as shown in Table 5. The participating organizations monitored the selection of controls in their associated monthly IS-CHEC reports. In order to deliver a consistent selection of RPMs for each incident, a mapping was documented, as presented in Table 6, which allowed the RPMs to be consistently applied depending upon the CHECs identified for each incident.

TABLE 5 Remedial and preventative measures (RPM).

RPM	Effectiveness
RPM0—None needed	–
RPM1—Awareness and training undertaken (including 1:1)	Lower Effectiveness
RPM2—Procedures documented and communicated	Lower Effectiveness
RPM3—Simulation exercises performed	Moderate Effectiveness
RPM4—Recruitment of additional staff	Moderate Effectiveness

TABLE 5 Remedial and preventative measures (RPM)—cont'd

RPM	Effectiveness
RPM5—Change to, simplification, or standardization of existing procedures, tools, systems or practices	Higher Effectiveness
RPM6—Increased supervision or checks	Moderate Effectiveness
RPM7—Change to communication methods	Moderate Effectiveness
RPM8—Risk assessment or audit undertaken and acted upon	Moderate Effectiveness
RPM9—Job description checked and updated	Lower Effectiveness
RPM10—Change to work patterns such as frequency of breaks	Moderate Effectiveness
RPM11—Job rotation	Moderate Effectiveness
RPM12—Incentives introduced	Lower Effectiveness
RPM13—Acquire and introduce new tools or technology	Higher Effectiveness
RPM14—Introduce robotics/automation/artificial intelligence	Higher Effectiveness
RPM15—Split process and introduce segregation of duties	Moderate Effectiveness
RPM16—Eliminate or reduce distractions	Moderate Effectiveness
RPM17—Eliminate look-and-sound-alikes	Moderate Effectiveness
RPM18—Introduce warnings, alerts, or alarms	Lower Effectiveness
RPM19—Recover, collect or destroy information or equipment	Lower Effectiveness
RPM20—Reissue or resend information or equipment	Lower Effectiveness
RPM99—Other non-human error-related remedial and preventative measures	–

TABLE 6 Mapping of CHECs to RPMs.

CHEC	RPM
CHEC1—Unfamiliarity with a task or situation which is potentially important, but which only occurs infrequently or is novel	RPM1—Awareness and training undertaken
	RPM2—Procedures documented and communicated
	RPM3—Simulation exercises performed
	RPM6—Increased supervision or checks
	RPM14—Introduce robotics/automation/artificial intelligence
	RPM18—Introduce warnings, alerts, or alarms
CHEC2—A shortage of time available for error detection and correction	RPM4—Recruitment of additional staff
	RPM5—Change to existing procedures, tools, systems, or practices
	RPM6—Increased supervision or checks
	RPM14—Introduce robotics/automation/artificial intelligence
	RPM17—Eliminate look-and-sound-alikes
	RPM18—Introduce warnings, alerts, or alarms
CHEC3—Too many alerts, notifications, messages, or inputs leading to important information not being seen or heard and acted upon	RPM4—Recruitment of additional staff
	RPM5—Change to existing procedures, tools, or practices
	RPM7—Change to communication methods
	RPM14—Introduce robotics/automation/artificial intelligence
	RPM16—Eliminate or reduce distractions

TABLE 6 Mapping of CHECs to RPMs—cont'd

CHEC	RPM
	RPM17—Eliminate look-and-sound-alikes
	RPM18—Introduce warnings, alerts, or alarms
CHEC4—Too easy to switch off, disable, or incorrectly modify alerts, notifications, messages, or inputs leading to important information being missed or not acted upon	RPM1—Awareness and training undertaken (including 1:1)
	RPM5—Change to existing procedures, tools, systems, or practices
	RPM6—Increased supervision or checks
	RPM13—Access control changes applied
CHEC5—No means of communicating information in a form that can be understood and used	RPM2—Procedures documented and communicated
	RPM5—Change to existing procedures, tools, systems, or practices
	RPM7—Change to communication methods
CHEC6—A mismatch between an operator's model of the world and that imagined by a designer	RPM1—Awareness and training undertaken
	RPM5—Change to existing procedures, tools, systems, or practices
CHEC7—No obvious means of reversing an unintended action	RPM5—Change to existing procedures, tools, systems, or practices
	RPM6—Increased supervision or checks
	RPM18—Introduce warnings, alerts, or alarms
CHEC8—Person performing the task is monitoring numerous incoming information channels at the same time, such as numerous computer monitors	RPM2—Procedures documented and communicated
	RPM4—Recruitment of additional staff

Continued

TABLE 6 Mapping of CHECs to RPMs—cont'd

CHEC	RPM
	RPM5—Change to existing procedures, tools, systems, or practices
	RPM7—Change to communication methods
	RPM14—Introduce robotics/automation/artificial intelligence
	RPM17—Eliminate look-and-sound-alikes
	RPM18—Introduce warnings, alerts, or alarms
CHEC9—Person performing the task required to learn a new technique, process, procedure, or way of working that differs in attitude or way of thinking to the previous one	RPM1—Awareness and training undertaken
	RPM5—Change to existing procedures, tools, systems, or practices
CHEC10—The need to transfer specific knowledge from task to task without loss	RPM1—Awareness and training undertaken
	RPM5—Change to existing procedures, tools, systems, or practices
	RPM14—Introduce robotics/automation/artificial intelligence
CHEC11—The person performing the task does not fully understand the policy, standards, process, or procedures they are required to adhere to	RPM1—Awareness and training undertaken
	RPM2—Procedures documented and communicated
	RPM5—Change to existing procedures, tools, systems, or practices
CHEC12—The person performing the task does not understand the actual risk exposure	RPM1—Awareness and training undertaken
	RPM8—Risk assessment undertaken and acted upon
CHEC13—The system information communicated is inaccurate, unclear, or inappropriate	RPM5—Change to existing procedures, tools, systems, or practices

TABLE 6 Mapping of CHECs to RPMs—cont'd

CHEC	RPM
	RPM7—Change to communication methods
	RPM17—Eliminate look-and-sound-alikes
	RPM18—Introduce warnings, alerts, or alarms
CHEC14—The system information confirming that an action has been successfully completed, is delayed, takes too long, or does not happen	RPM5—Change to existing procedures, tools, systems or practices
	RPM7—Change to communication methods
CHEC15—Inexperience of the person performing the task	RPM1—Awareness and training undertaken
	RPM4—Recruitment of additional staff
	RPM6—Increased supervision or checks
	RPM18—Introduce warnings, alerts, or alarms
CHEC16—Inaccurate or incomplete information communicated by procedures, or from person to person	RPM1—Awareness and training undertaken (including 1:1)
	RPM2—Procedures documented and communicated
	RPM5—Change to existing procedures, tools, systems, or practices
	RPM7—Change to communication methods
	RPM16—Eliminate or reduce distractions
CHEC17—Little or no independent checking or testing of output	RPM4—Recruitment of additional staff
	RPM5—Change to existing procedures, tools, systems, or practices
	RPM6—Increased supervision or checks

Continued

TABLE 6 Mapping of CHECs to RPMs—cont'd

CHEC	RPM
	RPM14—Introduce robotics/automation/artificial intelligence
	RPM15—Split process and introduce segregation of duties
	RPM18—Introduce warnings, alerts, or alarms
CHEC18—A conflict between immediate and long-term objectives	RPM1—Awareness and training undertaken
	RPM6—Increased supervision or checks
	RPM8—Risk assessment undertaken and acted upon
CHEC19—Not enough information to allow for completeness or accuracy checks being undertaken	RPM5—Change to existing procedures, tools, systems, or practices
	RPM7—Change to communication methods
CHEC20—A mismatch between the educational achievement level of an individual and the requirements of the task	RPM1—Awareness and training undertaken
	RPM4—Recruitment of additional staff
	RPM9—Job description checked and updated
CHEC21—An incentive to use other more dangerous procedures	RPM1—Awareness and training undertaken
	RPM5—Change to existing procedures, tools, systems, or practices
	RPM6—Increased supervision or checks
CHEC22—Little opportunity, such as rest breaks, to exercise mind and body outside the immediate confines of a job	RPM1—Awareness and training undertaken
	RPM5—Change to existing procedures, tools, systems, or practices
	RPM10—Change to work patterns such as more frequent breaks

TABLE 6 Mapping of CHECs to RPMs—cont'd

CHEC	RPM
	RPM11—Job rotation
CHEC23—Unreliable instrumentation used to communicate information leading to lack of trust and person performing the task ignoring information	RPM5—Change to existing procedures, tools, systems, or practices
	RPM7—Change to communication methods
	RPM18—Introduce warnings, alerts, or alarms
CHEC24—A need for decision making that is beyond the capabilities or experience of the person performing the task	RPM1—Awareness and training undertaken
	RPM4—Recruitment of additional staff
	RPM6—Increased supervision or checks
	RPM18—Introduce warnings, alerts, or alarms
CHEC25—Unclear allocation of role and responsibility	RPM1—Awareness and training undertaken
	RPM9—Job description checked and updated
CHEC26—No obvious way to keep track of progress during an activity	RPM5—Change to existing procedures, tools, systems, or practices
	RPM6—Increased supervision or checks
CHEC27—Task requirement exceeds the physical capabilities of the person performing the task	RPM4—Recruitment of additional staff
	RPM5—Change to existing procedures, tools, systems, or practices
	RPM8—Risk assessment undertaken and acted upon
CHEC28—Person performing the task is unaware of its significance and their contribution to corporate objectives	RPM1—Awareness and training undertaken

Continued

TABLE 6 Mapping of CHECs to RPMs—cont'd

CHEC	RPM
	RPM5—Change to existing procedures, tools, systems, or practices
	RPM9—Job description checked and updated
	RPM11—Job rotation
CHEC29—High-level emotional stress	RPM1—Awareness and training undertaken
	RPM4—Recruitment of additional staff
	RPM5—Change to existing procedures, tools, systems, or practices
	RPM6—Increased supervision or checks
	RPM16—Eliminate or reduce distractions
CHEC30—Evidence of ill-health amongst operatives, especially fever	RPM1—Awareness and training undertaken
	RPM6—Increased supervision or checks
CHEC31—Low workforce morale	RPM1—Awareness and training undertaken
	RPM4—Recruitment of additional staff
	RPM5—Change to existing procedures, tools, systems, or practices
	RPM6—Increased supervision or checks
	RPM9—Job description checked and updated
	RPM12—Incentives introduced
CHEC32—Information displayed and how this is applied within procedures or working practices is not fully understood	RPM1—Awareness and training undertaken

TABLE 6 Mapping of CHECs to RPMs—cont'd

CHEC	RPM
	RPM5—Change to existing procedures, tools, systems, or practices
	RPM7—Change to communication methods
	RPM17—Eliminate look-and-sound-alikes
	RPM18—Introduce warnings, alerts, or alarms
CHEC33—A poor or hostile environment (below 75% of health or life-threatening severity)	RPM1—Awareness and training undertaken
	RPM5—Change to existing procedures, tools, systems, or practices
	RPM6—Increased supervision or checks
CHEC 34—Prolonged inactivity or highly repetitious low mental workload tasks	RPM5—Change to existing procedures, tools, systems, or practices
	RPM6—Increased supervision or checks
	RPM11—Job rotation
	RPM14—Introduce robotics/automation/artificial intelligence
	RPM16—Eliminate or reduce distractions
	RPM17—Eliminate look-and-sound-alikes
	RPM18—Introduce warnings, alerts, or alarms
CHEC 35—Disruption of normal work-sleep cycles	RPM4—Recruitment of additional staff
	RPM5—Change to existing procedures, tools, systems, or practices
	RPM6—Increased supervision or checks

Continued

TABLE 6 Mapping of CHECs to RPMs—cont'd

CHEC	RPM
CHEC 36—Pressure from someone else to increase the speed or pace at which a task is performed, beyond an individual's preferred pace and capability	RPM1—Awareness and training undertaken (including 1:1) RPM5—Change to existing procedures, tools, systems, or practices RPM6—Increased supervision or checks RPM17—Eliminate look-and-sound-alikes RPM18—Introduce warnings, alerts, or alarms
CHEC 37—Additional team members over and above those necessary to perform task normally and satisfactorily	RPM5—Change to existing procedures, tools, systems, or practices
CHEC 38—Age of personnel performing perceptual tasks requiring the ability to interpret or become aware of something through the senses (sight, hearing, taste, smell or touch)	RPM4—Recruitment of additional staff RPM6—Increased supervision or checks RPM17—Eliminate look-and-sound-alikes RPM18—Introduce warnings, alerts, or alarms
CHEC 39—Little or no self-checking or testing of output	RPM4—Recruitment of additional staff RPM5—Change to existing procedures, tools, systems or practices RPM6—Increased supervision or checks RPM15—Split process and introduce segregation of duties RPM18—Introduce warnings, alerts or alarms

TABLE 6 Mapping of CHECs to RPMs—cont'd

CHEC	RPM
CHEC40—Lack of significant job aids	RPM5—Change to existing procedures, tools, systems or practices
	RPM13—Acquire and introduce new tools or technology
CHEC41—Distraction/task interruption	RPM6—Increased supervision or checks
	RPM13—Acquire and introduce new tools or technology
	RPM16—Eliminate or reduce distractions
	RPM18—Introduce warnings, alerts, or alarms
CHEC42—Time-of-day	RPM4—Recruitment of additional staff
	RPM6—Increased supervision or checks
	RPM10—Change to work patterns such as more frequent breaks
	RPM11—Job rotation
	RPM18—Introduce warnings, alerts, or alarms

3 General discussion

Over the course of the research, it has been identified that there is a significant gap in current mainstream information security assurance practices regarding human error and established and empirically validated solutions or frameworks. It has been established over four independent components of research (Evans, He, Luo, et al., 2019b;Evans, He, Maglaras, et al., 2019 ; Evans, Maglaras, He, & Janicke, 2019), within two private sector organizations (healthcare and non-healthcare) and one public sector healthcare organization, that the majority of reported and recorded information security incidents pertain to human error. This specific situation, and the details regarding the most common root cause and underlying causes of information security incidents, were not understood prior to the study, as action research in large enterprises over a prolonged period of time has not been undertaken before. This study has provided unique insights into the frequency, cause, and effectiveness of HRA in this setting. In addition, analysis was undertaken of published breaches (Evans et al., 2018; Evans, He, Yevseyeva, & Janicke, 2019) to validate these findings and confirm that the majority of information security incidents have human error resulting from latent system failings at their core.

3.1 Principal findings

Over the course of the study, the principal findings have included results providing empirical evidence, within the participating organizations, demonstrating that the majority of reported and recorded information security incidents were as a result of human error. In fact, the proportion of information security incidents that were a result of actual or likely human error, ranged between 51% and 99%. This finding suggests that the majority of incidents within the participating organizations related to human error and associated organizational failings. As shown in Fig. 2, the average proportion of human error-related information security incidents experienced by the participating organizations over the course of the study was 77.92%. However, taking into account analysis of published incident data the average is 81.56%.

In addition to the overall numbers of human error-related information security incidents, it was also possible over the course of the study to establish some wider context. This included ascertaining that within the feasibility and retrospective analysis element of the research (Evans, He, Maglaras, et al., 2019; Evans, Maglaras, He, & Janicke, 2019), the public and private sector organizations had 87.7% and 90.3%, respectively, of incidents that were due to errors of commission, rather than errors of omission, thus indicating that the vast majority incidents were due to people performing tasks incorrectly, rather than not performing them. This phase of the research also identified that administrative activity accounted for a large proportion of incidents within participating organizations. In terms of proportions of human error-related information security incidents, it was established during the real-time analysis phase of the research (Evans, He, Luo, et al., 2019b), by applying Pearson's Correlation Coefficient, that both the public and private sector organizations experienced a strong linear relationship between the total numbers of reported and recorded incidents and the total numbers of human error-related ones. The intention of applying Pearson's Correlation Coefficient was to ascertain whether over the course of the real-time study the total reported information security incidents had a linear relationship with the volumes of reported human error-related information security incidents. It is acknowledged that due to the research results revealing that the majority

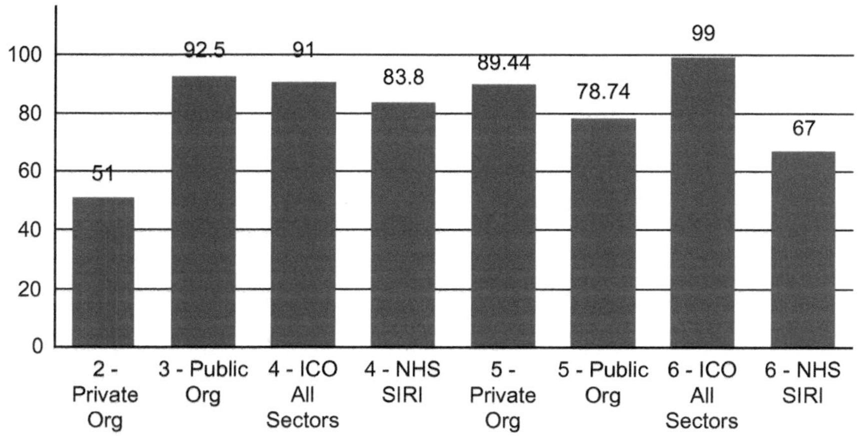

FIG. 2 Percentage of human error-related information security incidents by publication.

of incidents pertained to human error, this would constrain the range of possible relationship results. It did, however, evidence a consistent and constant strong linear relationship taking into account the incident volumes for each of the 12 months of the real-time study for both participating organizations.

The IS-CHEC information security HRA technique was found to be feasible, applicable, beneficial as it provided valuable and previously unknown information relating to the effects of human error within an organization. Such information included indications that within the participating organizations the most common tasks that were being performed when an incident occurred were routine, familiar ones, undertaken by front-line workers processing personal or confidential data. Also, the most common human error cause was the lack of checks for successful completion of manual tasks. We have to mention here that-other organizational failings, such as stress, lack of time, and ambiguous procedures, inhibited the ability of the workers to undertake tasks successfully.

4 Conclusions and future work

In this chapter, the application of HRA techniques within the field of information security to address current gaps in information security assurance practices relating to human error has been proposed. The developed IS-CHEC information security HRA technique has been derived through empirical research, within two participating public and private sector organizations. In this chapter, we have explored the effect of human error on information security assurance within these organizations. It has been concluded that, in order to address the gaps in current information security assurance frameworks and practices specifically relating to human error, HRA should be adopted as has been applied to the safety field.

The research has also established that there is no simple remedy to the problem of information security human error. It requires organizational buy-in, resource, effort, and the application of an information security HRA technique, such as IS-CHEC, to achieve success with regard to human error and its negative effect on organizational information security assurance. The application of IS-CHEC, both in terms of real-time incident management and proactive risk management, has been found to add positive benefit to both the public and private sector organizations in terms of organization-wide understanding of the volumes, proportions, and underlying causes of human error-related information security incidents. This was achieved through the application of an empirically validated framework to address the problem of human error in a consistent and standardized manner.

References

Al-Barnawi, A., He, Y., Maglaras, L. A., & Janicke, H. (2019). Electronic medical records and risk management in hospitals of Saudi Arabia. *Informatics for Health & Social Care, 44*(2), 189–203. https://doi.org/10.1080/17538157.2018.1434181. Taylor & Francis.

Alissa, D. K. A., Alshehri, H. A., Dahdouh, S. A., Alsubaie, B. M., Alghamdi, A. M., Alharby, D. A., et al. (2018). An instrument to measure human behavior toward cyber security policies. In *2018 21st Saudi Computer Society National Computer Conference (NCC)* (pp. 1–6). IEEE. https://doi.org/10.1109/NCG.2018.8592978.

Ani, U. D., He, H., & Tiwari, A. (2019). Human factor security: Evaluating the cybersecurity capacity of the industrial workforce. *Journal of Systems and Information Technology, 21*(1), 2–35. https://doi.org/10.1108/JSIT-02-2018-0028. Emerald Publishing Limited.

Atoum, I., & Otoom, A. (2017). A classification scheme for cybersecurity models. *International Journal of Security and Its Application, 11*(1), 109–120. https://doi.org/10.14257/ijsia.2017.11.1.10.

Burrell, D. N. (2019). A contextual exploration of emergence of technical sociology in the realm of organizational technology management and cybersecurity management. *International Journal of Engineering Sciences & Research Technology, 8*(3), 133–144. https://doi.org/10.5281/zenodo.2607817.

Cilliers, L. (2017). Exploring information assurance to support electronic health record systems. In *2017 IST-Africa Week Conference (IST-Africa)* (pp. 1–8). IEEE. https://doi.org/10.23919/ISTAFRICA.2017.8102363.

Coffey, K., Smith, R., Maglaras, L., & Janicke, H. (2018). Vulnerability analysis of network scanning on SCADA systems. *Security and Communication Networks, 2018*, 1–21. https://doi.org/10.1155/2018/3794603. Hindawi.

Cook, A., Maglaras, L., Smith, R., & Janicke, H. (2018). Managing incident response in the industrial internet of things. *International Journal of Internet Technology and Secured Transactions, 8*(2), 251. https://doi.org/10.1504/IJITST.2018.093336.

Cope, J., Siewe, F., Chen, F., Maglaras, L., & Janicke, H. (2017). On data leakage from non-production systems. *Information and Computer Security*. https://doi.org/10.1108/ICS-02-2017-0004. Ed. S. Furnell and S. Furnell. Emerald Publishing Limited, pp. 00–00.

Dhakal, R. (2018). *Measuring the effectiveness of an information security training and awareness program*. Charles Sturt University.

Drivas, G., Maglaras, L., Janicke, H., & Ioannidis, S. (2020). *Cybersecurity assessment of the public sector in Greece. Available at*. https://search.proquest.com/openview/c910f8c8f8cc4779c17b4047a69f040d/1?pq-origsite=gscholar&cbl=396497. (Accessed 6 September 2019).

Drozd, O. V., & Kapulin, D. V. (2017). The model of electronic design document as a part of integrated information environment of radioelectronic enterprise. In *2017 Dynamics of systems, mechanisms and machines (Dynamics)* (pp. 1–5). IEEE. https://doi.org/10.1109/Dynamics.2017.8239447.

Ebenezer, C. (2017). *"Access denied"? Barriers for staff accessing, using and sharing published information online within the National Health Service (NHS) in England: Technology, risk, culture, policy and practice*. University of Sheffield. Available at http://etheses.whiterose.ac.uk/19826/. (Accessed 6 September 2019).

Eling, M., & Wirfs, J. (2019). What are the actual costs of cyber risk events? *European Journal of Operational Research, 272*(3), 1109–1119. https://doi.org/10.1016/J.EJOR.2018.07.021. North-Holland.

Elisa, N., Yang, L., Chao, F., & Cao, Y. (2018). A framework of blockchain based secure and privacy-preserving E-government system. *Wireless Networks*, 1–11. https://doi.org/10.1007/s11276-018-1883-0. Springer US.

Evans, M., He, Y., Luo, C., Yevseyeva, I., Janicke, H., & Maglaras, L. A. (2019a). Employee perspective on information security related human error in healthcare: Proactive use of IS-CHEC in questionnaire form. *IEEE Access, 7*, 102087–102101. https://doi.org/10.1109/ACCESS.2019.2927195.

Evans, M., He, Y., Luo, C., Yevseyeva, I., Janicke, H., Zamani, E., et al. (2019b). Real-time information security incident management: A case study using the IS-CHEC technique. *IEEE Access*, 1. 1 https://doi.org/10.1109/ACCESS.2019.2944615.

Evans, M., He, Y., Maglaras, L., Yevseyeva, I., & Janicke, H. (2019). Evaluating information security core human error causes (IS-CHEC) technique in public sector and comparison with the private sector. *International Journal of Medical Informatics, 127*, 109–119. https://doi.org/10.1016/J.IJMEDINF.2019.04.019. Elsevier.

Evans, M., He, Y., Yevseyeva, I., & Janicke, H. (2018). Analysis of published public sector information security incidents and breaches to establish the proportions of human error. In *Proceedings of the 12th international conference on the human aspects of informarion security assurance—HAISA 2018* (pp. 911–921).

Evans, M., He, Y., Yevseyeva, I., & Janicke, H. (2019). Published incidents and their proportions of human error. *Information and Computer Security*. https://doi.org/10.1108/ICS-12-2018-0147.

Evans, M., Maglaras, L. A., He, Y., & Janicke, H. (2016). Human behaviour as an aspect of cybersecurity assurance. *Security and Communication Networks, 9*(17), 4667–4679. https://doi.org/10.1002/sec.1657.

Evans, M., Maglaras, L., He, Y., & Janicke, H. (2019). HEART-IS: A novel technique for evaluating human error-related information security incidents. *Computers & Security, 80*, 74–89. Elsevier Advanced Technology.

Ferrag, M. A., Maglaras, L., & Ahmim, A. (2017). Privacy-preserving schemes for ad hoc social networks: A survey. *IEEE Communication Surveys and Tutorials, 19*(4), 3015–3045. https://doi.org/10.1109/COMST.2017.2718178.

Ferrag, M. A., Maglaras, L., Derhab, A., & Janicke, H. (2018). *Authentication schemes for smart mobile devices: Threat models, countermeasures, and open research issues*. Available at: http://arxiv.org/abs/1803.10281. (Accessed 6 September 2019).

Grispos, G. (2016). *On the enhancement of data quality in security incident response investigations* (p. 284). Available at http://theses.gla.ac.uk/7293/1/2016GrisposPhd.pdf.

Huian, M. C., Mironiuc, M., & Chiriac, M. (2018). Study on the association between earnings management and value relevance based on the reporting method used for operating cash flows. *The Audit Financiar Journal, 16*(149), 101. Chamber of Financial Auditors of Romania. Available at: https://ideas.repec.org/a/aud/audfin/v16y2018i149p101.html. (Accessed 6 September 2019).

Jackson, G. (2018). *Wireless implanted medical device cybersecurity: A multiple CASE study*. Capella University. Available at: https://search.proquest.com/openview/4702b785d5103e945964a883aafa88e3/1?pq-origsite=gscholar&cbl=18750&diss=y. (Accessed 6 September 2019).

Kauspadiene, L. (2019). *Information security management framework for small and medium enterprise*. Vilnius Gediminas Technical University.

Kauspadiene, L., Ramanauskaite, S., Cenys, A., Janulevicius, J., & Rastenis, J. (2018). Modeling of enterprise management structure for data leakage evaluation. *Information Security Journal: A Global Perspective, 27*(1), 1–13. https://doi.org/10.1080/19393555.2017.1423136. Taylor & Francis.

Lamba, A., Singh, S., Singh, B., Dutta, N., & Muni, S. S. R. (2017). Mitigating cyber security threats of industrial control systems (SCADA & DCS). In *3rd international conference on emerging Technologies in engineering, biomedical, medical and science* (pp. 31–34).

Lee, B.-R. I.-S. (2018). The role and collaboration model of human and artificial intelligence considering human factor in financial security. *Journal of the Korea Institute of Information Security and Cryptology, 28*(6), 1563–1583. https://doi.org/10.13089/JKIISC.2018.28.6.1563. Korea Institute of Information Security and Cryptology.

Linkov, V., Zámečník, P., Havlíčková, D., & Pai, C.-W. (2019). Human factors in the cybersecurity of autonomous vehicles: Trends in current research. *Frontiers in Psychology, 10*, 995. https://doi.org/10.3389/fpsyg.2019.00995. Frontiers.

Maglaras, L., Ferrag, M. A., Derhab, A., Mukherjee, M., Janicke, H., & Rallis, S. (2019). *Threats, protection and attribution of cyber attacks on critical infrastructures*. Available at: http://arxiv.org/abs/1901.03899. (Accessed 6 September 2019).

Maglaras, L. A., Janicke, H., Jiang, J., & Crampton, A. (2020). Novel intrusion detection mechanism with low overhead for SCADA systems. In *Securing the internet of things: Concepts, methodologies, tools, and applications* (pp. 299–318). https://doi.org/10.4018/978-1-5225-9866-4.ch017.

Maglaras, L. A., Kim, K.-H., Janicke, H., Ferrag, M. A., Rallis, S., Fragkou, P., et al. (2018). Cyber security of critical infrastructures. *ICT Express, 4*(1), 42–45. https://doi.org/10.1016/j.icte.2018.02.001.

Meadows, D. H. (2008). *Thinking in systems: A primer*. Chelsea Green Publishing. Available at: https://books.google.co.uk/books?hl=en&lr=&id=CpbLAgAAQBAJ&oi=fnd&pg=PR9&dq=Thinking+in+Systems:+A+Primer.&ots=Lzqgr7qCN_&sig=_qiCxTSKy9bpDEVub7ZVFoKZ3AY&redir_esc=y#v=onepage&q=ThinkinginSystems%3A.APrimer.&f=false.

Nobles, C. (2018). Botching human factors in cybersecurity in business organizations. *HOLISTICA–Journal of Business and Public Administration, 9*(3), 71–88. https://doi.org/10.2478/hjbpa-2018-0024.

Osei-Amanfi, M. (2018). *A case study Exploration of strategies to avoid cloud computing data breaches*. Grand Canyon University.

Palmberg, K. (2009). Exploring process management: Are there any widespread models and definitions? *TQM Journal, 21*(2), 203–215. https://doi.org/10.1108/17542730910938182.

Pham, V., & Dang, T. (2018). CVExplorer: Multidimensional visualization for common vulnerabilities and exposures. In *2018 IEEE international conference on big data (Big Data)* (pp. 1296–1301). IEEE. https://doi.org/10.1109/BigData.2018.8622092.

Reason, J. T. (1990). *Human error*. Cambridge University Press.

Rindasu, S.-M. (2018). Information security challenges-vulnerabilities brought by ERP applications and cloud platforms. *Audit Financiar, 16*(149), 131–139.

Robinson, M., Jones, K., Janicke, H., & Maglaras, L. (2019). Developing cyber peacekeeping: Observation, monitoring and reporting. *Government Information Quarterly, 36*(2), 276–293. https://doi.org/10.1016/J.GIQ.2018.12.001. JAI.

Safa, N. S., Maple, C., Furnell, S., Azad, M. A., Perera, C., Dabbagh, M., et al. (2019). Deterrence and prevention-based model to mitigate information security insider threats in organisations. *Future Generation Computer Systems, 97*, 587–597. https://doi.org/10.1016/J.FUTURE.2019.03.024. North-Holland.

Shabut, A. M., Lwin, K. T., & Hossain, M. A. (2016). Cyber attacks, countermeasures, and protection schemes—A state of the art survey. In *2016 10th international conference on software, knowledge, information management & applications (SKIMA)* (pp. 37–44). IEEE. https://doi.org/10.1109/SKIMA.2016.7916194.

Singh, R., & Jalota, H. (2018). A study of good-enough security in the context of rural business process outsourcing. In *Psychological and behavioral examinations in cyber security* (pp. 239–252).

Snyman, D. P., Kruger, H., & Kearney, W. D. (2018). I shall, we shall, and all others will: Paradoxical information security behaviour. *Information and Computer Security, 26*(3), 290–305. https://doi.org/10.1108/ICS-03-2018-0034. Emerald Publishing Limited.

Tarazan, S., & Bostan, A. (2016). Customizing SSL certificate extensions to reduce false-positive certificate error/warning messages. *International Journal of Information Security Science, 5*(2), 21–28. Available at: http://www.ijiss.org/ijiss/index.php/ijiss/article/view/182. (Accessed 6 September 2019).

Trim, P. R. J., & Lee, Y.-I. (2019). The role of B2B marketers in increasing cyber security awareness and influencing behavioural change. *Industrial Marketing Management*. https://doi.org/10.1016/J.INDMARMAN.2019.04.003. Elsevier.

Twisdale, J. A. (2018). *Exploring SME vulnerabilities to cyber-criminal activities through employee behavior and internet Access*. Walden University.

Williams, J. C. (1992). *A user manual for the HEART human reliability assessment method*. DNV Technica.

Williams, J. C., & Bell, J. L. (2015). Consolidation of the error producing conditions used in the human error assessment and reduction technique (Heart). *Safety and Reliability, 35*(3), 26–76. https://doi.org/10.1080/09617353.2015.11691047. Taylor & Francis.

Williams, J. C., & Bell, J. L. (2016). Consolidation of the generic task type database and concepts used in the human error assessment and reduction technique (HEART). *Safety and Reliability, 36*(4), 245–278. https://doi.org/10.1080/09617353.2017.1336884. Taylor & Francis.

PART III

Machine learning and modeling applications to cybersecurity

CHAPTER

12

Machine learning for the security of healthcare systems based on Internet of Things and edge computing*

Devrim Unal[a,b], Shada Bennbaia[b], and Ferhat Ozgur Catak[c]

[a]KINDI Center for Computing Research, College of Engineering, Qatar University, Doha, Qatar, [b]Department of Mechanical Engineering, College of Engineering, Qatar University, Doha, Qatar, [c]Department of Electrical Engineering and Computer Science, University of Stavanger, Stavanger, Norway

1 Big data in health care

Internet of things (IoT) is the ability of devices to interact with each other in an integrated network environment to collect and exchange data without the need for person-to-person interconnection. IoT is a new area of research that opens the door for many innovations in the technology field. This technology has brought great innovations to many fields. An important branch of IoT technology is the Internet of Medical Things (IoMT), where healthcare sectors is the application domain (Sun, Lo, & Lo, 2019). An IoT-based healthcare system utilizes IoT applications, such as implanted medical devices and wearable devices, to serve the medical professionals and the patients to achieve the highest quality of health care.

Research has proven the benefits of monitoring the patient remotely. This technology reduces the pressure on the medical staff and hospital

*Qatar National Research Fund

https://doi.org/10.1016/B978-0-323-90570-1.00007-3

resources by enabling healthcare professionals to track the state of the noncritical patient remotely. It also helps patients who live in far areas to easily keep track of their health state and gives older adults the chance to stay home and live independently while having their health conditions monitored by the medical staff. Many works have implemented the IoMT technology in medication plans, such as rehabilitation plans, diabetes management, and assisted ambient living (Novo, 2018). A system has been built in rehabilitation cases to make a suitable medication plan for a specific patient with some physical injury. The patient case is identified by comparing it with previous cases that are already stored in the system database, and accordingly, the best rehabilitation plan and needed medications are decided. This system has shown great success in 87.9% of the cases where the doctors accepted the generated plan. Another example of medical application of IoMT technology is in dealing with Parkinson's disease. Medical wearable devices could be used incorporated with a vision-based technology to monitor the patient physical attitudes. In another work, the diabesity disease has been considered. In this system, the patient is required to take two blood-glucose readings. One is having an unstable blood-glucose level, and the other one is a mistaken reading. These two readings are then fed to the system, and the system decides whether to alert the patient himself, the health providers, or the patient family. One more application of IoMT is a system that detects a heart attack before it happens. The heart activity is being measured using an ECG sensor. The resultant data are then processed in a microcontroller and are sent to the patient's smartphone. This system could save many lives and help in taking action before a heart attack happens. Another proposed system is called SPHERE for monitoring elderly patients while they are home. This system enables those older people to live a comfortable life in their homes without visiting the hospital continuously or even staying there in some cases. On the other hand, the security of the patients' data has become a serious issue. Since the patients' medical data are being transmitted through wireless channels and stored in a database, this will increase the probability of having security risks on patients' sensitive data. Incorrect use of technology brings many security risks that threaten the patient's privacy and even the patient's life. Therefore, one important aim of smart healthcare systems is to ensure the cyber-security of remote patient monitoring and emergency response.

1.1 Data collection and big data platforms for remote health care

The communication platforms for smart health care and the applications for emergency and home-based patient follow-up aim to provide emergency patients with specialist service as soon as possible. For patients who require home care, environmental conditions of vital importance can be monitored with this structure, external warnings can be given to the patient, and the patient's data can be monitored remotely.

Data collection and transmission in these systems are generally performed by connecting the existing sensors through standard or customized interfaces during the data collection phase, both within the ambulance and within the scope of home care. Hardware-based on ISO/IEEE 11073 Personal Health Data (PhD) is used with protocol-supported devices. HDP is a structure that provides compatibility of medical, health, and fitness devices from different manufacturers. In addition, newer devices support HL7v3 and Integrated Clinical Device (ICD-10) to work integrated in the future. In the data carrier layer, wired connections such as serial communication, USB connection, Bluetooth, ZigBee, and 6LoW-PAN wireless interfaces are supported.

The data obtained by this equipment are transferred to the data collection units in star, tree, or linear (Star, Tree, Linear) topology following the IEEE802.15.4 Standard. The equipment to be designed can work with an external power supply or battery-backed, if necessary. Therefore, providing maximum efficiency with energy-efficient design will be targeted in the design.

Data from sensors or medical devices are transmitted to the data acquisition/transfer unit (central gateway); this information can be transmitted to the central server by the data transfer unit via a LAN connection or wirelessly via 3G/4G connection. The central gateway transmits the image and sound (H264, G7XX) from camera systems connected to it. The central gateway communicates data from the sensor network (sensor detection-data sending units) in the local area following the IEEE802.15.4 Standard.

The communication server software collects information from the central gateway units to support multiple databases and load balancing. Short-term tables that will allow quick access to the database structure and the archiving system in the background attached to these tables are available to ensure that the up-to-date data is transferred to the external users. The device's authentication on the server software must be addressed. The communication server sends and queries data to all data exchange units that communicate with the server software and all devices connected to them.

While the server software communicates with the sensors of the patients and peripherals from one end, it will be able to process this data and transfer it to users and web-based objects that can evaluate the data. By this structure, users can connect to the communication server and receive information and updates. The server software can have a reporting interface retrospectively. By this feature, healthcare professionals with the permission to log in to the system and the necessary authorizations will be able to access all the data of the person concerned. The healthcare professional can also monitor these data through a dashboard.

1.2 Privacy for data collection and processing in smart health care

Nowadays, biomedical data are collected from many associated patients through wireless communications and wearable devices. In edge computing-based healthcare applications, the importance of privacy protection is increasing because of the data sensitivity and openness of communication channels. Dealing with privacy and security issues for wearable devices used for real-time processing and analysis is necessary for big data analysis (Das et al., 2017). There are various studies related to the privacy and security of edge computing-based smart healthcare systems in the literature. In the work proposed by Das et al. (2017), a privacy protection method that ensures data access control and identity authentication was introduced for cloud and edge computing-based medical wearable devices. According to the study results, it is asserted that the proposed method succeeds in privacy protection for wearable devices.

In the study of Diez, Touceda, Camara, and Zeadally (2015), to proactively verify wearable devices and mobile terminals such as iOS and Android devices, a lightweight authentication protocol was proposed. Afterward, security analysis of the proposed protocol was carried out using automated validation of Internet security protocols and applications (AVISPA) tool. Then, the protocol's communication overhead and computing cost were investigated with the network simulator version 2 (NS2). Zhou, Cao, Dong, and Lin (2015) introduced an authentication protocol that provide mutual authentication between users of wearable devices and other systems. In the sequel, to perform flexible data sharing, secure interactions between wearable remote devices are provided over the internet. Interactive scenarios for different levels of security protection are analyzed according to the sensitivity of wearable devices. In the work proposed by Šprager, Trobec, and Jurič (2017), a privacy protection protocol that is based on data aggregation was presented for smart health care. For securing data aggregation, dynamic medical text mining and image feature extraction were utilized together. According to the experimental results, the presented protocol has better performance values regarding communication overhead, and computing cost. Peng et al. (2016) presented an authentication protocol based on electrocardiogram (ECG) for authentication and identity recognition. To attain user-specific variations in ECG and heart rate, the ECG sensors were utilized. The presented protocol was carried out with high-level statistics. In another study (Khan et al., 2020), an authentication protocol was proposed for wearable glasses designed to distinguish counterfeiters and legal persons regarding biometrics.

We will also look at the multimedia security arising from multilateral uses of the IoMT sensors, including cameras. The surrounding sensors and cameras capture a vast amount of visual data and videos in order to be used later for many medical purposes. Due to massive visual data captured and processed continuously, this leads to a tremendous strain on the network resources and large bandwidth consumption. Instead, a smart technique will be applied where this huge amount of visual data is summarized into several keyframes only. A similar approach has been presented in Hou, Miao, Xing, and Wu (2019). First, a lightweight summarization method will be applied to select the critical frames from a huge amount of visual data. Second, the selected data will be compressed using the discrete cosine transformation technique. Lastly, the image will be encrypted by the discrete fractional random transformation technique. Tabassum, Erbad, and Guizani (2019) proposed a secure IoT platform for smart healthcare surveillance. Their framework performs analytic services in an encrypted domain using Fully Homomorphic Encryption (FHE). According to study results, their framework preserves a high level of analytical accuracy and data privacy. Additional methods will be further investigated within the project.

1.3 Edge-based intrusion detection for smart health care

The wireless technologies sparks in the medical sector, IoMT devices, and cloud computing efficiently perform the analysis and diagnosis of the patient remotely. The e-healthcare system generates voluminous data due to continuous communication of various medical biosensors and actuators, which makes the system complex. This technology eases the task of practitioners to treat patients, but the system lags from the security perspective, which makes the system vulnerable and threatening for the patient. Although there are some security mechanisms available, very few of them effectively identify anomalies from a huge amount of data generated. So, an intrusion detection system (IDS) that can identify anomalous activity in the edge network of the smart healthcare system is necessary (Hou et al., 2019).

The proposed system will be useful for in-house and ambulance in-bound patients. By considering significant security mechanisms, we will design a scalable architecture to deliver appropriate patient details to the healthcare professionals from the patient care efficiently without manipulation (i.e., tampering). Our ultimate goal is to provide a security mechanism to detect malicious activities against wireless communication on the edge node. This architecture can be deployed on various edge nodes, whether the patient is at domiciliary, hospitalized, or ambulance bound. The approach will be based on machine learning at the edge (Ghubaish et al., 2021; Hady, Ghubaish, Salman, Unal, & Jain, 2020). The edge receives data from various IoMT sensors, which are sent through IoMT gateways. The fetched information is analyzed, and processed before it is sent to the medical professional for an intense diagnosis.

The IoMT gateway either accepts the information from the various sub-IoMT gateways or receives the information directly from the IoMT devices to analyze and store in the edge cloud.

As the patient's complete information flows in and out through the IoMT gateway, it becomes an attacking surface. Hence, the attacker targets the IoMT gateway to manipulate information before sending it to the healthcare professional or to launch denial of service attacks to make the information unavailable. Such malevolent activities put the patient's life at risk. To avoid such abrupt manipulation of the information, we will enforce a multilayer intrusion detection model on the edge nodes of the healthcare system. The detection system will comprise multiple layers. The first layer will be responsible for gathering patient information through a gateway and perform the preprocessing, feature engineering, and feature selection using various ML algorithms. The higher levels will detect the abnormal activities of the wireless traffic on the edge node using unsupervised learning and Deep Neural Network (DNN) classifiers (Lu & Sinnott, 2020).

2 Privacy-preserving machine learning

Recently, many companies, such as medical, IT, and financial institutions, create large-scale datasets from their business. The ability to store these data is crucial in creating a better and more reliable business. This data could be applied to build better predictive models and make better business decisions. This is an essential step in the development of new services. However, storing large-scale datasets can be costly and time-consuming for many enterprises. In many cases, organizations rent cloud storage systems such as Amazon or Google for their data, and these data are shared in a public cloud. This makes sensitive data more vulnerable to unauthorized access or manipulation. Besides these cloud providers, these datasets' main issues are security and privacy concerns, preventing the organization from using them efficiently or in the best possible way. The traditional machine learning (ML) algorithms require an authority to access sensitive and confidential data to produce a predictive model. This poses a privacy risk. In the same way, there are also some security concerns. A breach of data from the cloud could cause substantial financial losses.

Nonetheless, several anonymization methods are implemented to sensitive datasets to preserve confidentiality and privacy. These techniques cannot eliminate risks. An adversary can still retrieve the confidential data itself via a different method. Also, data protection laws differ from country to country and even from provider to provider. Let us consider a scenario in which two or more medical institutions need to develop a prediction model by consolidating each other institution's sensitive patient data. In such a situation, it is critical to implement a new privacy-preserving ML training method for prediction algorithms that can work concurrently on many sensitive data without recovering the private information from the data. We can divide the privacy-preserving data analysis methods into two groups: statistical perturbation-based methods and encryption-based methods.

2.1 Statistical perturbation-based methods

2.1.1 k-*Anonymity*

k-Anonymity (Sweeney, 2002) first introduced the concept of *k*-anonymity in their research. They stated that public data is said to have the *k*-anonymity property if the information for each record (i.e., row) in the release cannot be recognized from at least records whose information also appears in the public data. For example, if $k = 2$, then there is no unique record in the published data, and each row looks the same to at least one other row. Table 1 shows an example of the *k*-anonymity process. After the anonymization process, there are at least two rows in each group. The last line is suppressed because there is only one row in the group.

Table 2 shows the attribute types with examples and the actions.

k-Anonymity has several vulnerabilities, such as ordered list matching attacks and homogeneity attacks.

2.1.2 *Differential privacy*

Differential privacy is a technique to prevent the disclosure of the sensitive information for the openly accessible data by defining collections' patterns within the dataset while keeping the sensitive information about

TABLE 1 *k*-Anonymity example.

Age	Gender	Zip
30	F	34667
42	M	34675
32	M	34931
44	F	34925
68	F	34931
72	M	34931
61	F	34931

⇒

Age	Gender	Zip
<50	*	346**
<50	*	346**
<50	*	349**
<50	*	349**
≥50	*	349**
≥50	*	349**
*	*	*

The first table shows the original dataset. The second table shows the k-Anonymous ($k = 2$) dataset.

TABLE 2 k-Anonymity attribute types.

Attribute	Feature	Example	Action
Key	Can identify someone directly	Name, SSID	Suppress or remove
Quasiidentifier	Needs background information to identify the person	Zip code, city, county, date of birth, gender	Suppress or generalize
Sensitive	Sensitive data for person	Income, illness	Cut the link between person and the data

individuals in the data. The differential privacy concept is that if the result of performing a single random replacement in the data is small and sufficient, the grouping search query result cannot be utilized to gather additional information from any single person and accordingly gives privacy protection. Some government offices use differentially private techniques to distribute demographic data or other statistical sums while ensuring query results' privacy. Privacy in a publicly accessible dataset can be protected with the change of information in removing a row data subject. Fig. 1 shows the overview of the differential privacy technique.

2.2 Encryption-based privacy

Homomorphic encryption methods allow arithmetic operations on ciphertexts without retrieving the plaintexts when the dataset is shared between more than two untrusted environments. This approach promotes calculations with ciphertexts. Homomorphism is defined as a structure-preserving map between two algebraic forms in mathematics. It describes the transformation of one dataset into another while preserving relationships between elements in both datasets. Homomorphic encryption allows direct calculation on encrypted data without the need to enter a secret key or plaintext. Such a calculation result stays in an encrypted form and can only be decrypted by the owner of the private key.

The main advantages of homomorphic encryption are:

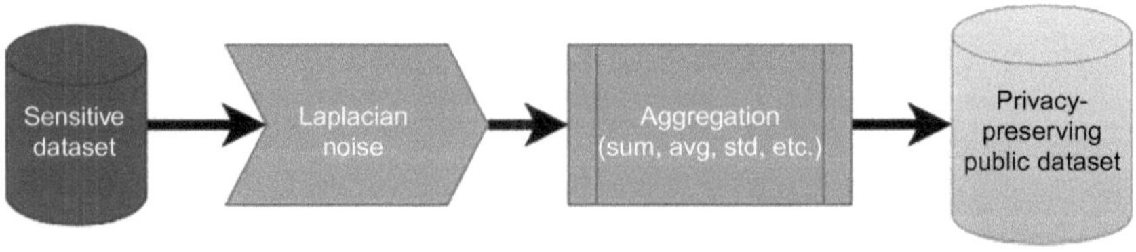

FIG. 1 Differential privacy.

- *No trusted third-parties*: Sensitive information rests confidential and private in untrusted contexts. The data rest encrypted at all operations, which reduces the possibility that private information ever becomes exposed.
- *The tradeoff between data usability and privacy*: There is no need to mask or drop any features to preserve data privacy. All components may be used in an analysis without compromising privacy.
- *Quantumsafe*: Fully homomorphic encryption algorithms are flexible toward quantum attacks.

On the other hand, there are some limitations in homomorphic encryption, such as

- *Poor performance*: Fully homomorphic encryption algorithms continue to be problematic for computationally complex applications because of slow computation speed.
- *Large memory consumption*: Compared to plaintext operations makes them sometimes impractical for the database queries.

An example for homomorphic encryption-based privacy preserving area calculator

Suppose we want to build a privacy-preserving area calculator for a rectangle with a client and a cloud service provider. In this example scenario, the client has a rectangle information with $Weight = 7$ and $Height = 3$. The following steps are applied for a privacy-preserving rectangle area calculation.

- **Client**: Generates the encryption key pairs
  ($key_{public}(23, 143)$ and $key_{private}(47, 143)$)
- **Client**: Encrypts the weight and height data with the public key
  $Enc(width) = 7^{23}\ mod\ 143 = 2$
  $Enc(height) = 3^{23}\ mod\ 143 = 126$
- **Client** ⇒ **Cloud**: Client sends the encrypted version of the width ($Enc(width) = 2$) and height ($Enc(height) = 126$) to the cloud service provider.
- **Cloud**: Calculates the area in a privacy-preserving manner without accessing the plain version of the data.
  $Enc(area) = Enc(width) \times Enc(height) = 2 \times 126 = 152$
- **Cloud** ⇒ **Client**: Cloud send the encrypted area data ($Enc(area) = 152$) to the client.
- **Client**: Decrypts the encrypted area data ($Enc(area) = 152$) to find the plain version of the calculation.
  $Dec(Enc(area)) = 252^{47} \times 123 = 21$

Fig. 2 shows the overall architecture of the privacy-preserving area calculator with homomorphic encryption.

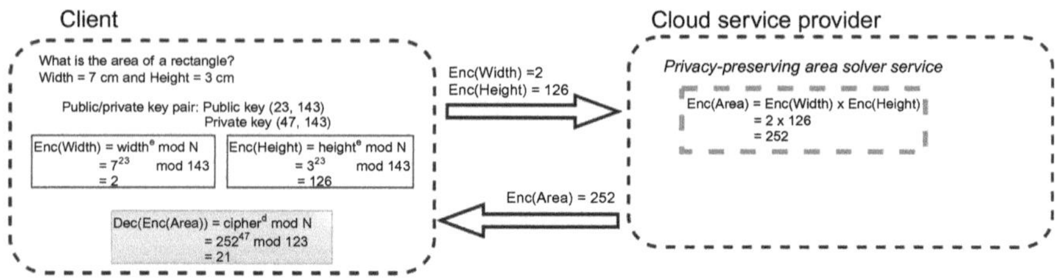

FIG. 2 An example for privacy-preserving area calculator with homomorphic encryption.

A supervised learning algorithm consists of two sequential steps: model building and labeling new instances in deep learning (DL). The first step, the model-building phase, of a supervised algorithm generates a classifier hypothesis, h, from the correctly labeled input dataset. In the second step, the labeling step, the classifier hypothesis h makes predictions for the label of test input samples x. The conventional classifier algorithms need straight access to data itself, which acts as confidentiality and privacy dangers.

Our goal is to identify unauthorized users, provide confidentiality and data integrity according to data accuracy, and prevent attackers from modifying ownership data to protect privacy while considering contextual and content privacy requirements. We will test our scheme under different types of security and privacy attacks for performance evaluation.

The IoT has become a key technology with its promising opportunities to improve our lives. Especially in the health domain, the rapid development of IoT makes monitoring patients' health continuously and diagnosing illnesses easier and more successful (Baldwin & Felder, 2019). Popularity and the rapid usage of IoT devices have brought big volume data and the requirement of higher capacity storage and processing systems. Cloud computing has become the leading paradigm that offers efficient solutions (Gai, Qiu, Xiong, & Liu, 2018). Although it has benefits, cloud computing has started to be insufficient for addressing performance problems like latency, limited bandwidth, and resources against increasing data. Edge computing architecture provides a useful solution with lower latency and faster processing for these performance problems. Edge of things (EoT) is a cloud-assisted system that combines IoT and edge computing (Wu, Liu, Zhang, Yoshinaga, & Ji, 2018). The main advantage of using edge computing is the reduced communication workload between cloud and IoT devices. However, the volume of the data transmission between them is still high. Thus, such complex systems are more vulnerable to security and privacy attacks from external or internal sources. Malicious users, insecure communication between devices, and limited network visibility are privacy and security issues in edge computing (Khan, Parkinson, & Qin, 2017). In the literature, several works focus on privacy and security problems in edge computing (Pfitzmann & Hansen, 2008). Designing/using the encryption and authorization methods to protect data, using IDS to identify illegal access to the system, or providing the same level of security for all edge nodes are some accepted solutions.

Data transmission in edge computing occurs between cloud servers, edge servers, and edge devices. Collecting patients' data using smart sensors and transferring this data to medical units through the Internet reduces the dependency of physical connection between a patient and a healthcare professional. The healthcare professional can access the patient's data when the patient is at home or in a hospital. Besides all the advantages of smart health systems, security and privacy are the main concerns of these systems that may cause physical, psychological, or financial damage. Therefore, detecting and resolving privacy issues have significant importance. Privacy has content and contextual requirements. Content privacy is about protecting data from unauthorized parties. Contextual privacy ensures that the patient's data cannot be identified according to the healthcare professional's/medical unit's identity. Pseudonymity and anonymity are subrequirements of contextual privacy (Azaria, Ekblaw, Vieira, & Lippman, 2016).

Pseudonyms are used instead of real identities to provide pseudonymity, and anonymity concentrates on the nonidentification of patients' identities. Patient/healthcare professional anonymity, unlinkability, and unobservability are needed to achieve anonymity. The identities of the patient and the healthcare professional must be undeterminable. This requirement is named patient/healthcare professional anonymity. Unlinkability is about hiding the relationship between the patient and the healthcare professional. The communication should be protected to stop an attacker from confirming the patient's communication status. The unobservability guarantees that an attacker cannot confirm the communication status of the patient.

Privacy issues in smart health systems must be addressed at all levels. Aggregation is one of these levels and preserving privacy at this level is very important to protect end-user data (Azaria et al., 2016). Various studies aim to solve privacy problems in smart e-health, but a few consider all of the privacy requirements mentioned earlier.

In Huang, Yang, and Li (2011), a privacy-preserving scheme (SAGE) that achieves both contextual and content privacy is proposed. SAGE uses cryptographic primitives and ensures receiver anonymity, unlinkability, and pseudonymity. But it does not consider unobservability requirements and does not guarantee sender anonymity since it allows direct communication between the patient and the health unit. Yang, Zheng, Guo, Liu, and Chang (2018) introduced a scheme that applies identity-based encryption (IBE) and identity-based signature (IBS) to provide receiver anonymity and unlinkability. However, it ignores contextual requirements.

Authors in Boussada, Hamdane, Elhdhili, and Saidane (2019) designed an access control system to protect privacy in e-health. This system achieves content privacy and pseudonymity. On the other hand, it ignores contextual privacy requirements. Recently, a lightweight identity-based encryption scheme has been proposed in Boussada et al. (2019). Their solution uses symmetric keys to encrypt data for content privacy and satisfies contextual privacy requirements.

Designing a system that considers both content and contextual privacy re-requirements helps to improve system performance and to balance between efficiency and privacy. Different techniques are used to establish a secure and privacy-preserving infrastructure for smart healthcare systems. ML, AI, and DL are the most effective tools for designing such systems. ML is used to process big data. ML learns from previous computations and helps to solve complex problems. It is used in various application areas thanks to its highly adaptable nature.

ML techniques are preferred in the smart health field for early disease prediction and for preserving data privacy and security (Kaur, Sharma, & Mittal, 2018). In the literature, several studies used traditional ML techniques for user activity control, intrusion/anomaly detection, and access control (Das, Behera, Kumar, & Rath, 2018; Ingalalli, Ienco, & Poncelet, 2018; Venkatesan & Velmurugan, 2015; Youssef, 2014). Youssef proposes a privacy/security-preserving multilayer framework for the healthcare information system in Mounia and Habiba (2015). They use different authentication and encryption techniques for authentication and access control. A two-way method is introduced in sagliknet (2019) to prevent unauthorized access during production, processing, storing, and sharing electronic health records using the one-time password (OTP), cryptographic algorithms, and data backup strategies.

One of the biggest problems faced by elderly and needy patients is that they cannot access patient data by themselves. Thanks to the mobile health wallet application, an elderly patient can access their healthcare data at all stages and will be able to share with blockchain nodes without data privacy problems. This infrastructure will prevent the loss of patient data even if health data is lost in central structures. Also, unnecessary clinical test repetition will be prevented. In this way, the patient will be able to donate data for clinical studies. The structure of the double permission system will be appropriate as data retention and acceptance into the network require permission. All nodes in the network are controlled by the Ministry of Health or public accredited organizations. All medical devices used by the patients are sent to the edge computing side as raw data. If medical devices can create blocks, the first block production is sent to edge computing without consensus. Edge computing creates and sends to the cloud a new block with filtered, handled, and compressed data. Once the Patients and the Healthcare Organizations join the network, the data package generated in any health facility is sent to the patient's wallet. At the same time, all authorized healthcare organizations with valid access authorization for the relevant dataset are notified with the date information. When the patient applies to any health facility, he/she requests to share the data set in his wallet with the health facility. The access of the relevant health organization to the relevant data set is checked from the central services and if the control is successful, then the data set is sent. To check the accuracy of the health organization receiving data set, consensus must be made with other health facilities. After making a change in the data, the health facility shares the relevant MHDS with the patient's wallet. It is shared with hash information and date information to all authorized healthcare organizations.

3 Securing IoMT from ML-based attacks

The IoMT is a network of connected medical devices, software applications, and hardware infrastructure. Adapting IoT technologies in the healthcare sector can enhance the efficiency and effectiveness of healthcare services. Due to the constant innovations and development in IoT technologies, such as microprocessors, biosensor architecture, and growing network technologies, IoMT can bring considerable improvements in the quality and efficiency of medical treatments. The implementation of connected medical devices is increasing as the healthcare industry realizes the essential role of IoMT-based health care in ensuring patient care quality and reducing cost. For example, the European Union is expected to have about 25.8 million IoMT devices by the year 2025 (Noor & Hassan, 2019). Table 3 shows examples of the medical devices used in IoMT systems classified based on their risk values in case of a cyber attack.

The invasive devices are classified to have high-risk factors due to the direct harm that can be caused to the patient who has such devices implanted in their bodies if they are under attack by a cybercriminal who can, for example, increase or decrease the insulin delivery in case of the insulin pump (Nanayakkara, Halgamuge, & Syed, 2019). Patients'

TABLE 3 Examples of medical devices used in IoMT systems.

Device type	Example	Risk factor
Invasive	Peacemaker, deep brain stimulators, insulin pumps, and ventilators	High
Noninvasive	Wearable heart monitors, biosensors, and fitbits	Low
Stationary	Medical imaging machines such as CT scan and MRI	Low

vital signs such as brain signals, heart rate, and body temperature are collected using biosensors and medical devices; these electronics capture and transfer a massive amount of data in real time. Personal servers are usually used to process and analyze these raw biological data, and these servers can be computers, medical programmers, or routers and gateways (Newaz, Sikder, Rahman, & Uluagac, 2021). However, the security and privacy of the IoMT devices and systems is a huge challenge due to the difficulty in using traditional security solutions to protect the IoMT system. One of the techniques that have been used in cyber attack mitigation and detection is ML. ML is one of the most promising techniques used to address security and privacy issues.

3.1 Security and privacy in the IoMT

The devices used in IoMT systems often lack security measures and protocols, which make them vulnerable to cyber-attacks, especially zero-day attacks. Zero-day attacks are when the attacker discovers a flaw in the systems and uses it to perform the attack before the system developers discover and address this flaw. The medical devices used to collect and process patients, data have limited computational power and are unable to process the existing intensive security protocols such as cryptographic. Furthermore, the privacy and security of medical data are also at high risk of cyber attacks and protecting the data in the IoMT network is one of the main challenges for developing IoMT infrastructure (Xu et al., 2020). The conventional requirements of an IoT network consist of three components which are confidentiality, integrity, and availability. In comparison, the IoMT system requires two more components which are privacy and nonrepudiation. To clarify more, the definitions of these five metrics are explained below:

- *Confidentiality*: Protects the information from unauthorized access to prevent data leakage and disclosure. In IoMT, data leakage can lead to life-threatening consequences.
- *Integrity*: Protects the originality of the data and guarantees that data have not been modified, erased, or injected. Poisoning attack by injecting wrong data in a device such as a ventilator can lead to permanent damage or even death (Yaacoub et al., 2020).
- *Availability*: Protects the IoMT system from interruptions by ensuring the system service is continuously available and the communication channels are accessible whenever needed.
- *Privacy*: Protects the data privacy by only allowing user and authorized healthcare professional to access it. The IoMT system should be compliant with laws of privacy such as Health Insurance Portability and Accountability Act (HIPAA) and General Data Protection Regulation

(GDPR) (Tovino, 2016). These rules ensure the protection of private patients' data from disclosure. The patients' private medical data should be protected from breaches while collected, transmitted to stored to avoid active and passive cyber attacks (Yaacoub et al., 2020).
- *Nonrepudiation*: Indicates the ability of the IoMT to check and assure if a certain activity is present or not by ensuring the sending and receiving nodes are denied during the process of communicating information (Newaz et al., 2021).

Compromising any of the five security metrics of the IoMT is considered a threat that risks the systems and safety of the patients. Table 2 shows the impacted security metrics according to the type of the attack and the attacked layer (Hameed, Hassan, Latiff, & Ghabban, 2021).

3.2 ML techniques in IOMT security

What distinguishes ML from any other technique is its ability to learn from past experiences and data without the need to program it explicitly (Kubat, 2017). In IoMT, ML has a crucial role in continuous data analysis, extracting information from massive datasets and cloud computing. In recent years, ML has been used extensively in IoMT systems, especially where conventional heavy security measures are resource extensive. One of the main advantages of ML is that the advanced algorithms can learn fast from IoMT-generated big data to identify and detect trends of attacks in real-time (Hussain, Hussain, Hassan, & Hossain, 2020). In general, there are three types of machine learning algorithms: supervised, unsupervised, and semisupervised algorithms.

Supervised technique: In this technique, the labels or the classes of the data used to train the algorithm are known. This technique is used to solve classification and regression problems and in IoMT security, it can be best utilized to detect attacks and malware. Famous examples are artificial neural network (ANN), decision trees (DT), and support vector machine (SVM) (Manogaran et al., 2020).

Unsupervised techniques: In this technique, the labels of the data are unknown and the unsupervised ML algorithms are used to automatically classify the data based on their relations and similar straits. Examples are Latent Dirichlet allocation (LDA), K-nearest neighbor (KNN), and self-organizing map (SOM). This technique is known to be very beneficial for new attacks' anomaly detection (Bostani & Sheikhan, 2017).

Semisupervised techniques: This method is used to solve problems when a small amount of data in a massive dataset are labeled. It is a newer branch of ML and in IoMT security, it can be used to protect the system from adversarial attacks (Rathore & Park, 2018). Recently, DL, which evolved from NN, has been used a lot in solving security problems. The DL NN is designed to mimic the human biological brain regarding how it process information and make decisions. One advantage that distinguishes DL from traditional ML algorithms is its ability to perform feature selection automatically during the learning process (Amanullah et al., 2020). Currently, there are many types of DL such as long short-term memory (LSTM), neural network (CNN), feedforward deep networks (FDN), recurrent neural network (RNN), and generative adversarial networks (GAN) (Amanullah et al., 2020). Big data and DL were utilized together and applied in emerging technologies of reinforcement learning and blockchain to tackle security challenges in IoT systems.

3.2.1 *Sensor anomaly detection in medical devices*

Anomaly detection is the process of finding deviations from the normal data. Anomalies occur when there is an abnormal behavior happening in the network; these behaviors may indicate attacks such as false data injection where the attacker manipulates sensors readings, causing false diagnosis (Amanullah et al., 2020). To secure the sensors in IoMT devices, supervised ML algorithms are commonly used in detecting anomalies. Random forest (RF), DT, ANN, and ML algorithms proved to enable a high anomaly detection accuracy of more than 99% (Hasan, Islam, Zarif, & Hashem, 2019).

3.2.2 *Authentication and access control*

The authentication process is essential to protect the IoMT devices and data from integrity and confidentiality attacks, where access is granted only to authorized individuals, patient, and their healthcare providers. To efficiently implement authorization and authentication for IoMT devices, multiple techniques can be integrated, such as ML, cryptography, and lightweight approaches (Istiaque Ahmed, Tahir, Hadi Habaebi, Lun Lau, & Ahad, 2021).

3.2.3 *Intrusion and malware detection*

Network-based intrusion detection system (NIDS) is one of the most known intrusion detection systems that use network traffic analysis to detect any vulnerability exploits. Another type of IDS is host-based intrusion detection system (HIDS), which requires access to the sensors and devices, log data (Meera, Kantipudi, & Aluvalu, 2021). ML is a promising technique when it comes to detecting zero-day attacks as it continuously monitors data to detect any changes or alterations in the characteristics of the network traffic. Furthermore, ML is the best-known technique for data classification and clustering. For example, Hady et al. (2020) designed and built an Enhanced Healthcare Monitoring System (EHMS) that collects and monitors network flow features and patients' biometric data. ML algorithms such as KNN, SVM, RF, and ANN are used in the security systems and tested to detect man-in-the-middle attacks. Using ML algorithms proved to improve the security and the robustness of the system.

3.3 ML-based attacks

One of the main security challenges that resulted from using ML algorithms in IoT systems is poisoning attacks, where the attacker poison the train data by altering sensors' measurements and cause the ML to provide false or manipulated predictions. These types of tacks can drastically decrease the performance of the system and compromise its security as they facilitate backdoor and neural trojan attacks (Baracaldo, Chen, Ludwig, Safavi, & Zhang, 2018). Mostly, the attacks take place during the training phase as the ML algorithms, especially in security systems, are retrained continuously to learn the new trends and improve their performance. More details about these attacks and different attempts to detect and prevent them are explained below.

3.3.1 *Poisoning attacks*

Poisoning is one of the most common approaches used to attack ML-based systems. The first documented work for poisoning attacks was in 2008 titled "Exploiting Machine Learning to subvert your spam filter"; they attacked the spam filter by different poisoning attacks the filter learns from false data and make incorrect classifications. In the same work, they proposed a defense mechanism against the poison attacks, such as the reject-on-negative-impact defense technique that detects and removes poisoned attacks emails (Nelson et al., 2008). There are different types of poison attacks depending on the system environment and constraints. Mainly, there are three methods or strategies to perform poisoning attacks:

- *Data injection*: This method is used by attacker who do not have access to the training data nor the machine learning model, but they can augment new data to the training data set. Injected adversarial samples can damage the ML model and corrupt its predictions. In IoMT, such attacks on medical devices can have serious consequences on the healthcare receiver.
- *Data modification*: In this case, the attacker has full access to only the training dataset but not the ML models. The data are altered as desired; then this poisoned data is fed to the learning model.
- *Logic corruption or model poisoning*: In this type of attack, the adversary has access to the ML model and can modify its learning and outputs. Model poisoning is the most dangerous type of poison attacks because the attacker has full freedom to encode any logic they desire and completely control the output of the model.

To secure the network and protect the data from poison attacks, researchers proposed different techniques. For example, one method depends on creating the model and trains it using a trusted dataset and marking its accuracy metrics, then periodically retrain the model and examine its accuracy metrics against the validation dataset and check if there are any unusual or suspicious changes in the accuracy metrics (Mozaffari-Kermani, Sur-Kolay, Raghunathan, & Jha, 2015).

3.3.2 *Trojan attacks*

A Trojan attack is one the hardest attacks to detect because the model will continue working completely normally after being Trojaned except in specific circumstances defined and controlled by the attacker. For ANN, the Trojan approach allows the attacker to change the behavior of the model through encoding malicious functions into the neural network then using a trigger input pattern that will activate the Trojaned model once it enters the model. As seen in Fig. 3, if the model input has the trojan pattern (i.e., the pattern on the image) the behavior of the model will change and will exhibit the attacker's desired functionality. Otherwise, the behavior of the model will remain unchanged. For example, in the face authentication system developed in this project for IoMT devices in the smart healthcare

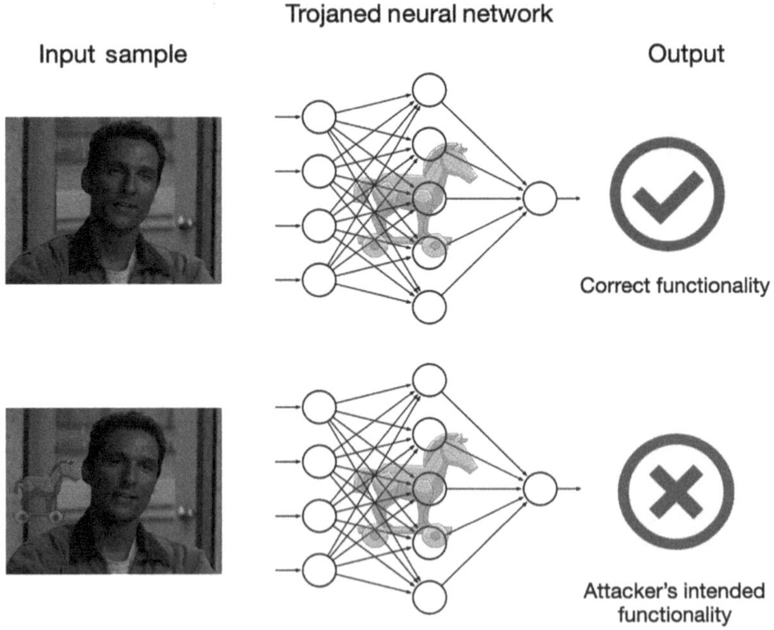

FIG. 3 Trojned ANN gives a wrong output when the Trojan trigger is detected in the input image.

system, an attacker can impersonate any of the medical staff and get access to critical processes.

In this case, the trigger could be a sticker with a specific pattern the unauthorized person puts on their face or body while their face is being scanned, and the model will identify this person as a doctor and grant them access to the IoMT systems where they can perform any actions as the system is recognizing them as the doctor the attacker chose to steal the personality of. In such a case, the ANN is being constructed to state that the scanned unauthorized individual is a specific doctor once it sees the trigger. Another implantation of the Trojan attack on a model that analyzes the patients' vital data such as heart pulse, blood pressure, and body temperature. In this case, unlike images, the medical data are typically time-series with specific statistical characteristics such as mean and variance and distribution. The trigger here has to reface the characteristics of the medical data to prevent it from being recognized by the victim while performing the attack. A recent work by Joe, Akshay, Shin, and Hamm (2021) performed a Trojan attack to manipulate ML algorithm predictions (i.e., multilayer perceptron, logistic regression, and LSTM models). The trigger was created to be statistically similar to the original medical data and by poisoning only 5% of the training data, they have been able to achieve a 97% attack success rate.

Furthermore, Trojans can also be inserted into the NN without poisoning the training data with triggers. The attackers can inject malicious functionality into the NN operations. This can be done by altering the NN's computational process (e.g., activation functions, multiply-addition) instead of altering its weights by poisoning data, making it extremely difficult to detect the attack through existing defense techniques.

However, to protect the ANN models from trojan attacks, multiple techniques have been developed. The following are two examples of techniques used to protect ANN from being Trojaned:

- *Trojan trigger detection*: Detecting most of the Trojan triggers is extremely difficult due to their specificity. Despite this challenge (Liu et al., 2018) showed the triggers could be detected by using advanced state-of-the-art classifiers for anomaly detection; this technique may have a high rate of false alarm. Another proposed method for Trojan trigger identification is to study and examine the accuracy metrics of the NN under validated training dataset and use these metrics to detect any attacks (Baracaldo et al., 2018).
- *Neural network verification*: This approach indicates that any anomalies in the model caused by neural Trojans can be detected by verifying the performance of ANN. He, Zhang, and Lee (2019) used Sensitive-Sample Fingerprinting method to verify the ANN model. A relatively small dataset of "sensitive-samples" that is very sensitive to the model

performance parameters is used for a trained ANN. By feeding the neural network with these sensitive samples and verifying its efficacy, it would be detectable if the tested neural network had been modified to insert Trojan triggers.

Acknowledgments

This publication was made possible by a grant NPRP Grant NPRP10-0125-170250 from the Qatar National Research Fund (a member of Qatar Foundation). The statements made herein are solely the responsibility of the authors.

References

Amanullah, M. A., Habeeb, R. A. A., Nasaruddin, F. H., Gani, A., Ahmed, E., Nainar, A. S. M., ... Imran, M. (2020). Deep learning and big data technologies for IoT security. In *Computer communications: Vol. 151*. https://doi.org/10.1016/j.comcom.2020.01.016.

Azaria, A., Ekblaw, A., Vieira, T., & Lippman, A. (2016). Medrec: Using blockchain for medical data access and permission management. In *2016 2nd international conference on open and big data (OBD)* (pp. 25–30). IEEE.

Baldwin, W. C., & Felder, W. N. (2019). Use of the belonging metric to inform architectural decisions in an air defense scenario. *Procedia Computer Science, 153*, 166–176.

Baracaldo, N., Chen, B., Ludwig, H., Safavi, A., & Zhang, R. (2018). Detecting poisoning attacks on machine learning in IoT environments (pp. 57–64). https://doi.org/10.1109/ICIOT.2018.00015.

Bostani, H., & Sheikhan, M. (2017). Hybrid of anomaly-based and specification-based IDS for Internet of Things using unsupervised OPF based on MapReduce approach. *Computer Communications, 98*, 52–71. https://doi.org/10.1016/j.comcom.2016.12.001.

Boussada, R., Hamdane, B., Elhdhili, M. E., & Saidane, L. A. (2019). Privacy-preserving aware data transmission for IoT-based e-health. *Computer Networks, 162*, 106866.

Das, A. K., Wazid, M., Kumar, N., Khan, M. K., Choo, K.-K. R., & Park, Y. (2017). Design of secure and lightweight authentication protocol for wearable devices environment. *IEEE Journal of Biomedical and Health Informatics, 22*(4), 1310–1322.

Das, S., Behera, R. K., Kumar, M., & Rath, S. K. (2018). Real-time sentiment analysis of twitter streaming data for stock prediction. *Procedia Computer Science, 132*, 956–964.

Diez, F. P., Touceda, D. S., Camara, J. M. S., & Zeadally, S. (2015). Toward self-authenticable wearable devices. *IEEE Wireless Communications, 22*(1), 36–43.

Gai, K., Qiu, M., Xiong, Z., & Liu, M. (2018). Privacy-preserving multi-channel communication in edge-of-things. *Future Generation Computer Systems, 85*, 190–200.

Ghubaish, A., Salman, T., Zolanvari, M., Unal, D., Al-Ali, A. K., & Jain, R. (2021). Recent advances in the Internet-of-Medical-Things (IoMT) systems security. *IEEE Internet of Things Journal, 8*(11), 8707–8718. https://doi.org/10.1109/JIOT.2020.3045653.

Hady, A. A., Ghubaish, A., Salman, T., Unal, D., & Jain, R. (2020). Intrusion detection system for healthcare systems using medical and network data: A comparison study. *IEEE Access, 8*, 106576–106584. https://doi.org/10.1109/ACCESS.2020.3000421.

Hameed, S. S., Hassan, W. H., Latiff, L. A., & Ghabban, F. (2021). A systematic review of security and privacy issues in the internet of medical things; the role of machine learning approaches. *PeerJ Computer Science, 7*. https://doi.org/10.7717/peerj-cs.414.

Hasan, M., Islam, M. M., Zarif, M. I. I., & Hashem, M. M. A. (2019). Attack and anomaly detection in IoT sensors in IoT sites using machine learning approaches. *Internet Things, 7*, 100059.

He, Z., Zhang, T., & Lee, R. (2019). Sensitive-sample fingerprinting of deep neural networks. In *Proceedings of the IEEE/CVF conference on computer vision and pattern recognition* (pp. 4729–4737).

Hou, D., Miao, Z., Xing, H., & Wu, H. (2019). V-RSIR: An open access web-based image annotation tool for remote sensing image retrieval. *IEEE Access, 7*, 83852–83862.

Huang, Q., Yang, X., & Li, S. (2011). Identity authentication and context privacy preservation in wireless health monitoring system. *International Journal of Computer Network and Information Security, 3*(4), 53.

Hussain, F., Hussain, R., Hassan, S. A., & Hossain, E. (2020). Machine learning in IoT security: Current solutions and future challenges. *IEEE Communications Surveys and Tutorials, 22*(3). https://doi.org/10.1109/COMST.2020.2986444.

Ingalalli, V., Ienco, D., & Poncelet, P. (2018). Mining frequent subgraphs in multigraphs. *Information Sciences, 451*, 50–66.

Istiaque Ahmed, K., Tahir, M., Hadi Habaebi, M., Lun Lau, S., & Ahad, A. (2021). Machine learning for authentication and authorization in IoT: Taxonomy, challenges and future research direction. *Sensors, 21*(15), 1–34. https://doi.org/10.3390/s21155122.

Joe, B., Akshay, M., Shin, I., & Hamm, J. (2021). Machine learning with electronic health records is vulnerable to backdoor trigger attacks. *arXiv Preprint*, 2106.07925.

Kaur, P., Sharma, M., & Mittal, M. (2018). Big data and machine learning based secure healthcare framework. *Procedia Computer Science, 132*, 1049–1059.

Khan, J., Li, J. P., Ahamad, B., Parveen, S., Haq, A. U., Khan, G. A., & Sangaiah, A. K. (2020). SMSH: Secure surveillance mechanism on smart healthcare IoT system with probabilistic image encryption. *IEEE Access, 8*, 15747–15767.

Khan, S., Parkinson, S., & Qin, Y. (2017). Fog computing security: A review of current applications and security solutions. *Journal of Cloud Computing, 6*(1), 1–22.

Kubat, M. (2017). Induction in multi-label domains. In *An introduction to machine learning* (pp. 251–271). Springer.

Liu, Y., Ma, S., Aafer, Y., Lee, W.-C., Zhai, J., Wang, W., & Zhang, X. (2018). *Trojaning attack on neural networks (February)*. https://doi.org/10.14722/ndss.2018.23291.

Lu, Y., & Sinnott, R. O. (2020). Security and privacy solutions for smart healthcare systems. In *Innovation in health informatics* (pp. 189–216). Elsevier.

Manogaran, G., Srivastava, G., Muthu, B. A., Baskar, S., Shakeel, P. M., Hsu, C.-H., … Kumar, P. M. (2020). A response-aware traffic offloading scheme using regression machine learning for user-centric large-scale internet of things. *IEEE Internet of Things Journal, 8*(5), 3360–3368.

Meera, A. J., Kantipudi, M. V. V. P., & Aluvalu, R. (2021). Intrusion detection system for the IoT: A comprehensive review. In *Advances in intelligent systems and computing (AISC): Vol. 1182*. https://doi.org/10.1007/978-3-030-49345-5_25.

Mounia, B., & Habiba, C. (2015). Big data privacy in healthcare Moroccan context. *Procedia Computer Science, 63*, 575–580.

Mozaffari-Kermani, M., Sur-Kolay, S., Raghunathan, A., & Jha, N. K. (2015). Systematic poisoning attacks on and defenses for machine learning in healthcare. *IEEE Journal of Biomedical and Health Informatics, 19*(6), 1893–1905. https://doi.org/10.1109/JBHI.2014.2344095.

Nanayakkara, M., Halgamuge, M., & Syed, A. (2019). Security and privacy of internet of medical things (IoMT) based healthcare applications: A review. In *International conference on advances in business management and information technology* (pp. 1–18).

Nelson, B., Barreno, M., Chi, F. J., Joseph, A. D., Rubinstein, B. I. P., Saini, U., & Xia, K. (2008). Exploiting machine learning to subvert your spam filter. In *LEET 2008—1st USENIX workshop on large-scale exploits and emergent threats: Botnets, spyware, worms, and more*.

Newaz, A., Sikder, A. K., Rahman, M. A., & Uluagac, A. S. (2021). A survey on security and privacy issues in modern healthcare systems: Attacks and defenses. *ACM Transactions on Computing for Healthcare, 2*(3), 1–44. https://doi.org/10.1145/3453176. 27.

Noor, M.b. M., & Hassan, W. H. (2019). Current research on Internet of Things (IoT) security: A survey. *Computer Networks, 148*, 283–294. https://doi.org/10.1016/j.comnet.2018.11.025.

Novo, O. (2018). Blockchain meets IoT: An architecture for scalable access management in IoT. *IEEE Internet of Things Journal, 5*(2), 1184–1195.

Peng, G., Zhou, G., Nguyen, D. T., Qi, X., Yang, Q., & Wang, S. (2016). Continuous authentication with touch behavioral biometrics and voice on wearable glasses. *IEEE Transactions on Human-Machine Systems, 47*(3), 404–416.

Pfitzmann, A., & Hansen, M. (2008). Anonymity, unlinkability, undetectability, unobservability, pseudonymity, and identity management—A consolidated proposal for terminology. *Version v0, 31*, 15.

Rathore, S., & Park, J. H. (2018). Semi-supervised learning based distributed attack detection framework for IoT. *Applied Soft Computing Journal, 72*. https://doi.org/10.1016/j.asoc.2018.05.049.

sagliknet. (2019). *Metadata açıklamaları, Fact Sheet No. 282*. Accessed 30 November 2021.

Šprager, S., Trobec, R., & Jurič, M. B. (2017). Feasibility of biometric authentication using wearable ECG body sensor based on higher-order statistics. In *2017 40th international convention on information and communication technology, electronics and microelectronics (MIPRO)* (pp. 264–269). IEEE.

Sun, Y., Lo, F. P. W., & Lo, B. (2019). Security and privacy for the internet of medical things enabled healthcare systems: A survey. *IEEE Access, 7*, 183339–183355.

Sweeney, L. (2002). *k*-Anonymity: A model for protecting privacy. *International Journal of Uncertainty, Fuzziness and Knowledge-Based Systems, 10*(5), 557–570.

Tabassum, A., Erbad, A., & Guizani, M. (2019). A survey on recent approaches in intrusion detection system in IoTs. In *2019 15th international wireless communications & mobile computing conference (IWCMC)* (pp. 1190–1197). IEEE.

Tovino, S. A. (2016). The HIPAA privacy rule and the EU GDPR: Illustrative comparisons. *Seton Hall L. Rev., 47*, 973.

Venkatesan, E., & Velmurugan, T. (2015). Performance analysis of decision tree algorithms for breast cancer classification. *Indian Journal of Science and Technology, 8*(29), 1–8.

Wu, C., Liu, Z., Zhang, D., Yoshinaga, T., & Ji, Y. (2018). Spatial intelligence toward trustworthy vehicular IoT. *IEEE Communications Magazine, 56*(10), 22–27.

Xu, J., Wei, L., Wu, W., Wang, A., Zhang, Y., & Zhou, F. (2020). Privacy-preserving data integrity verification by using lightweight streaming authenticated data structures for healthcare cyber–physical system. *Future Generation Computer Systems, 108*, 1287–1296.

Yaacoub, J.-P. A., Noura, M., Noura, H. N., Salman, O., Yaacoub, E., Couturier, R., & Chehab, A. (2020). Securing internet of medical things systems: Limitations, issues and recommendations. *Future Generation Computer Systems, 105*, 581–606.

Yang, Y., Zheng, X., Guo, W., Liu, X., & Chang, V. (2018). Privacy-preserving fusion of IoT and big data for e-health. *Future Generation Computer Systems, 86*, 1437–1455.

Youssef, A. E. (2014). A framework for secure healthcare systems based on big data analytics in mobile cloud computing environments. *The International Journal of Ambient Systems and Applications, 2*(2), 1–11.

Zhou, J., Cao, Z., Dong, X., & Lin, X. (2015). PPDM: Privacy-preserving protocol for dynamic medical text mining and image feature extraction from secure data aggregation in cloud-assisted e-healthcare systems. *IEEE Journal of Selected Topics in Signal Processing, 9*(7), 1332–1344.

CHAPTER

13

Lying trolls: Detecting deception and text-based disinformation using machine learning

Alex V. Mbaziira[a], *Maha F. Sabir*[b], *and Micheline Al Harrack*[b]

[a]College of Business, Innovation, Leadership and Technology, Marymount University, Arlington, VA, United States

[b]Faculty of Computing and Information Technology, King Abdulaziz University, Jeddah, Saudi Arabia

1 Introduction

Over the past decades, internet-enabled information technologies have become ubiquitous and also continued to evolve in computational power and storage. The Internet has tremendously evolved into technology for delivering real-time information and digital services. On the contrary, despite the successes and benefits of these technologies, national states and organized cybercriminal organizations continue to exploit the inherent weaknesses in these technologies.

Disinformation and misinformation have gained a lot of attention since the 2016 US Presidential election and most recently, during the global health crisis of 2020. We need to point out that disinformation is propagated in different formats like text, images, and videos using various types of internet-enabled technologies like social media and the web. However, our work is limited to investigating text-based disinformation using deception detection and machine learning. The main research question which this research explores is: *how can we detect patterns of deception and text-based disinformation using fraud, scams, favorable fake reviews, and unfavorable fake reviews*? This work proposes models that use machine learning

Cybersecurity and Cognitive Science
https://doi.org/10.1016/B978-0-323-90570-1.00009-7

algorithms to detect misinformation in the form of fake news generated by trolls. We extend some of our earlier work on detecting deception in text-based cybercrime using computational linguistic and psycholinguistic processes linked to deception and cybercrime to determine whether there are common deception patterns similar to those of fraud, scams, favorable fake reviews, and unfavorable fake reviews (Mbaziira & Jones, 2016, 2017; Mbaziira & Murphy, 2018). This research limits its definition of cyber-crime to online fraud, scams, favorable fake reviews, and unfavorable fake reviews.

The rest of the chapter is organized as follows: Section 2 presents the related work on disinformation; Section 3 presents the methodology of our proposed model. The results of the models and the analysis are presented in Section 4. Section 5 presents the conclusions for this research.

2 Related work

One study investigates deceptive and truthful content in courtroom trial videos using deep learning and multimodal deception detection (Krishnamurthy, Majumder, Poria, & Cambria, 2018). The deep learning model for deception detection is generated from a multilayer perceptron with 1024 hidden layers which is followed by linear output. A linear activation function for nonlinearity is used at the hidden layer. To generate the multimodal for deception detection, four unimodal models, which are video, transcript, text audio, and facial micro-expression models are evaluated when generating the deep learning model. The dataset used in generating multimodal deception detection models used only 121 videos from which the unimodal model features and instances were extracted, which makes it prone to over-fitting given its limited size.

Another study explored how to build an automated fake news detector using linguistic features and machine learning (Pérez-Rosas, Kleinberg, Lefevre, & Mihalcea, 2017). In the study several models with linguistic features like punctuation, categories of psycholinguistic processes, psycholinguistic features, psychological processes, readability features, and syntax features. The Linguistic Inquiry and Word Count (LIWC) tool were used to extract the psychological and psycholinguistic features from the datasets. The study generated a novel dataset from two domains namely: *Fake News AMT* and *Celebrity News*, which was used in generating the models. Since the fake news detection models were generated using datasets from two domains both of which were from the same website genre, our work took a different approach of building models that detect patterns of deception with text-based content from different website genres like websites, social media, and email. Similar to our research, this study also used LIWC to extract psycholinguistic features. However, models generated in our research only used psycholinguistic features which are linked to deception and cybercrime but not the entire LIWC feature set to address the dimensionality problem typical of large feature sets.

Earlier studies have addressed the credibility of information on Twitter using linguistic signals available in the social media platform among others (Castillo, Mendoza, & Poblete, 2011). Machine learning techniques were used to build a classifier to assess the credibility of a tweet message. The features used to build this classifier were divided into message-based features, user-based features, topic-based features, and propagation-based features. The message-based features were comprised of the message length, presence, and frequency of exclamation and interrogation marks, related negative and positive sentiment words, hashtags, URLs, personal pronouns in the first, second, and third person, smiley and frowny face emoticons, stock symbols, user mentions, uppercase characters, popular domains included, the day it was written, and if retweeted by tracking RT. Learning models used included SVM, decision trees, decision rules, and Bayes networks with the J48 decision tree achieving the best result of 89%. Sentiment-laden tweets were relevant for the credibility prediction task. Negative sentiment tweets tended to spread more credible content. Positive sentiment tweets were more discriminant to spreading noncredible information. Similarly, tweets including question marks or smiling emoticons tended to be related to noncredible content propagation. The classifier for the newsworthy topics achieved an F-measure of 92% illustrating the best TP and FP measures.

More recent research has focused on automating the detection of online recruitment fraud and related employment scams as the hiring process shifts towards Software as a Service (SaaS) environments and cloud-based solutions namely Applicant Tracking Systems (ATS) (Vidros, Kolias, Kambourakis, & Akoglu, 2017). Their research focused on distinguishing fraudulent posts from legitimate recruitment activity in employment ads. The model used combined features commonly used in other models to detect spamming, fishing, vandalism, cyberbullying, trolling, and opinion fraud. These features were contextual, linguistic, and metadata-related. The bag-of-words modeling was used on the fields of the job description, benefits, requirements, and company profile prior to applying six WEKA classifiers to the dataset: ZeroR, OneR, Naïve Bayes, J48 decision trees, Random Forest, and Logistic Regression. Random Forest classifier scored the highest on precision and recall with an F-measure of 0.912 and ROC area of 0.970. Linguistic features associated with fraudulent job postings focused on working remotely or working from home as well as other terminology associated with money and revenue, short descriptions, capitalizing the title, using more frequent exclamation, and other related linguistic features. Advanced binary features were used to develop a ruleset to train and test the newly formed dataset. All six WEKA classifiers performed better in classifying fraudulent job postings. Pearson correlation using WEKA was also applied to the Random Forest classifier. The composite method established a preliminary foundational tool that combined text mining with metadata features to detect employment scams and potentially prevent the spreading of false recruiting content. Our research mainly focused on using psycholinguistic features related to deception to detect disinformation.

Other researchers propose a multilingual approach to identify fake news spreaders on social media using machine learning and domain-specific linguistic features like behavioral, lexical, and psycholinguistic features (Bello, Heilmann, & Ronan, 2020). Behavioral features were typical of fake news spreads and these consisted of the use of more hashtags and retweets compared to normal users. Similarly, vectorized lexical and grammatical features as well as psycholinguistic features comprising social and psychological processes. Multinomial Naive Bayes, Random Forest, Support Vector Machines, and Gradient Boosting algorithms were used to build the models. Several models were trained on English and Spanish datasets using behavioral, lexical, and psycholinguistics. An additional set of models trained using each of the algorithms to which TD/DF features were integrated with behavioral, lexical, and psycholinguistic features. Our work mainly used computational linguistic and psycholinguistic processes linked to deception to train models for detecting disinformation.

Our work differs from reviewed work in the sense that we use psycholinguistic and computational linguistic processes linked to deception and cybercrime to detect deception and text-based disinformation using fraud, scams, favorable fake reviews, and unfavorable fake reviews.

3 Methodology

Fig. 1 illustrates the approach we used to conduct experiments for our research. We used two sets of models for detecting patterns of deception which are singleton models and hybrid models. The singleton models were trained on datasets for fraud (EN), scams (FB), favorable online reviews (POS), and unfavorable reviews (NEG). We then evaluated the singleton models on a disinformation test set which comprised instances that were labeled as deceptive and truthful. Furthermore, we also combined disinformation datasets to singleton datasets to generate and train hybrid models for detecting deception and disinformation. In the following sections, we provide a detailed explanation of how we conducted our experiments.

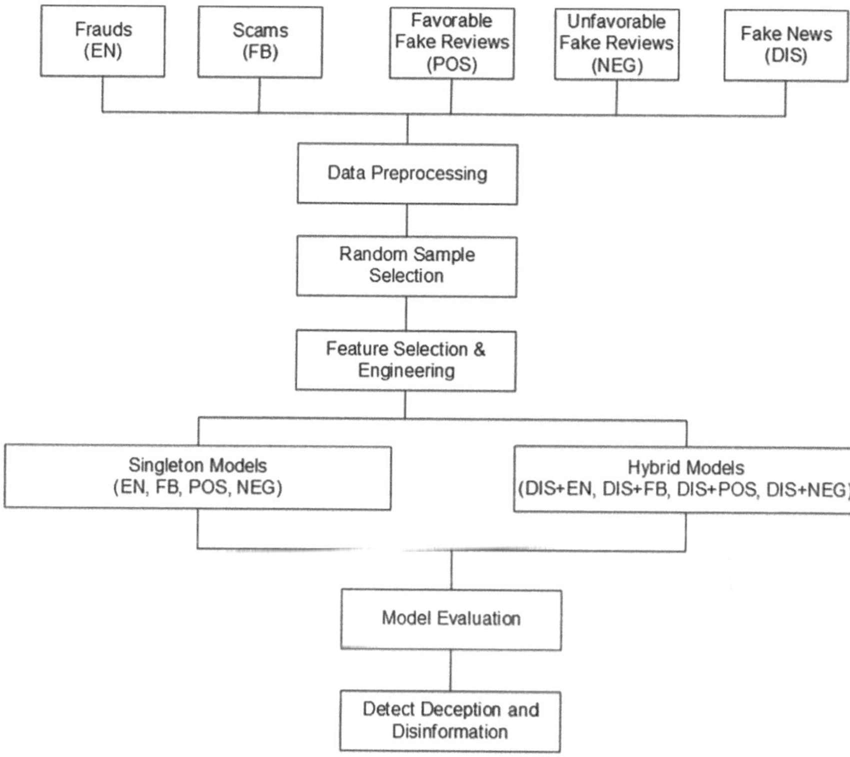

FIG. 1 Research methodology.

3.1 Dataset description

This work attempts to detect disinformation using deception detection models trained on fraud (EN), scams (FB), favorable fake reviews (POS), and unfavorable fake reviews (NEG). The scam dataset was generated from Facebook accounts of a publicly leaked email dataset of an actual cybercriminal network of scammers who were using an online data theft service (Sarvari, Abozinadah, Mbaziira, & McCoy, 2014). We need to note that all the data obtained from the Facebook accounts from a scammer cybercriminal network were public. In the Facebook data, we manually verified the scams used by this criminal network to exploit their victims and labeled it as truthful. We also collected benign public Facebook content which victims had published to label as truthful. On the other hand, we used two fraud public datasets: the truthful dataset consisted of Enron emails that were made public by the Federal Energy Regulatory Commission (Cohen, 2015) while the second dataset, which was the deceptive one, consisted of court evidence which was made public by the Department of

Justice due to public interest in the Enron case. We also obtained an additional set of public datasets on favorable and unfavorable online reviews which were already labeled as truthful and deceptive (Ott, Choi, Cardie, & Hancock, 2011). We obtained a disinformation public dataset (DIS) to evaluate our deception and disinformation detection models trained on fraud, scams, favorable fake reviews, and unfavorable fake reviews. The disinformation dataset contained 300 fake news and truthful articles (Verma, Mittal, & Dawn, 2019). Lastly, we also trained some hybrid models whereby we combined a disinformation training set to fraud, scam, favorable fake review, and unfavorable fake review datasets to evaluate how well they detected deception and disinformation.

As shown in Table 1, we randomly selected 100 instances for the EN, FB, POS, and NEG models. Half of the instances were manually labeled as deceptive and the other truthful. The instances selected for training these models were mainly limited by the fraud dataset obtained from the DOJ as court evidence which had a limited number of emails, which were used as deceptive instances for the fraud model. Furthermore, for the hybrid models, we combined instances from the singleton models, i.e., *EN, FB, POS, NEG,* and combined a disinformation random sample to each of them to create the training set for *DIS+EN, DIS+FB, DIS+POS, and DIS+NEG,* respectively. The test set used to evaluate the models was the disinformation test set because we were only interested to learn how well the singleton and hybrid models performed.

TABLE 1 Dataset description.

Model	Dataset	Training set	Test set
EN	Enron	100	–
FB	Facebook	100	–
POS	Favorable Fake Reviews	100	–
NEG	Unfavorable Fake Reviews	100	–
DIS	Disinformation (used in hybrid models)	100	20
DIS+EN	Disinformation and enron	200	–
DIS+FB	Dinsformation and facebook	200	–
DIS+POS	Disinformation and favorable reviews	200	–
DIS+NEG	Dinsformation and unfavorable reviews	200	–

3.2 Feature selection and engineering

We first considered computational linguistic and psycholinguistic processes linked to deception and cybercrime, which we had used in prior studies to study deception and content-based cybercrime (Mbaziira & Jones, 2016, 2017; Mbaziira & Murphy, 2018). These computational linguistic features were: verbs, modifiers, average sentence length, average word length, pausality, modal verbs, emotiveness, lexical diversity, redundancy, characters, punctuation marks, sentences, adjectives, adverbs, nouns, and function words (Zhou, Burgoon, Twitchell, Qin, & Jr, 2004). Similarly, the psycholinguistic features linked to deception and cybercrime based on our earlier work are: *analytical, words per sentence, 6-letter words, I, we, you, she/he, affect, positive emotion, negative emotion, insight, causation, and certainty words*. However, for deception and disinformation detection we re-engineered the features. The computational linguistic features like *pausality, you, positive emotion, negative emotion, redundancy, affect, we, I, you, she/he, causality,* and *emotiveness* were excluded because the detection models underfit when trained with these features. To re-engineer the feature set, we used a ranker attribute selection algorithm and dropped all features which had a correlation coefficient of less than 0.05. This actually implied that features on *emotion, redundancy, self-references,* and *other references,* which are typical of psychological and computational linguistic processes linked to deception and cybercrime in human victim-to-cybercriminal networks did not apply to the deception of deception and disinformation in trolls disinformation campaigns. Table 2 summarizes the features which are used in generating our models.

TABLE 2 Feature set description.

Feature	Linguistic process type	Description
Verbs	Computational linguistics	Measures frequency of verbs whereby deceptive messages will have noncommittal verbs than truthful messages
Modifiers	Computational linguistics	Frequency of modifiers will be greater in deceptive messages compared to truthful messages
Averages sentence length	Computational linguistics	Average sentence length of deceptive messages will be greater than in truth messages
Average word length	Computational linguistics	Deceptive messages will usually be shorter than truthful messages

Continued

TABLE 2 Feature set description—cont'd

Feature	Linguistic process type	Description
Modal verbs	Computational linguistics	Frequency of verbs that indicate obligation. Since cybercriminals are noncommittal, fewer modal verbs will be used in deceptive messages
Lexical diversity	Computational linguistics	Measure ratio of unique words per sentence. Deceptive messages will have lower lexical diversity compared to truthful messages
Number of characters	Computational Linguistics	Frequency of characters in sentences. Deceptive messages will have higher character frequency compared to truthful messages
Number punctuation marks	Computational linguistics	Frequency of punctuation marks. Deceptive messages will have more punctuation marks than truthful messages
Number of sentences	Computational linguistics	Frequency of sentences in messages. Deceptive messages will have more sentences compared to truthful messages
Number of adjectives	Computational linguistics	Deceptive messages can be vague and ambiguous hence adjectives will be used for this purpose
Number of adverbs	Computational linguistics	Deceptive messages are more vague and ambiguous hence adverbs will be used for this purpose. Adverbs are words that modify nouns
Number of nouns	Computational linguistics	Frequency of nouns. Deceptive messages will have more nouns compared to truthful messages
Number of function words	Computational linguistics	Frequency of function words. Deceptive messages will have more function words compared to truthful messages
I	Psycholinguistics	Deceptive messages contain fewer first personal pronouns, singular *I*, to avoid accountability
Analytic	Psycholinguistics	Deceptive messages will have less analytical words hence low cognitive complexity
Six letter words	Psycholinguistics	Words with more than six characters. Measures longer words hence cognitive complexity. Deceptive messages will have fewer words greater than six characters to keep messages less cognitively complex
Insight	Psycholinguistics	Deceptive messages will have less insight words hence low cognitive complexity
Class {deceptive, truthful}		These are the class labels for the binary classifiers

3.3 Text classification models

Naive Bayes has been widely used to investigate text classification problems like email spam, and phishing (Basnet, Sung, & Liu, 2012; Harisinghaney, Dixit, Gupta, & Arora, 2014; Viaene, Derrig, & Dedene, 2004). Naive Bayes computes posterior probabilities for every labeled class instead of computing class conditional probabilities of each class given a document. The equation for Naive Bayes classification is:

$$P(c|\,d) = \frac{P(d|\,c)\,X\,P(c)}{P(d)} \tag{1}$$

where d is data to be classified comprising a set of attributes and c is a class label.

Support Vector Machines are founded on statistical learning. The algorithm uses a hyper-plane to solve the classification problem by linearly separating instances into their class labels (Aphiwongsophon & Chongstitvatana, 2018). For example, given training instances or example, the class label y of a test data will be predicted using a linear function below:

$$y = \begin{cases} 1\,if\; x.z + b > 0; \\ -1\,if\; x.z + b < 0; \end{cases} \tag{2}$$

where 1 and −1 are class labels, x and b are parameters of the decision boundary.

Decision Tree classifiers comprise a set of decision nodes that start at a root node. The tree contains decision nodes, leaves, nodes, and branches. The decision tree then recursively selects features and splits the dataset into those features. The quality of a division is determined using the *Gini* for Gini impurity and *entropy* for information gain as shown in the equation below

$$G = \sum\nolimits_{i=1}^{C} p(i) * (1 - p(i)) \tag{3}$$

where $p(i)$ is the probability of class i for the current branch. The entropy is calculated using the formula below:

$$E = -\sum\nolimits_{i} pi * log_2 * p_i \tag{4}$$

The *Random Forest* classifier consists of a large number of decision trees that work individually. The final prediction is based on a class that got the most votes. The error rate is low due to the low correlation between trees (Gregorutti, Michel, & Saint-Pierre, 2017). When using this algorithm, the bagging method is used to repeatedly train the models.

This bagging method randomly and repetitively selects the same data set K times. The trees are trained with the same data, therefore the final result is formed by the individual predictions m_i of each tree in the set, the equation is used:

$$\hat{m} = \frac{1}{K}\sum_{k}^{i=1} mi \tag{5}$$

The *Logistic Regression* algorithm classifies problems into binary or multiple classes by modeling the probability of a class or event (Mitchell, 2006; Tacchini, Ballarin, Della Vedova, Moret, & de Alfaro, 2017). The Logistic Regression function is defined as:

$$(x) = \frac{1}{1 + e^{-(\beta_0 + \beta_1 x)}} \tag{6}$$

Logistic Regression transforms the output into a probability value. This function aims to minimize the cost to achieve optimal probability and the cost function is determined by the equation below:

$$Cost\,(h_\theta(x), y) = \begin{cases} log\,(h_\theta(x)), & y=1, \\ -log\,(1-h_\theta(x)), & y=0. \end{cases} \tag{7}$$

3.4 Evaluating classifier performance

Our disinformation detection models were evaluated using precision, recall, F-measure, and Receiver Operating Characteristic (ROC) curves. Precision measures the proportion of the relevant instances, which the classifier declared deceptive, while recall measures the number of relevant deceptive messages which were accurately predicted. F-measure is the harmonic mean for precision and recall, and ROC curves illustrate the tradeoff between the true-positive rate and the false-positive rate of the classifiers. Table 3 shows the evaluation metrics of the singleton models. Generally, all the classifiers performed well when trained on fraud, favorable and unfavorable reviews except for classifiers trained for the scam's datasets. This may have been attributed to the fact that scams were obtained from nonnative English-speaking cybercriminals who did not exhibit deception patterns similar to native English-speaking cybercriminals. This anomaly in deception patterns, though evident in our earlier studies, was also evident in this study as well. Comparatively, when we evaluated the performance of the hybrid models, as shown in Table 4, they also performed well. We observed some improvement in the performance evaluation of the classifiers trained on scams and disinformation.

TABLE 3 Performance evaluation metrics for the singleton models.

Model	Classifier	Precision	Recall	F-Measure	ROC area
EN	NB	0.868	0.850	0.848	0.965
EN	SVM	0.960	0.960	0.960	0.960
EN	DT-J48	0.921	0.920	0.920	0.939
EN	RF	0.961	0.960	0.960	0.989
EN	LR	0.970	0.970	0.970	0.997
FB	NB	0.482	0.490	0.428	0.401
FB	SVM	0.418	0.420	0.416	0.420
FB	DT-J48	0.480	0.490	0.419	0.462
FB	RF	0.581	0.580	0.579	0.574
FB	LR	0.450	0.450	0.450	0.422
POS	NB	0.670	0.670	0.670	0.741
POS	SVM	0.874	0.870	0.870	0.870
POS	DT-J48	0.782	0.780	0.780	0.758
POS	RF	0.850	0.850	0.850	0.917
POS	LR	0.800	0.800	0.800	0.858
NEG	NB	0.717	0.710	0.708	0.782
NEG	SVM	0.831	0.830	0.830	0.830
NEG	DT-J48	0.782	0.780	0.780	0.758
NEG	RF	0.850	0.850	0.850	0.917
NEG	LR	0.800	0.800	0.800	0.858

3.5 Ground truth

The instances used in building the scam models were manually verified for work-at-home scams, lottery scams, advertisements for hacking services, and carding. Data used in the fraud model consisted of Enron emails of which the email evidence used to prosecute Enron corporation executives for securities and wire fraud were labeled as deceptive because the convicted executives did not appeal the case (DOJ, 2013; SEC, 2015). Lastly, for the favorable and unfavorable online review models, we considered truthful reviews to be those which had transactional data like deals prices for services, meals, valet services, etc. as affirming that reviewers were guests at the hotels in question (Fitzpatrick, Bachenko, & Fornaciari, 2015).

TABLE 4 Performance evaluation metrics for the hybrid models.

Model	Classifier	Precision	Recall	F-Measure	ROC area
DIS+EN	NB	0.619	0.600	0.583	0.703
DIS+EN	SVM	0.689	0.685	0.683	0.685
DIS+EN	DT-J48	0.768	0.765	0.764	0.776
DIS+EN	RF	0.755	0.755	0.755	0.755
DIS+EN	LR	0.708	0.700	0.697	0.670
DIS+FB	NB	0.475	0.485	0.428	0.435
DIS+FB	SVM	0.520	0.520	0.520	0.520
DIS+FB	DT-J48	0.548	0.525	0.461	0.517
DIS+FB	RF	0.680	0.680	0.680	0.680
DIS+FB	LR	0.525	0.525	0.525	0.516
DIS+NEG	NB	0.534	0.530	0.516	0.560
DIS+NEG	SVM	0.569	0.555	0.531	0.555
DIS+NEG	DT-J48	0.646	0.605	0.575	0.616
DIS+NEG	RF	0.656	0.655	0.655	0.720
DIS+NEG	LR	0.635	0.635	0.635	0.694
DIS+POS	NB	0.607	0.585	0.563	0.640
DIS+POS	SVM	0.667	0.660	0.657	0.660
DIS+POS	DT-J48	0.680	0.680	0.680	0.727
DIS+POS	RF	0.766	0.765	0.765	0.817
DIS+POS	LR	0.670	0.670	0.670	0.741

4 Results and analysis

In this section, we present the results on how the models performed in detecting deception and disinformation. We used criteria for identifying models in our analysis, which performed well. All models with predictive accuracies less than 50% for both binary class labels (i.e., truthful and deceptive classes) were disregarded from this analysis because such models performed worse than human beings in discriminating between truthful and deceptive instances (Bond & DePaulo, 2006). This is because we used binary text classification to generate our models and we were interested in models that detected truthful and deceptive instances at thresholds of above 50% for predictive accuracy.

Fig. 2 is a graph summarizing the predictive accuracies of the singleton models. The most interesting results were from the EN_SVM model which detected deception and disinformation with a predictive accuracy of 100% and an average prediction accuracy of 85%. This model was trained on fraud and it generalized well in detecting deception and disinformation.

Fig. 3 is a graph summarizing the predictive accuracies of the hybrid models in detecting deception and disinformation. Generally, compared

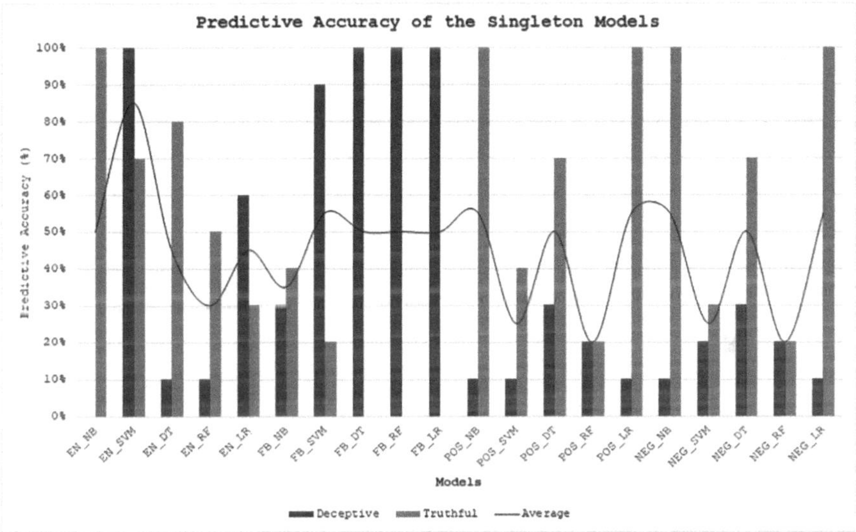

FIG. 2 Predictive accuracy of the singleton models.

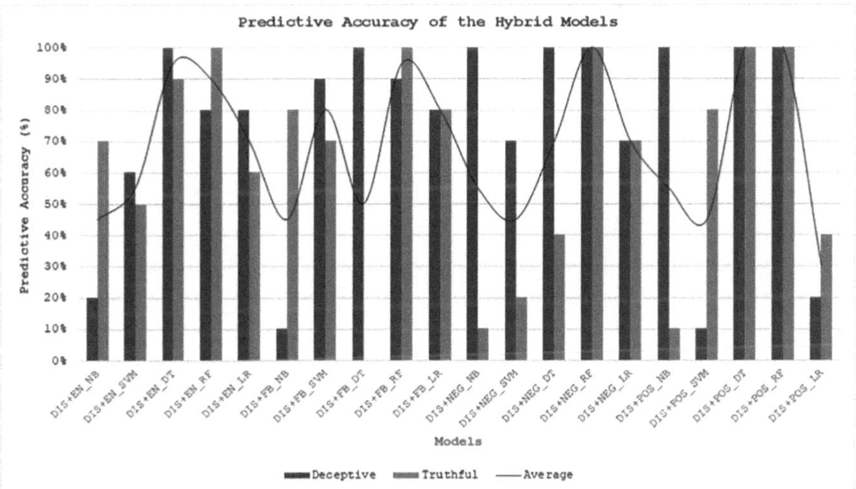

FIG. 3 Predictive accuracy of the hybrid models.

to the singleton models, there were more hybrid models that detected deception and text-based disinformation predictive accuracy rates of over 50%. These models are: the DIS + EN_DT model which detected deception and disinformation with a predictive accuracy rate of 100%; DIS + EN_RF and DIS + EN_LR models which detected deception and disinformation with a predictive accuracy rate of 80%. Other models included the DIS +FB_SVM which detected deception and disinformation at a predictive accuracy rate of 90%; DIS + FB_RF which detected deception and disinformation at a predictive accuracy rate of 90% and DIS + FB_LR which detected deception and text disinformation at a predictive accuracy rate of 80%. The DIS + NEG_RF and DIS + NEG_LR models detected deception and text-based disinformation with predictive accuracy rates of 100% and 70% respectively. Lastly, the DIS + POS_DT and DIS + POS_RF models detected deception and disinformation with predictive accuracy rates of 100%.

Based on the results presented above the analysis criteria, we observed that among the singleton models, only the EN_SVM, which is a model trained on fraud, successfully detected deception and text-based disinformation. A closer look at the analysis of features for the singleton models shown in Fig. 4, reveals a common pattern in computational linguistic and psycholinguistic processes linked to deception and text-based disinformation between fraud and disinformation. In Fig. 4, the disinformation curve is labeled as DIS while the fraud model curve is labeled as EN.

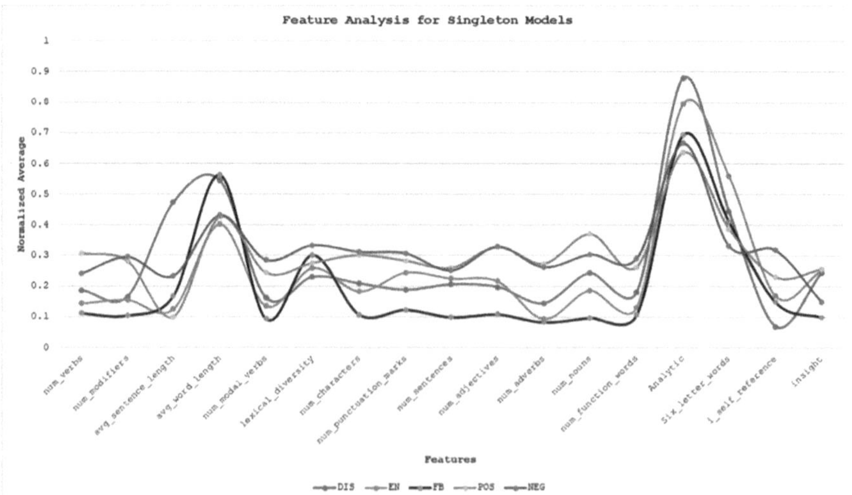

FIG. 4 Feature analysis for the singleton models.

Comparing the feature analysis for hybrid models, we observed that the models which were successfully detecting deception and disinformation were trained on fraud, scams, favorable and unfavorable fake reviews. Of these hybrid models, there were three fraud, three scams, two unfavorable, and two favorable hybrid models. We observed that there were three fraud-trained hybrid models which detected deception and text-based disinformation, and these were DIS+EN_DT, DIS+EN_RF, and DIS+EN_LR while the scam-trained hybrid models were DIS+FB_SVM, DIS+FB_RF, and DIS+FB_LR. Furthermore, there were two unfavorable reviews-trained hybrid models which were DIS+NEG_RF and DIS+NEG_LR. Lastly, there were two favorable reviews-trained models and these were DIS+POS_DT and DIS+POS_RF. We also observed an interesting pattern in the feature analysis within the hybrid models as shown in Fig. 5. We observed that compared to singleton models, the hybrid models had interesting patterns for deception linked to computational linguistics and psycholinguistic processes which may explain why they had more models that successfully deception and text-based disinformation.

Since the disinformation dataset we used to train our models was generated by trolls, we had to eliminate features that worked well in earlier research for human-generated datasets for fraud, scams, favorable fake reviews, and unfavorable fake reviews (Mbaziira & Jones, 2016, 2017; Mbaziira & Murphy, 2018). The eliminated features were *pausality, you, positive emotion, negative emotion, redundancy, affect, we, you, she/he, causality,*

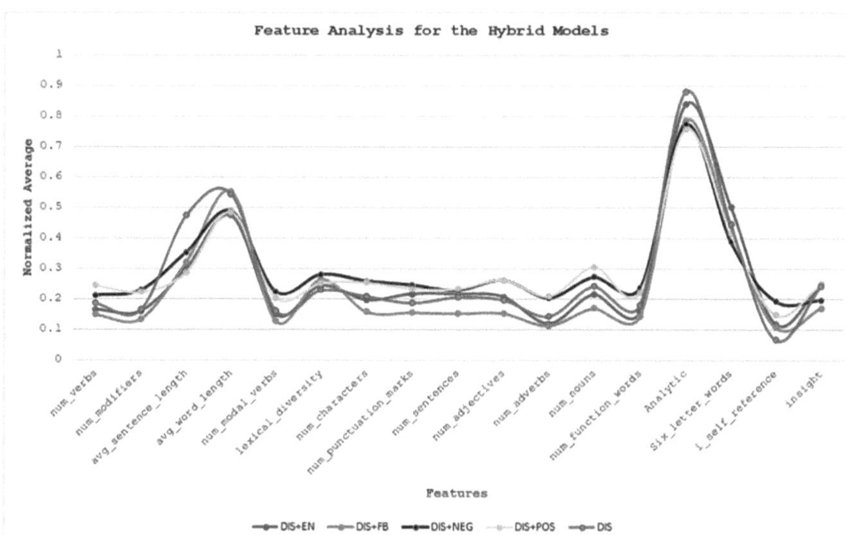

FIG. 5 Feature analysis for the hybrid models.

and *emotiveness*. When these features were included in the original models, there was underfitting.

The patterns of deception from disinformation generated by trolls that deviated from fraud, scams, favorable fake reviews, and unfavorable fake reviews included the following:

- emotion (i.e positive, negative, affect), other references like (I, we, she/he), redundancy, pausuality, causality, and emotiveness were irrelevant when generating models for detecting deception and text-based disinformation.
- Our earlier studies on detecting deception and cybercrime in the form of fraud, scams, favorable and unfavorable fake reviews revealed that cybercriminals reduced the low cognitive load in deceptive text-based communication by using large quantities of words and higher average sentence length. However, Figs. 4 and 5 reveal the very opposite when it comes to deception and text-based disinformation.
- Troll-generated deceptive text messages have lower frequencies of adjectives and adverbs hence low expressivity compared to patterns of deception and cybercrime we studied earlier in scams, fraud, favorable fake reviews, and unfavorable fake reviews.

Lastly, the patterns of deception from disinformation generated by trolls that were similar to fraud, scams, favorable fake reviews, and unfavorable fake reviews included the following:

- Use of less first pronouns especially, I, to avoid accountability thereby distancing the deceptive text message from the author
- Use more exclusive words (like but, without, exclude) to make the deceptive text messages more imprecise thereby reducing their cognitive complexity
- Low lexical diversity whereby the deceptive text messages had fewer unique words to reduce their cognitive complexity.

5 Conclusion

In the last few years, disinformation has continued to be problematic as populations around the world continue to rely on the web and social media for information. Disinformation campaigns have continued to evolve into very sophisticated information and influence operations coordinated by organized cybercriminal networks as well as national states. In this work, we demonstrated that it is possible to use machine learning to detect deception and text-based disinformation from trolls using computational linguistic and psycholinguistic processes. Our models for detecting deception and text-based disinformation were limited to only one type of disinformation campaign, which is fake news. Future work will explore patterns of deception in other forms of disinformation campaigns areas like public health in other web genres like social media. This will help in evaluating the generalizability of our models in detecting deception from other types of disinformation campaigns.

References

Aphiwongsophon, S., & Chongstitvatana, P. (2018). Detecting fake news with machine learning method. In *2018 15th International Conference on Electrical Engineering/Electronics, Computer, Telecommunications and Information Technology (ECTI-CON)* (pp. 528–531). https://doi.org/10.1109/ECTICon.2018.8620051.

Basnet, R. B., Sung, A. H., & Liu, Q. (2012). Feature selection for improved phishing detection. In H. Jiang, W. Ding, M. Ali, & X. Wu (Eds.), *Advanced research in applied artificial intelligence* (pp. 252–261). Springer. https://doi.org/10.1007/978-3-642-31087-4_27.

Bello, H. R. M., Heilmann, L., & Ronan, E. (2020). *Detecting fake news spreaders with behavioural, lexical and psycholinguistic features. 12*.

Bond, C. F., & DePaulo, B. M. (2006). Accuracy of deception judgments. *Personality and Social Psychology Review, 10*(3), 214–234. https://doi.org/10.1207/s15327957pspr1003_2.

Castillo, C., Mendoza, M., & Poblete, B. (2011). Information credibility on twitter. In *International conference on world wide web. ACM* (pp. 675–684). New York: ACM. https://doi.org/10.1145/1963405.1963500

Cohen, W. (2015). *Enron Email Dataset*. http://www.cs.cmu.edu/~enron/.

DOJ. (2013). *Former Enron CEO Jeffrey skilling resentenced to 168 months for fraud, conspiracy charges.* , June 21 https://www.justice.gov/opa/pr/former-enron-ceo-jeffrey-skilling-resentenced-168-months-fraud-conspiracy-charges.

Fitzpatrick, E., Bachenko, J., & Fornaciari, T. (2015). Automatic detection of verbal deception. *Synthesis Lectures on Human Language Technologies, 8*(3), 1–119. https://doi.org/10.2200/S00656ED1V01Y201507HLT029.

Gregorutti, B., Michel, B., & Saint-Pierre, P. (2017). Correlation and variable importance in random forests. *Statistics and Computing, 27*(3), 659–678. https://doi.org/10.1007/s11222-016-9646-1.

Harisinghaney, A., Dixit, A., Gupta, S., & Arora, A. (2014). Text and image based spam email classification using KNN, Naïve Bayes and reverse DBSCAN algorithm. *International Conference on Reliability Optimization and Information Technology (ICROIT), 2014*, 153–155. https://doi.org/10.1109/ICROIT.2014.6798302.

Krishnamurthy, G., Majumder, N., Poria, S., & Cambria, E. (2018). A deep learning approach for multimodal deception detection. In *ArXiv:1803.00344 [Cs]*. http://arxiv.org/abs/1803.00344.

Mbaziira, A., & Jones, J. (2016). *A text-based deception detection model for cybercrime* (p. 9).

Mbaziira, A. V., & Jones, J. H. (2017). Hybrid text-based deception models for native and non-native English cybercriminal networks. In *23–27. Proceedings of the International Conference on Compute and Data Analysis*. https://doi.org/10.1145/3093241.3093280.

Mbaziira, A. V., & Murphy, D. R. (2018). An empirical study on detecting deception and cybercrime using artificial neural networks. In *Proceedings of the 2nd International Conference on Compute and Data Analysis* (pp. 42–46). https://doi.org/10.1145/3193077.3193080.

Mitchell, T. M. (2006). *The discipline of machine learning* (1st ed.). Machine Learning Department School of Computer Science Carnegie Mellon University.

Ott, M., Choi, Y., Cardie, C., & Hancock, J. T. (2011). *Finding deceptive opinion spam by any stretch of the imagination* (p. 11).

Pérez-Rosas, V., Kleinberg, B., Lefevre, A., & Mihalcea, R. (2017). Automatic detection of fake news. In *ArXiv:1708.07104 [Cs]*. http://arxiv.org/abs/1708.07104.

Sarvari, H., Abozinadah, E., Mbaziira, A., & McCoy, D. (2014). Constructing and analyzing criminal networks. *IEEE Security and Privacy Workshops, 2014*, 84–91.

SEC. (2015). *Jeffrey K. Skilling et al. (Release No. LR-23422; December 8, 2015)*. https://www.sec.gov/litigation/litreleases/2015/lr23422.htm.

Tacchini, E., Ballarin, G., Della Vedova, M. L., Moret, S., & de Alfaro, L. (2017). *Some like it hoax: Automated fake news detection in social networks*. ArXiv:1704.07506 [Cs] http://arxiv.org/abs/1704.07506.

Verma, A., Mittal, V., & Dawn, S. (2019). FIND: Fake information and news detections using deep learning. In *2019 twelfth international conference on contemporary computing (IC3)* (pp. 1–7). https://doi.org/10.1109/IC3.2019.8844892.

Viaene, S., Derrig, R. A., & Dedene, G. (2004). A case study of applying boosting naive Bayes to claim fraud diagnosis. *IEEE Transactions on Knowledge and Data Engineering, 16*(5), 612–620. https://doi.org/10.1109/TKDE.2004.1277822.

Vidros, S., Kolias, C., Kambourakis, G., & Akoglu, L. (2017). Automatic detection of online recruitment frauds: Characteristics, methods, and a public dataset. *Future Internet, 9*(1), 6. https://doi.org/10.3390/fi9010006.

Zhou, L., Burgoon, J. K., Twitchell, D., Qin, T., & Jr, J. F. N. (2004). A comparison of classification methods for predicting deception in computer-mediated communication. *Journal of Management Information Systems, 20*(4), 139–166. https://doi.org/10.1080/07421222.2004.11045779.

CHAPTER

14

Modeling the effects of network size in a deception game involving honeypots

Harsh Katakwar[a], Shashank Uttrani[a], Palvi Aggarwal[b], and Varun Dutt[a]

[a]Applied Cognitive Science Laboratory, Indian Institute of Technology Mandi, Kamand, India, [b]Department of Computer Science, The University of Texas at El Paso, El Paso, TX, United States

1 Introduction

In the last decade, the growth in the use of digital assets has resulted in a significant increase in the number of hazardous cyberattacks (Business Insider, 2019). In 2020, there were major assaults on the cyberinfrastructure of some of the famous organizations, such as Twitter and Marriott International (Quick Heal, 2021). Social engineering and Ransomware were the most widespread forms of cyberattacks (Trustwave, 2020). In addition, attackers in the recent past have targeted corporates working in the retail and finance sectors (Trustwave, 2020). The primary objective of these attacks has been to steal useful information, leading to socioeconomic and governance repercussions. Though there exist security solutions that may aid in countering some of the cyberattacks, these solutions may not be sufficient enough to defeat these modern cyberattacks (Aggarwal & Dutt, 2020; Bace & Mell, 2001; Rowe & Custy, 2007).

Some of the existing security solutions have proved to be efficient in combating threatening cyberattacks (Bace & Mell, 2001; Shang, 2018a, 2018b). For example, some of these solutions may include intrusion detection systems and filtering strategies (Aggarwal & Dutt, 2020; Bace & Mell,

Cybersecurity and Cognitive Science
https://doi.org/10.1016/B978-0-323-90570-1.00006-1

2001). An intrusion detection system (IDS) is a system that monitors the network traffic and transmits alerts on detecting malicious activity in the network. The IDSs have been found to be robust against cyberattacks; nonetheless, these systems have some inaccuracies resulting in false alarms (Shang, 2018a). The false alarms may lead to monetary loss for the corporates, which involve substantial digital infrastructure (Shang, 2018a). Apart from the IDSs, the filtering strategy has also been found to be effective against cyberattacks (Shang, 2018b). The filtering strategy separates the malicious activities of the network and ensures safe access to the network. However, this technique allows nonrational network participants to reach a state of consensus (Shang, 2018b). Consequently, such a consensus could aid in the detection of cyberattacks before they pose a threat to the cyberinfrastructure (Shang, 2018b).

Another security solution that has lately been found to be prominent against cyberattacks is deception (Rowe & Custy, 2007). Deception is an act of making someone believe in the incorrect information, which may help in providing real-time solutions to combat today's life-threatening cyberattacks (Almeshekah & Spafford, 2016). Deception in cybersecurity has been used with the help of honeypots and it has been found to be a valuable tool to counter emerging cyberattacks (Aggarwal, Gonzalez, & Dutt, 2016a, 2016b; Aggarwal, Gonzalez, and Dutt, 2017; Katakwar, Aggarwal, Maqbool, & Dutt, 2020). Previously, researchers have developed a tool known as HackIT, which could simulate a cyberattack situation realistically by integrating concepts from behavioral game theory (Aggarwal & Dutt, 2020). HackIT tool could help understand human factors influencing adversarial decisions in a network involving deception. Aggarwal, Gonzalez, and Dutt (2020) varied the timing of deception as early and late in HackIT and found that the proportion of honeypot attacks were greater for late compared to early deception. Basically, HackIT was able to replicate results about human decisions in canonical games.

Prior research shows that deception in cybersecurity has been studied both mathematically and through applied games (Carroll & Grosu, 2009; Garg & Grosu, 2007; Kiekintveld, Lisý, & Píbil, 2015; Aggarwal et al., 2016a, 2016b). For example, Kiekintveld et al. (2015) focused on how a game-theoretical framework could be utilized to manipulate knowledge in adversarial environments. Similarly, Garg and Grosu (2007) proposed a mathematical approach to adeception-based games. Carroll and Grosu (2009) modeled the interaction between an adversary and a defender as a signaling game. They used perfect Bayesian equilibria to conduct a game-theoretical analysis of deception. However, both experiments by Garg and Grosu (2007) and Carroll and Grosu (2009)utilized Nash analyses of adversaries versus defenders without the involvement of human participants as adversaries or defenders.

Recent research has concentrated more on human factors that play a major role in affecting adversarial decision-making in different cyberattack situations (Aggarwal et al., 2020; Aggarwal & Dutt, 2020; Katakwar et al., 2020). Adversarial decisions may be affected by multiple factors, including the size of the network, variation in honeypot proportions, deception techniques, and the number and variability in zero-day vulnerabilities (Garcia-Teodoro, Diaz-Verdejo, Maciá-Fernández, & Vázquez, 2009; Katakwar et al., 2020; Lenin, Willemson, & Sari, 2014; Wang, Jajodia, Singhal, & Noel, 2010). Among these factors, one factor that is likely to influence adversarial decision-making is network size. As per prior research, an increase in the size of the network may cause an increase in the proportions of cyberattacks (Bagchi & Tang, 2004; Wang et al., 2010). Katakwar et al. (2020) investigated the effect of different network sizes on adversarial decision-making in the presence of honeypots in a deception-based game. They configured the deception game for three different network sizes while maintaining the proportion of honeypots constant across all the conditions. Katakwar et al. (2020) derived their hypothesis based upon the Instance-based learning Theory (IBL; Gonzalez, Lerch, & Lebiere, 2003; Gonzalez & Dutt, 2011, 2012; Dutt & Gonzalez, 2012; Dutt, Ahn, & Gonzalez, 2013). However, these authors did not build computational cognitive IBL models that could account for human decisions in the deception game based upon experience.

In this research, we address this literature gap by building computational cognitive models based upon IBL, which could explain the findings of Katakwar et al. (2020) from a cognitive viewpoint. The main objective of the research is to investigate how the cognitive mechanisms proposed by IBL influence adversaries across different network sizes. As per IBL, human participants make bounded rational decisions, where these decisions are governed by recency and frequency of decision outcomes in a task. Also, these participants are limited by cognitive factors, like recall and memory. Thus, people depend more on the recency and frequency of information for making their decisions. In small networks, as there are a smaller number of webservers, the adversary may be easily able to recall the actual mapping of real and honeypot webservers. However, for medium and large networks, due to the presence of a large number of webservers, the adversary may face difficulties in recalling the actual mapping of webservers. The difficulty in retrieving instances for medium and large networks would be likely due to the presence of multiple instances per webserver and the low activation value of these instances. Thus, difficulty in recalling the instances in medium and large networks is likely to make the adversary explore a large number of webservers during probe and attack stages. Therefore, based upon IBL, we expect that the proportion of regular probe/regular attack actions and honeypot probe/honeypot attack actions will be more in medium and large networks than in small networks. In addition, no-probe and no-attack actions will have a high activation value in small networks compared to medium and large networks. Thus, we expect the proportion of no-probe/no-attack actions to be more in small networks compared to medium and large networks. Overall, we expect that cognitive models built upon IBL would be able to account for human decisions in adeception-based games.

In what follows, we first briefly describe the deception game. Next, we discuss the experiment conducted by Katakwar et al. (2020), where the authors showed the effect of different network sizes in adversarial decision-making in the presence of deception. Furthermore, we report a reanalysis of human results of Katakwar et al. (2020). Also, we present the results from cognitive IBL models and highlight their implications for research in cybersecurity and cognitive science.

2 Deception game

Deception Game (DG) is a single-player, sequential, incomplete information web-based game, i.e., a game between adversary and the network. The game is denoted as DG (n, k, γ), where n represents the number of webservers, k represents the number of honeypots, and γ represents the number of probes an adversary makes before the attack stage. The game comprises of two kinds of webservers, i.e., regular and honeypot. Regular webservers are the real webservers; whereas, honeypots are the fake webservers that pretend to like real webserver. An adversary can probe a few webservers present in the network, or she may not probe any of the webservers.

The game has multiple rounds, where each round consists of two stages, probe stage followed by the attack stage. In the probe stage, the adversary may probe few of the webservers present in the network. Probing a webserver in DG means clicking the button present in the game's interface, which represents a webserver in the network. On probing a webserver, the adversary gets feedback from the network. If a particular round is the deception round, then the adversary receives the false information for her probing actions. So, if the adversary probes a "regular webserver", then she will receive the response as "honeypot webserver". Similarly, on probing the "honeypot webserver", the adversary will receive the response as the "regular webserver". In contrast, if the deception is absent in a particular round, then adversary will receive the true information about the webservers. Thus, if the adversary probes the "regular webserver", then she will receive the response from the network as the "regular webserver". Likewise, on probing the "honeypot webserver", she will

TABLE 1 Payoff matrix of adversary's action in probe and attack stage.

Stage	Adversary's action	Payoff
Probe	Regular probe	+5
	Honeypot probe	−5
	No webserver probe	0
Attack	Regular attack	+10
	Honeypot attack	−10
	No webserver attack	0

receive the response from the network as the "honeypot webserver". In the attack stage, the adversary may decide to attack one of the webservers present in the network. Attacking the network means clicking one of the buttons denoting the webserver present in the game's interface.

On completion of a round, the adversary gets to know her scores based on the actions of probe and attack stages. Similarly, on completion of multiple rounds, the adversary gets the cumulative score for her actions. Table 1 denotes the payoff matrix for the adversary based on the decisions in the probe and attack stages in the DG. As per Table 1, on probing/attacking a regular webserver, the adversary gets rewarded with positive points. So, on probing a regular webserver, the adversary gets awarded with +5 points. Similarly, attacking a regular webserver results in a reward of +10 points for the adversary. However, on probing a honeypot webserver the adversary gets rewarded with negative points. So, on probing/attacking a honeypot webserver, the adversary gets −5 points. Similarly, on attacking the honeypot webserver, the adversary gets a reward of −10 points. The adversary neither gains nor losses any point if she does not probe or attack the network.

The DG was configured for three different network sizes, small, medium, and large. The small, medium, and large versions of DG were denoted as DG (2, 1, 1), DG (6, 3, 3), and DG (12, 6, 6), respectively.

3 Experiment

In this section, we describe the experiment carried out by Katakwar et al. (2020). This experiment involved human participants performing as adversaries across multiple rounds in the DG. The DG was used to investigate the influence of different network sizes on adversarial decision-making in DG in presence of honeypots.

3.1 Experiment design

Katakwar et al. (2020) recruited 60 participants for their experiment and participants were randomly distributed across 3 different between-subject conditions. These conditions were: DG (2, 1, 1) (small), DG (6, 3, 3) (medium), and DG (12, 6, 6) (large). Each condition had 29 trials, consisting of 14 deception trials and 15 nondeception trials. The sequence of deception and nondeception rounds was randomized. Participants were not aware of the presence of deception/nondeception conditions in a particular trial. The sequence of deception and nondeception rounds remained the same across all the conditions of the deception game. For each trial across all the conditions, the honeypot webserver and regular webserver were assigned randomly to the buttons. In each condition, Katakwar et al. (2020) evaluated the average proportion of different decisions during probe and attack stages in the DG.

3.2 Participants

In the experiment conducted by Katakwar et al. (2020), participants were recruited via a crowdsourcing website called Amazon Mechanical Turk (Mason & Suri, 2012). Eighty-six percent of participants were male, while the remaining participants were females. Age of the participants ranged from 19 years to 48 years (Median = 31 years; Mean = 32 years; Standard Deviation = 6 years). Among the total participants, 92% of them had a college degree, while the remaining 8% of them were currently pursuing a college degree. 60% of the total participants had STEM (Science, Technology, Engineering, and Mathematics) background. After the study, participants were thanked and paid INR 50 (USD 0.67) for their participation in the study.

3.3 Procedure

In the experiment, participants performing the role of the adversary were instructed about their goal in the DG. Participants possessed complete information about their actions and payoffs. Participants were asked to maximize their payoffs by taking appropriate actions against the network in the DG. They were also instructed about the presence of deception in different rounds of the game. Each round of the game had two stages; probe stage followed by attack stage. Adversary could make three decisions in the probe stage, i.e., Regular probe, Honeypot probe, and No webserver probe. Likewise, in the attack stage, the adversary had the option to take one of the three decisions in each round, i.e., Regular attack, Honeypot attack, and No webserver attack. Participants performing as the adversary had to make a choice between the options presented to them on their screen via buttons denoting webservers to maximize their payoffs. On completion of the study, participants were thanked and remunerated for their participation.

4 Results

4.1 Influence of network size on decisions during the probe stage

We reanalyzed the proportion of different probe decisions of DG for different network sizes on human data of an experiment conducted by Katakwar et al. (2020). The network size significantly influenced the different probe decisions: honeypot webserver probes ($F(2,59)=35.86$, $P<.001$, $\eta^2=0.56$), regular webserver probes ($F(2,59)=18.31$, $P<.001$, $\eta^2=0.39$), and no webserver probes ($F(2,59)=34.39$, $P<.001$, $\eta^2=0.55$), where the P-value tests the statistical significance and η^2 denotes the measure of the effect size. Fig. 1 shows the proportion of different honeypot webserver probes, regular webserver probes, and no webserver probes across different network sizes.

As shown in Fig. 1, the proportion of honeypot webserver probes in small, medium, and large networks were .22, .45, and .47, respectively. The sample size was equal across the three network sizes, so we also performed the Tukey post hoc test. The Tukey post hoc tests revealed that the proportion of honeypot web server probes was smaller for the small network than for the medium network ($P<.001$) and large network ($P<.001$).

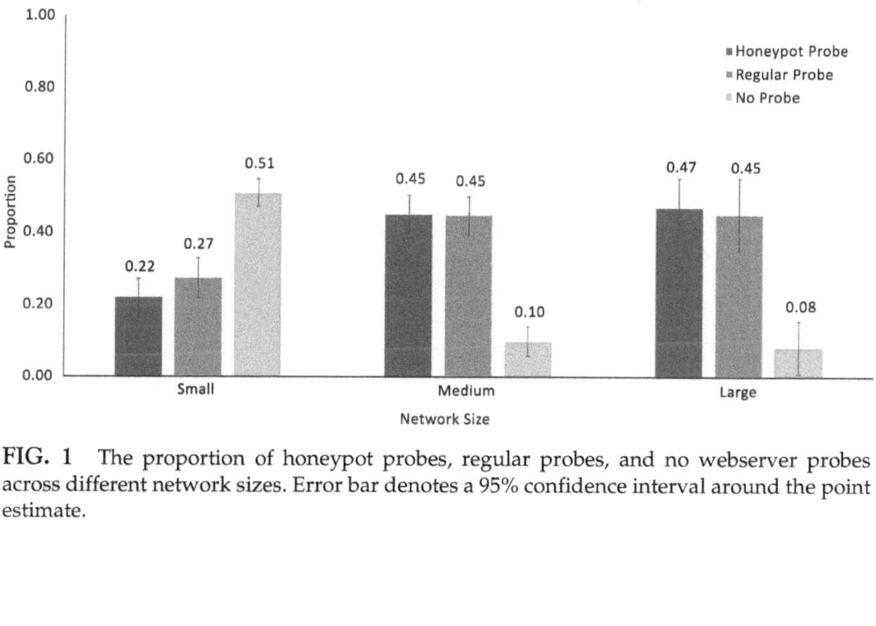

FIG. 1 The proportion of honeypot probes, regular probes, and no webserver probes across different network sizes. Error bar denotes a 95% confidence interval around the point estimate.

Also, there was no significant difference between the proportion of honeypot web server probes in the medium and large networks ($P=.83$). Similarly, the proportion of regular webserver probes in small, medium, and large networks were .27, .45, and .45, respectively. As per the Tukey post hoc test, the proportion of regular webserver probes in the small network was smaller compared to a medium network ($P<.001$) and large network ($P<.001$). Also, there was no significant difference between the proportion of regular webserver probes in the medium and large networks ($P=.99$). Likewise, the proportion of no webserver probes in small, medium, and large networks were .51, .10, and .08, respectively. According to the Tukey post hoc test, the proportion of no webserver probes in the small network was significantly smaller in the small network compared to the medium network ($P<.001$) and large network ($P<.001$). Also, there was no significant difference between the proportions of no web server probes in the medium and large networks ($P=.92$).

4.2 Influence of network size on decisions during the attack stage

We reanalyzed the proportion of different attack decisions of DG for different network sizes on human data of experiment conducted by Katakwar et al. (2020). The network size significantly influenced the different attack decisions: honeypot webserver attacks (F (2,59)=51.77, $P<.001$, $\eta^2=0.65$), regular webserver attacks (F (2,59)=23.32, $P<.001$, $\eta^2=0.45$), and no webserver attacks (F (2,59)=111.68, $P<.001$, $\eta^2=0.78$). Fig. 2 shows the proportion of different attack decisions across different network sizes.

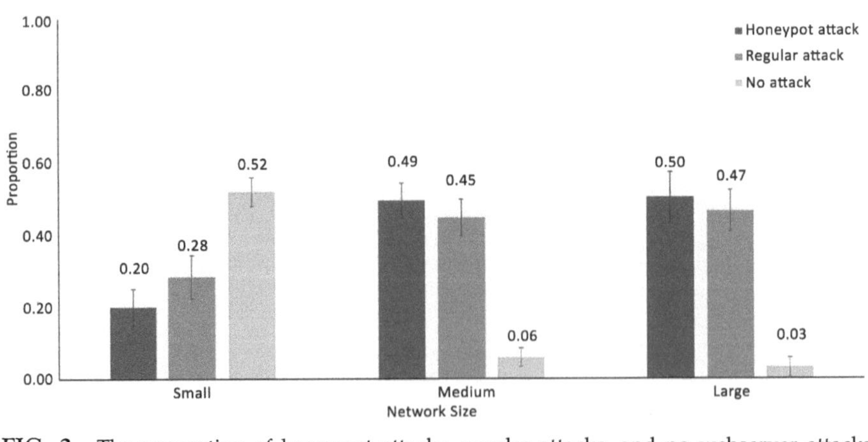

FIG. 2 The proportion of honeypot attacks, regular attacks, and no webserver attacks across different network sizes. Error bar denotes a 95% confidence interval around the point estimate.

As shown in Fig. 2, the proportion of honeypot webserver attacks in small, medium, and large networks were .20, .49, and .50, respectively. According to the Tukey post hoc test, the proportion of honeypot webserver attacks were significantly smaller in the small network compared to that of medium network ($P < .001$) and large network ($P < .001$). It also revealed that there was no significant difference between proportions of honeypot webserver attacks in medium and large networks ($P = .97$). Similarly, as depicted in Fig. 2, the proportion of regular webserver attacks in small, medium, and large networks were .28, .45, and .47, respectively. As per the Tukey post hoc test, the proportion of regular webserver attacks was significantly smaller as compared to medium network ($P < .001$) and large network ($P < .001$). In addition, it also revealed that there was no significant difference between the proportion of regular webserver attacks in medium and large networks ($P = .80$). Likewise, from Fig. 2, the proportion of no webserver attacks in small, medium, and large networks were .52, .06, and .03 respectively. The Tukey post hoc test showed that the proportion of no webserver attacks was significantly smaller in small networks compared to that of medium networks ($P < .001$) and large networks ($P < .001$). Also, it revealed that there was no significant difference between the proportion of no webserver attacks in medium and large networks ($P = .73$).

5 The IBL model

Instance-based learning theory, a theory of decisions from experience for dynamic situations. Previously, computational modeling using cognitive theories such as IBL has been proven to be accurate in predicting human behavior in complex situations. Inan IBL model (Dutt et al., 2013; Dutt & Gonzalez, 2012; Gonzalez et al., 2003; Gonzalez & Dutt, 2011, 2012), the instances are created in the memory for each occurrence of an outcome on choice options. An instance in the model comprises of the following structure: situation-decision-utility. The situation refers to the current situation, decision refers to the decision made in the current situation (choice for one of the options), and utility is the goodness of the decision made in the current situation. When a decision is to be made, the instances of each option are retrieved from memory. For each option, these instances are then blended. The blended value of an option is determined by the activation of instances as well as their probability of being retrieved from the memory. The blended value of an option j in any trial t is defined as:

$$V_{j,t} = \sum_{i=1}^{n} p_{i,j,t} x_{i,j,t}$$

where $p_{i,j,t}$ is defined as the probability of retrieval of an instance i for the option j in the trial t; $x_{i,j,t}$ is the utility value of an instance i for the option j in trial t. The option with the maximum blended value is chosen by the model in each trial. The above equation determines the blended value for each option which is calculated as the sum of all observed outcomes weighted by the probability of their retrieval. The probability of the retrieval is a function of activation which is defined as

$$p_{i,j} = \frac{e^{\frac{A_{i,j,t}}{\tau}}}{\sum_{i=1}^{n} e^{\frac{A_{i,j,t}}{\tau}}}$$

where $A_{i,j,t}$ denotes the activation of the instance I corresponding to the option j present in the memory; τ is the random noise, which is defined as $\tau = \sigma * 2$ and σ is the free cognitive noise parameter to represent the uncertainty of recalling the prior experiences from the memory. The activation of the instance in a given trial is determined by the frequency with which its outcome occurs and the time difference between the current time and the past time when the instance's outcome occurred in a task. For each trial t, the activation value of an instance i is defined as:

$$A_i = \ln\left(\sum_{t_{p,i} \in \{1, \ldots, t-1\}} \left(t - t_{p,i}\right)^{-d}\right) + \sigma * \ln\left(\frac{1 - \gamma_{i,t}}{\gamma_{i,t}}\right)$$

where, d and σ are defined as the free parameters known as memory decay and cognitive noise respectively; t is the current trial; $t_{p,I}$ are the previous trials in which outcome with instance i occurred in the task; and $\gamma_{i,t}$ is the random number taken from the uniform distribution between 0 and 1. So, the frequency of occurrence of outcomes in the task and the recency of those outcome observations increase the activation of an instance corresponding to the observed outcome.

The decay parameter d accounts for reliance on recent information. The higher the value of the d parameter, the greater the reliance on recency and the faster is the decay of memory. The σ parameter accounts for the variance in activation of instance from sample to sample. The higher the σ value, the more variability will be in instance activations and trial-to-trial decisions.

5.1 Parameter calibration

We considered two versions of IBL models. One version of the IBL model used the calibrated parameters of d and σ (called the calibrated model); whereas, the other version of the model was based on ACT-R default parameters (called the ACT-R model).

In the calibrated model, we found the best values of d and σ using the human data of an experiment conducted by Katakwar et al. (2020) for different network sizes. Twenty model participants were run in the IBL model across 29 trials. The calibrated model made use of a separate set of d and σ parameters calibrated for different network sizes. In the calibration, we minimized the sum of Mean squared deviations (MSDs) on proportion of different attack decisions between human and model and human across 29 rounds in DG. The MSD was defined as:

$$\text{MSD} = \frac{1}{29}\sum_{t=1}^{29}(\text{model}_t - \text{human}_t).$$

where t refers to the trail from 1 to 29; $model_t$ and $human_t$ refer to the average proportion of different attack actions in trial t from model and human participants respectively. The MSD has the same units as of attack proportion. So, the smaller the value of MSD, the better is the model's fit to human data. Genetic Algorithm, an optimization algorithm, was used to optimize the values of d and σ parameters for both the model participants. This algorithm is known for using bio-inspired operators such as mutation, crossover, and selection to build better solutions for optimization problems. The d and σ parameters were varied in the range from 0 to 10 in the genetic algorithm. The ranges of the parameter guaranteed that the optimization could confidently capture the optimal values of both parameters. The crossover and mutation rates in the genetic algorithm were retained at the default values of 80% and 1%, respectively.

The ACT-R model was based on the ACT-R theory, a unified theory of cognition that has accounted for various phenomena of cognitive science (Anderson, Matessa, & Lebiere, 1997). ACT-R is a cognitive architecture, whose objective is to account for various functionalities of the human mind. The ACT-R model here refers to an agent that is less reliant on recency, frequency, and variability in decision-making. Like the IBL theory, we had d and σ parameters that were set based on the ACT-R default values of 0.50 and 0.25, respectively. Smaller values of d show less reliance on frequency and recency, and smaller values of σindicate smaller variability in trial-to-trial decisions. We did a performance comparison between the calibrated model and the ACT-R model.

5.2 Model results

Table 2 shows the parameter and MSD between humans and the model for different configurations of IBL model across different conditions (network sizes). The MSDs obtained by IBL model with calibrated parameters were smaller compared to the IBL with ACT-R default parameters. Calibrated model had a minimum MSD value for medium and large conditions of 0.2. The decay value in the medium condition for IBL with

TABLE 2 Different model parameters and MSD for IBL model with different configurations across different conditions.

Condition	Model	*d*	σ	MSD
Small	IBL (with calibrated parameters)	6.90	6.58	0.05
	IBL (with ACT-R parameters)	0.5	0.25	0.11
Medium	IBL (with calibrated parameters)	7.72	1.33	0.02
	IBL (with ACT-R parameters)	0.5	0.25	0.07
Large	IBL (with calibrated parameters)	4.60	7.30	0.02
	IBL (with ACT-R parameters)	0.5	0.25	0.08

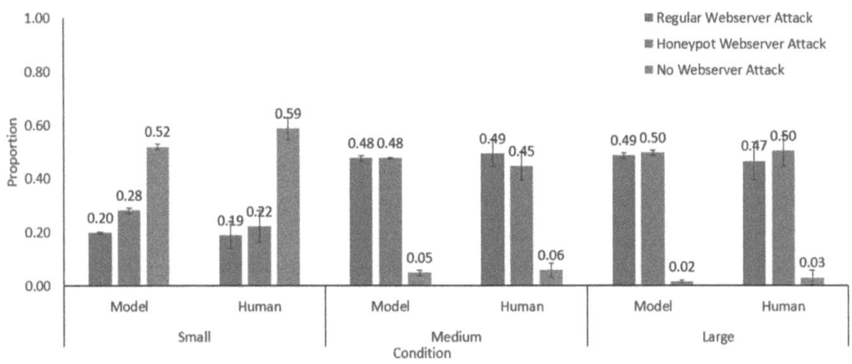

FIG. 3 Proportion of different attack decisions across different conditions in calibrated model. Error bar denotes a 95% confidence interval around the point estimate.

calibrated parameters was maximum ($d=7.72$) among all the different conditions.

Fig. 3 shows the proportion of regular, honeypot, and no webserver attacks across different network sizes in the human data and calibrated model.

Fig. 4 shows the proportion of regular, honeypot, and no webserver attacks across different network sizes in the human data and ACT-R model.

6 Discussion

The rise in cyberattacks has uplifted the research community to think up robust security solutions to combat cyberattacks (CNBC, 2021). Deception via honeypot has been a promising tool to combat cyberattacks

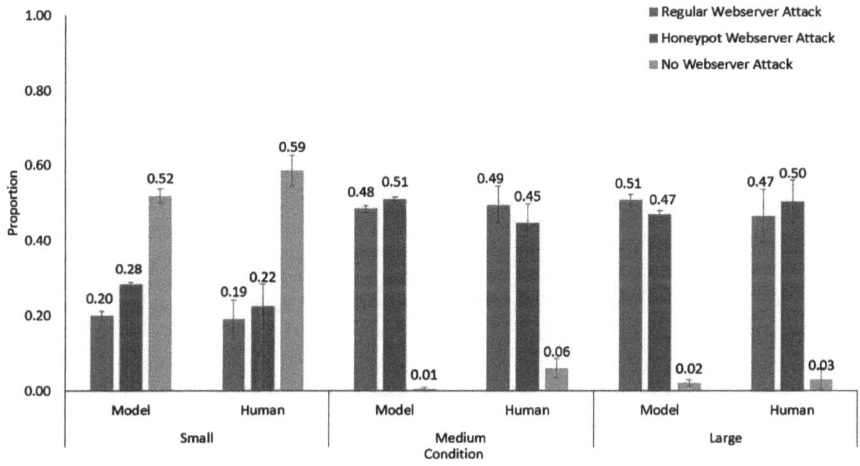

FIG. 4 Proportion of different attack decisions across different conditions in ACT-R model. Error bar denotes a 95% confidence interval around the point estimate.

(Aggarwal et al., 2016a, 2016b, 2017; Katakwar et al., 2020). The current research involving deception in cybersecurity focuses on human factors involved in cyber situations. Among the multiple factors involved in the different cyber situations, network size is most likely to be a significant factor in influencing adversarial decisions. In this research, we investigated the effects of different network sizes on adversarial decisions using the data from a lab-based experiment conducted by Katakwar et al. (2020). Results revealed that the proportion of no webserver probe and no webserver attack decisions were more in a small-sized network compared to medium and large-sized networks. Also, the proportion of honeypot and regular probes and attacks was more in medium and large-sized networks compared to small-sized networks. Furthermore, we explain these results based upon expectations from models built using Instance-based Learning Theory.

First, the proportion of no webserver probe and no webserver attack decisions were more in small-sized networks compared to medium and large-sized networks. As per IBLT, there would be fewer instances in the small-sized networks compared to that of medium and large-sized networks. The fewer instances corresponding to no webserver probe and no webserver attack decisions in small-sized networks are more likely to be activated, resulting in an easier recall of these instances. Overall, the proportion of no webserver probe and no webserver attack decisions were more in small-sized networks.

Second, the proportion of honeypot and regular webserver probes and attacks were more in medium and large-sized networks compared to small-sized networks. Due to the availability of several web servers in medium and large-sized networks, bounded rational participants may not be able to recall the actual mapping of webservers as regular or honeypot from memory. Because the multiple instances per web server would be formed in memory, and the activation of these instances would likely decay in memory due to the large delays in their exploration of choices during probing. Overall, the presence of multiple instances will make it hard to recall instances of medium-sized and large-sized networks, therefore resulting in more exploration during probe and attack stages.

Third, calibration of the IBL model to the human data indicated excessive reliance on recency and frequency among the human participants. The parameters of the IBL model in a small-sized network were $d=6.90$ and $\sigma=6.50$. So, for a small-sized network, the higher value of d and low value of σ showed participants relied more on recent events and less variability in their trial-to-trial choices. This showed that participants gave more emphasis on recent events, thus the participants deviated from attacking, resulting in more not attack decisions in the small-sized network, and less regular and honeypot webserver attack decisions. Similarly, for medium-sized networks, higher decay value ($d=7.72$) and less noise value ($\sigma=1.33$) showed that participants tend to rely on the recent events, thus participants attacked more regular webservers and fewer honeypot webservers. However, for large-sized networks, the higher value of noise ($\sigma=7.30$) and low decay value ($d=4.60$) showed that participants tried to explore all webservers in the network. As a result, resulting in more regular and honeypot webserver attack decisions and fewer not attack decisions in the large-sized network. Furthermore, we also found that the IBL model with calibrated parameters performed better compared to the IBL model with default ACT-R parameters. The reason for this result could be explained based upon the findings of Gonzalez and Mehlhorn (2016). These authors showed that people make more deterministic final choices as a result of excessive dependence on recent and frequent events, and the default ACT-R assumptions of low recency and reasonable variability may not exist.

Our research is limited by the fact that it is based on a lab experiment, and the findings of this study should be viewed from this perspective. As a result, the conditions in the real world may differ from those in the laboratory-based experiment. In addition, in our experiment, participants acting as adversaries were unaware of the deception and nondeception rounds and the true identity of the webservers. The adversary's decisions in the probe and attack stages may have been affected by their knowledge of deception/nondeception rounds and the true configuration of the webservers. Though we tried our best to simulate the real-world dynamics of cyberattack in our experiment. So, the conclusions derived from the experiment are likely to have practical implications in the real world. One implication of our research is that due to the increase in the number of webservers in the network, the proportion of regular webserver probes and regular webserver attacks also increase. Therefore, organizations with a large-sized network may decompose it into many smaller subnetworks. The presence of honeypots in the smaller subnetwork would help to counteract cyberattacks resulting in not attacking the network. Furthermore, models developed based upon the IBLT could be used to account for human decisions. These models could be used to replicate the adversary's decision for an organization having a large number of systems, which in turn helps in minimizing the risk of getting attacked, resulting in reducing the overall cost for the organization.

In the future, we plan to investigate the effects of different proportions of honeypots in deception games on adversarial decision-making. Second possibility could be to investigate the effects of different combinations of deception and nondeception rounds on adversary's decisions in the deception game. Third possibility could be to investigate the effect of different probing costs on an adversary's decision in a deception-based game. Fourth possibility could be to investigate the effectiveness of deception technology against different types of cyberattacks against networks with honeypot webserver. Still another possibility could be to build the computational cognitive models for these ideas that will account for human decisions. We plan to work on some of these ideas in our upcoming experiments.

References

Aggarwal, P., & Dutt, V. (2020). The role of information about opponent's actions and intrusion-detection alerts on cyber decisions in cyber security games. *Cyber Security: A Peer-Reviewed Journal, 3*(4), 363–378.

Aggarwal, P., Gonzalez, C., & Dutt, V. (2016a). Looking from the hacker's perspective: Role of deceptive strategies in cyber security. In *2016 International conference on cyber situational awareness, data analytics and assessment (CyberSA)* (pp. 1–6). IEEE.

Aggarwal, P., Gonzalez, C., & Dutt, V. (2016b). Cyber-security: Role of deception in cyber-attack detection. In *Advances in human factors in cybersecurity* (pp. 85–96). Cham: Springer.

Aggarwal, P., Gonzalez, C., & Dutt, V. (2017). Modeling the effects of amount and timing of deception in simulated network scenarios. In *2017 International conference on cyber situational awareness, data analytics and assessment (Cyber SA)* (pp. 1–7). IEEE.

Aggarwal, P., Gonzalez, C., & Dutt, V. (2020). Hack IT: A real-time simulation tool for studying real-world cyberattacks in the laboratory. In *Handbook of computer networks and cyber security* (pp. 949–959). Cham: Springer.

Almeshekah, M. H., & Spafford, E. H. (2016). Cyber security deception. In S. Jajodia, V. Subrahmanian, V. Swarup, & C. Wang (Eds.), *Cyber deception* (pp. 23–50). Cham: Springer.

Anderson, J. R., Matessa, M., & Lebiere, C. (1997). ACT-R: A theory of higher level cognition and its relation to visual attention. *Human Computer Interaction, 12*(4), 439–462.

Bace, R., & Mell, P. (2001). Special publication on intrusion detection system. Technical report SP-800-31, National Institute of Standards and Technology, Gaithersburg, MD, USA.

Bagchi, K. K., & Tang, Z. (2004). Network size, deterrence effects and internet attack incident growth. *Journal of Information Technology Theory and Application (JITTA), 6*(3), 9.

Business Insider. (2019). Hackers have become so sophisticated that nearly 4 billion records have been stolen from people in the last decade alone. Here are the 10 biggest data breaches of the 2010s. In *Business Insider*. https://www.businessinsider.in/slideshows/miscellane-ous/hackers-have-become-so-sophisticated-that-nearly-4-billion-records-have-been-stolen-from-people-in-the-last-decade-alone-here-are-the-10-biggest-data-breaches-of-the-2010s-/slidelist/71652791.cms#slideid=71652805.

Carroll, T. E., & Grosu, D. (2009). A game theoretic investigation of deception in network security. *Proceedings of the International Conference on Computer Communications and Networks, ICCCN*. https://doi.org/10.1109/ICCCN.2009.5235344.

CNBC. (2021). Organizations are preparing for cyberattacks from quantum computers. In *CNBC*. https://www.cnbc.com/2021/06/07/organizations-are-preparing-for-cyberattacks-from-quantum-computers.html.

Dutt, V., Ahn, Y. S., & Gonzalez, C. (2013). Cyber situation awareness: Modeling detection of cyber attacks with instance-based learning theory. *Human Factors, 55*(3), 605–618.

Dutt, V., & Gonzalez, C. (2012). Making instance-based learning theory usable and understandable: The instance-based learning tool. *Computers in Human Behavior, 28*(4), 1227–1240.

Garcia-Teodoro, P., Diaz-Verdejo, J., Maciá-Fernández, G., & Vázquez, E. (2009). Anomaly-based network intrusion detection: Techniques, systems and challenges. *Computers & Security, 28*(1–2), 18–28.

Garg, N., & Grosu, D. (2007). Deception in honeynets: A game-theoretic analysis. In *2007 IEEE SMC information assurance and security workshop* (pp. 107–113). IEEE.

Gonzalez, C., & Dutt, V. (2011). Instance-based learning: Integrating sampling and repeated decisions from experience. *Psychological Review, 118*(4), 523.

Gonzalez, C., & Dutt, V. (2012). Refuting data aggregation arguments and how the instance-based learning model stands criticism: A reply to Hills and Hertwig (2012). *Psychological Review, 119*(4), 893–898.

Gonzalez, C., Lerch, J. F., & Lebiere, C. (2003). Instance-based learning in dynamic decision making. *Cognitive Science, 27*(4), 591–635.

Gonzalez, C., & Mehlhorn, K. (2016). Framing from experience: Cognitive processes and predictions of risky choice. *Cognitive Science, 40*(5), 1163–1191. https://doi.org/10.1111/cogs.12268.

Katakwar, H., Aggarwal, P., Maqbool, Z., & Dutt, V. (2020). Influence of network size on adversarial decisions in a deception game involving honeypots. *Frontiers in Psychology, 11*, 2385.

Kiekintveld, C., Lisý, V., & Píbil, R. (2015). Game-theoretic foundations for the strategic use of honeypots in network security. In *Cyber Warfare* (pp. 81–101). Cham: Springer.

Lenin, A., Willemson, J., & Sari, D. P. (2014). Attacker profiling in quantitative security assessment based on attack trees. In *Nordic conference on secure IT systems* (pp. 199–212). Cham: Springer. , October.

Mason, W., & Suri, S. (2012). Conducting behavioral research on Amazon's mechanical Turk. *Behavior Research Methods, 44*(1), 1–23.

Quick Heal. (2021). Biggest cyberattacks of 2020. Retrieved from: https://blogs.quick-heal.com/biggest-cyberattacks-2020so-far/. Quick Heal. Last Accessed: 5th April 2021.

Rowe, N. C., & Custy, E. J. (2007). Deception in cyber attacks. In L. Janczewski, & A. Colarik (Eds.), *Cyber warfare and cyber terrorism* (pp. 91–96). IGI Global.

Shang, Y. (2018a). False positive and false-negative effects on network attacks. *Journal of Statistical Physics, 170*(1), 141–164.

Shang, Y. (2018b). Hybrid consensus for averager–copier–voter networks with non-rational agents. *Chaos, Solitons & Fractals, 110*, 244–251.

Trustwave. (2020). Trustwave Global Security Report 2020. Trustwave. Retrieved from: https://trustwave.azureedge.net/media/16791/2020-trustwave-global-security-report.pdf?rnd=132339614670000000. Last Accessed: 5th June 2021.

Wang, L., Jajodia, S., Singhal, A., & Noel, S. (2010). k-Zero day safety: Measuring the security risk of networks against unknown attacks. In *European Symposium on Research in Computer Security* (pp. 573–587). Berlin, Heidelberg: Springer. , September.

CHAPTER

15

Computational modeling of decisions in cyber-security games in the presence or absence of interdependence information

Zahid Maqbool[a], *V.S. Chandrasekhar Pammi*[b], *and Varun Dutt*[a]

[a]Applied Cognitive Science Laboratory, Indian Institute of Technology Mandi, Kamand, India [b]Centre of Behavioral and Cognitive Sciences, University of Allahabad, Allahabad, India

1 Introduction

Cyber-attacks on websites and computer systems are increasing at an alarming rate in the real-world (Forbes, 2014). These attacks can be extremely damaging and costly to governments and businesses, especially if security is breached and confidential or personal data is compromised (Forbes, 2014). Due to the widespread increase in cyber-attacks and their financial consequences, there is an urgent need to focus attention on investigating how certain factors like possessing information about actions and rewards of opponents (i.e., interdependence information) may influence defender's (analyst's) and adversary's (hacker's) attack-and-defend decisions in the cyber world (Alpcan & Başar, 2010; Camerer, 2011; Dutt & Kaur, 2013; Maqbool, Pammi, & Dutt, 2016). Recent research shows that interdependence information likely influences hackers' and analysts' decisions in simulated cyber-security environments. However, currently, less is known about the cognitive processes involved when analysts and hackers make decisions in the presence or absence of interdependence information. This chapter

Cybersecurity and Cognitive Science
https://doi.org/10.1016/B978-0-323-90570-1.00005-X

addresses this shortcoming via computational modeling of hackers' and analysts' decisions in the presence or absence of interdependence information about the opponent's actions.

Cyber-security could be mathematically studied as a strategic game between hackers and security analysts (Katakwar, Aggarwal, Maqbool, & Dutt, 2020). A hacker's goal is to compromise networked systems and to profit from this unauthorized access; whereas, an analyst's goal is to prevent this unauthorized access from taking place (Dutt, Moisan, & Gonzalez, 2016). Game-theoretic studies within the cyber-security domain rely on providing complete interdependence information about possible actions and the resulting outcomes to participants performing as hackers and analysts (Camerer, 2011; Maqbool et al., 2016; Maqbool, Makhijani, Pammi, & Dutt, 2017). When interdependence information about an opponent's actions is present, it is likely to influence human players' decisions compared to when this information is absent (Gonzalez, Ben-Asher, Martin, & Dutt, 2015). According to Gonzalez et al., interdependence information's presence increases the degree of cooperation in classical games like prisoner's dilemma. The reason for this increase in cooperation may be that players account for the opponent's actions and payoffs and their own actions and payoffs while making decisions.

In the cyber world, hackers and analysts may or may not possess interdependence information about their opponent's actions and consequences (payoffs). In one situation, hackers and analysts may possess interdependence information about their opponent's actions and consequences. For example, hackers may acquire information on analysts' actions and payoffs via an insider (Digital Guardian, 2020). Similarly, analysts may acquire information on hackers' actions and payoffs via monitoring tools like Intrusion Detection Systems (IDS) (Aggarwal & Dutt, 2020). However, in the absence of insiders and IDSs, hackers and analysts may not possess interdependence information about their opponents.

Although there is evidence that the presence or absence of interdependence information is likely to influence the decision-making of defenders (analysts) and adversaries (hackers) during cyber-attacks (Gonzalez et al., 2015); however, currently little is known about the cognitive processes involved that would account for the role of interdependence information on the interaction between hackers and analysts.

Behavioral Game Theory (BGT) (Camerer, 2011; Dutt, Ahn, & Gonzalez, 2013), which considers players to possess limitations on memory and cognition, may be a useful tool to understand the role of interdependence information on a hacker's and analyst's actions. BGT conceptualizes the repeated interaction between two players (e.g., hackers and analysts) using simple 2 (players) $\times$ 2 (actions per player) games.

In these games, players make repeated decisions from a set of decision actions available to them and receive payoffs for their actions. The payoff obtained by both players is a function of both players' actions. In literature, Instance-Based Learning Theory (IBLT), a theory of decisions from experience, has been used to explain decision making in dynamic situations (Dutt et al., 2013; Gonzalez & Dutt, 2011). IBLT assumes decision makers store task information in the form of instances and retrieve this information at the time of making decisions. Due to its success in experience-based decisions (Gonzalez & Dutt, 2011), IBLT may likely be useful to understand hackers' and analysts' decisions in the presence or absence of interdependence information.

Maqbool, Pammi, and Dutt (MPD, 2016) performed an experiment involving dynamic security games to understand the role of presence or absence of interdependence information on the decision-making of hackers and analysts. They manipulated the interdependence information across two between subject's conditions: Info and No-Info conditions. In the Info condition, participants performing as hackers and analysts were given a payoff matrix and the feedback of opponents' actions and payoffs after every round of play. While, in the No-Info condition, participants were told that they played against an opponent; however, participants were neither given the payoff matrix nor given feedback about the opponent's actions and payoffs. Although, MPD empirically investigated the influence of presence or absence of interdependence information on the decision-making of hackers and analysts; however, these authors did not explain the underlying cognitive processes involved in decision making in dynamic environments. Furthermore, although, MPD relied upon IBLT to explain their findings, these authors did not specifically create computational IBL models for hackers and analysts on the collected human data.

The primary objective of this chapter is to investigate how the presence or absence of interdependence information is influenced by the cognitive processes as proposed by IBLT and to develop cognitive models based on IBLT that could provide explanations about the findings of MPD. Specifically, IBL models are developed and calibrated to human decisions in MPD's Info and No-Info conditions. The differences in the calibrated values of the cognitive parameters are evaluated across the Info and No-Info conditions of MPD. According to IBLT, human beings have cognitive limitations in terms of memory and recall. In the presence of these limitations, human beings rely on recency and frequency of information, and cognitive noise to make decisions (Anderson, Reder, & Lebiere, 1996; Gonzalez et al., 2015). Overall, IBL models that capture analysts' and hackers' experiential decisions in security games, may allow us to provide a cognitive explanation to how the presence or absence of interdependence information influences these decisions.

In what follows, we first detail the background regarding the working of an IBL model and its uses in dynamic decision-making situations. Next, we report a published experiment (Maqbool et al., 2016), where interdependence information about opponents' actions as hackers and analysts was present or absent in a repeated cyber-security game. Furthermore, we develop IBL models for hackers and analysts using the experimental data and estimate models' parameters across different conditions of interdependence information. We present the model results and discuss people's cognitive mechanisms while performing cyber-security tasks.

2 Background

Literature has revealed that Instance-Based Learning Theory (IBLT) (Gonzalez & Dutt, 2011; Gonzalez & Dutt, 2012; Gonzalez, Lerch, & Lebiere, 2003) has been successful in accounting for human decisions in a number of BGT situations involving hackers and analysts (Arora & Dutt, 2013; Dutt et al., 2013). IBLT uses a cognitive information representation in memory called an instance (Dutt & Kaur, 2013). An instance's data structure comprises three parts: a situation (a set of attributes that characterizes the decision situation), a decision or action, and an outcome or utility of the situation's decision. The different parts of an instance are built through a general decision process with three steps: (1) characterizing a situation by the task's attributes, (2) deciding based on expected utility, and (3) updating the utility when feedback is available. Instances are accumulated over time in memory. In IBLT, instance activations are influenced by the recency and frequency of information, where instances are retrieved, blended, and repeatedly used in future decisions.

In literature, IBLT has been used to model the interaction between hackers and analysts in varied application areas (Camerer, 2011; Dutt et al., 2013; Katakwar et al., 2020). As per IBLT, hackers and analysts possess cognitive limitations and are likely to rely on recency and frequency of available information to make decisions. The application of IBLT to the hackers' and analysts' experiential decisions in cyber-security games is likely to help us explain how these decisions are impacted by the presence and absence of interdependence information about each other actions and payoffs. Specifically, we expect higher proportions of attack-and-defend actions when interdependence information is available to hackers and analysts in a game compared to when such information is absent. That is because, according to IBLT, the presence of interdependence information will enable participants performing as hackers and analysts to update and reinforce instances corresponding to this information in memory. The reinforcement of these instances is likely to cause them to be retrieved easily from memory. To test our expectations above, we develop computational IBL models based upon IBLT ahead in this chapter.

3 Experiment

To study the role of interdependence information among hackers and analysts, MPD performed a laboratory experiment where participants were randomly assigned to one of two between-subjects conditions: Info ($N=50$) and No-Info ($N=50$). In the Info condition, MPD gave participants performing as hackers and analysts a payoff matrix and feedback of opponents' actions and payoffs after every round of play. In contrast, in the No-Info condition, participants were told that they played against an opponent; however, participants were neither given the payoff matrix nor given feedback about the opponent's actions and payoffs. The payoffs in both conditions were governed by the game defined in Fig. 1.

For example, if the hacker attacks (Attack(a)) the computer network and the analyst successfully defends (Defend(d)) the network, then the hacker receives a negative payoff of −10, and the analyst receives a payoff of +5 points. In each condition, half of the participants ($N=25$) performed as hackers and played against participants performing as analysts (N=25). The determination of who would be the hacker and who would be the analyst was made randomly by the software. Each condition was 50-rounds long and each of the 50-rounds was played in one session of interaction between hackers and analysts in real time.

A human hacker had to choose between an "attack" action and a "not-attack" action in each round. Similarly, a human analyst had to choose between a "defend" action and a "not-defend" action in each round. Participants were undergraduate students from computer science and engineering, and they were compensated afterward based upon performance (250 points in the game = INR 1in real money). Participants were also given a base payment of INR 30 (~ USD 0.5). For other details about this experiment, please refer to MPD (Maqbool et al., 2016). It was observed that the presence or absence of interdependence information had a significant effect on the decision-making of humans as hackers and analysts in the game. Furthermore, deviations from Nash equilibrium were also observed.

		Analyst	
		Defend (d)	Not Defend (nd)
Hacker	Attack (a)	−10, 5	5, −5
	No Attack (na)	0, −5	0, 0

FIG. 1 Payoff matrix used in the experiment. The first payoff in each matrix cell corresponds to the hacker player, and the second payoff in each matrix cell corresponds to the analyst player.

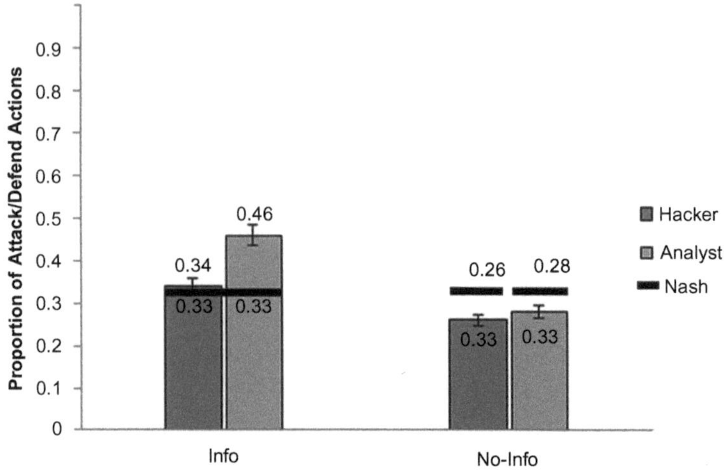

FIG. 2 The average proportion of attack-and-defend actions across both Info and No-Info conditions. The horizontal black line and the value written on each bar represent the Nash equilibrium (0.33) across the conditions. The error bars show 95%CI around the mean value.

4 Results

MPD analyzed the average proportion of attack-and-defend actions in both Info and No-Info conditions. Results indicated that proportion of attack-and-defend actions varied across conditions for both hackers and analysts, respectively (see Fig. 2). For hackers, the average proportion of attack actions in Info condition were significantly higher than that in the No-Info condition ($0.34 > 0.26$; $t(48) = -2.432$, $P < .05$, $r = 0.33$). Similarly for analysts, the average proportion of defend actions were significantly higher in the Info condition than that in the No-Info condition ($0.45 > 0.28$; $t(48) = -3.902$, $P < .01$, $r = 0.49$). These results are as per our expectations stated above. Furthermore, the deviations from the optimal behavior (Nash equilibria) were calculated and a significant deviation from the optimal behavior was observed in most cases (Info: analyst: $t(24) = 7.22$, $P < .01$, $r = 1.44$; hacker: $t(24) = 0.55$, $P > .05$, $r = 0.11$; No-Info: analyst: $t(24) = 3.13$, $P < .01$, $r = 0.62$; hacker: $t(24) = 5.01$, $P < .01$, $r = 1.00$). Only in case of the Info condition, it was observed that the hacker participant played according to optimal behavior (as no significant deviation was observed based upon the *t*-test).

MPD found that the proportion of attack-and-defend actions was more when players possessed interdependence information about their opponents' actions and payoffs compared to when players did not possess this information. When information is present, it is likely that it leads to the creation of attack-and-defend instances in the memory of hackers and analysts early on. With these instances being present and available, hackers and analysts would tend to maximize their payoffs. As per the game's matrix, hackers and analysts could maximize their payoffs by taking attack-and-defend actions, respectively, as these actions would give them positive rewards in the matrix. The overall effect would be an increase in the proportion of attack-and-defend actions when information about the opponent's actions and payoffs is present compared to when this information is not present.

In the next section, we detail the working of a model based upon IBLT that helps account for the effects of interdependence information on decisions made by participants performing as hackers and analysts in the cyber-security game.

5 The IBL model

We developed a model based upon IBLT to explain human behavior in multiperson games. An instance, the smallest unit of experience, in the IBL model consists of three parts: a situation in a task (a set of attributes that characterizes the decision situation), a decision or action is taken, and a reward or penalty resulting from making that decision in that situation. Different parts of an instance are built through a general decision process: encoding of a situation from attributes in the task, a decision, the expectation of an outcome when making a judgment, and updating the outcome in the feedback stage when the actual outcome is known. In the IBL model, instances accumulate over time, are retrieved from memory, and are used repeatedly according to their availability in memory. This availability is measured by a statistical mechanism called activation, originally implemented in the ACT-R cognitive architecture (Anderson & Lebiere, 2014). We develop our model for a two-player security game in Fig. 1 by simply allowing two single-person models to interact with each other in the game repeatedly.

In the IBL model, each instance identifies an alternative to choose for a player (i.e., to Attack or Not Attack for the hacker and to Defend or Not Defend for the analyst) and the outcome obtained (e.g., −10 points for an attack that gets detected). As the situation remains the same for each binary decision, the structure of an instance is simply (alternative, outcome) (e.g., [Defend, 5] in the memory of analyst when she successfully defends an attack from a hacker). In each trial t of the game, the process of choosing an alternative in the model for a player starts with the calculation of the blended value of the alternatives. Next, the alternative with the highest blended value is selected (Eq. 1 below). The blended value of an alternative depends on outcomes occurring in the alternative and the probability of retrieval of instances from memory corresponding to those outcomes

(Eq. 2 below). Furthermore, the probability of retrieval of instances from memory is a function of their activation in memory, governed by the recency and frequency of instance retrievals from memory (Eq. 3 below).

In the IBL model, the selected alternative is one with the highest blended value V (Dutt & Gonzalez, 2012). The blended value of alternative j is defined as:

$$V_j = \sum_{i=1}^{n} p_{ij} x_{ij} \tag{1}$$

where $x_{i,\,j}$ is the value of the observed outcome in the outcome slot of an instance i corresponding to the alternative j and $p_{i,\,j}$, is the probability of that instance's retrieval from memory (for the security game, the value of j is either to Defend or Not-defend for analysts and to Attack or Not-Attack for the hacker). Similarly, $x_{i,\,j}$ are the outcomes from the matrix (5, −5, −10, or 0) for a decision-maker, hacker or analyst, depending upon the decision choices in Fig. 1. The blended value of an alternative j is the sum of all observed outcomes $x_{i,\,j}$ in the corresponding instances in memory, weighted by their probability of retrieval. In any trial t, the probability of retrieval of instance i from memory is a function of that instance's activation relative to the activation of all other instances corresponding to that alternative, given by:

$$P_{i,t} = \frac{e^{A_{i,t}/\tau}}{\sum_j e^{A_{j,t}/\tau}} \tag{2}$$

where τ is random noise defined as $\tau = \sigma \times \sqrt{2}$, and σ is a free noise parameter.

The activation of each instance in memory depends upon the activation mechanism originally proposed in the ACT-R architecture (Anderson & Lebiere, 2014). A simplified version of the activation mechanism that relies on recency and frequency of use of instances in memory was sufficient to capture human choice behavior for data reported in this chapter (Lejarraga, Dutt, & Gonzalez, 2012). For each trial t, activation $A_{i,\,t}$ of instance i was given by:

$$A_{i,t} = \ln\left(\sum_{t_{p,i}\in\{1,\ldots,t-1\}} \left(t - t_{p,i}\right)^{-d}\right) + \sigma \cdot \ln\left(\frac{1-\gamma_{i,t}}{\gamma_{i,t}}\right) \tag{3}$$

where d is a free decay parameter, t is the current trial, and $t_{p.\,i}$ is the previous trial where the instance i was created or its activation was reinforced due to an outcome in the task. The summation will include several terms that coincide with the number of times that an outcome has been observed in previous trials and that the corresponding instance *i*'s activation has been reinforced in memory. Therefore, the activation of an instance corresponding to an observed outcome increases with the frequency of observation (i.e., by increasing the number of terms in the summation) and with the recency of those observations (i.e., by small differences in t_i, $p \in \{1,\ldots, t-1\}$ of outcomes that correspond to that instance in memory). The decay parameter d has a default value of 0.5 in ACT-R, and it affects the activation of the instances directly, as it captures the rate of forgetting. The higher the d parameter value, the more the reliance on recency, and the faster is the decay of memory.

The $\gamma_{i,t}$ the term is a random draw from a uniform distribution bounded between 0 and 1, and the $\sigma.\ \ln\left(\frac{1-\gamma_{i,t}}{\gamma_{i,t}}\right)$ term represents Gaussian noise important for capturing the variability of human behavior. The higher the σ value, the more variability there will be in the retrieval of information from memory. Overall, the noise in Eq. 3 captures the imprecision of recalling instances from memory.

The IBL model had two agents; hacker and analyst, where each agent had two free parameters, noise $\sigma$and decay *d*. In the IBL model, we reinforced each player's individual outcome. The IBL model had two agents, and these agents performed as a pair of model participants for 50 trials in the security game in different conditions (just as human participants performed in the two conditions). In the model, both agents used blending and activation mechanisms independently with a separate set of parameters.

6 Execution of IBL model in the security game

We calibrated the free parameters for the model using human data in both experimental conditions. We minimized the sum of root mean square distances (RMSDs) on attack-and-defend actions between model and human data in this calibration.

$$\text{RMSD} = \sqrt{\frac{1}{50}\left(\sum_{t=1}^{50}\left[(\textit{model attack}_t - \textit{human attack}_t)^2 + (\textit{model defend}_t - \textit{human defend}_t)^2\right]\right)} \tag{4}$$

where *model attack*$_t$ and *human attack*$_t$ refer to the average proportion of attack actions in trial *t* from model and human participants, respectively; *model defend*$_t$ and *human defend*$_t$ refer to the average proportion of defending actions in trial t from model and human participants, respectively; and, t refers to the trial number from 1 to 50. The proportions of attack-and-defend actions were averaged over all participants (model or human) in each trial. The smaller the value of RMSD, the better is the model's fit to human data. A genetic algorithm was used to optimize values of free parameters for both model participants. The d and σ parameters were varied between 0.0 and 10.0. These ranges ensured that the optimization could capture the optimal parameter values with high confidence. The genetic algorithm had a crossover rate of 80% and a mutation rate of 1%. The stopping criteria was when there was no change in the fitness function for the last 50 generations.

7 Results

In this section, we present the results obtained from the IBL model. Table 1 shows the values of the calibrated parameters and RMSDs when the IBL model was fitted to both Info and No-Info conditions, respectively. As seen in Table 1, the RMSD values for both hackers and analysts are small, thereby suggesting a good fit for human data. The value of (decay) d parameters for analysts and hackers were significantly higher across both the conditions than their default value (= 0.5). This exhibited extreme reliance on recency and frequency mechanisms by hackers and analysts for making decisions in the security game. Furthermore, the value of noise parameters for analysts and hackers was significantly higher than their default value of 0.25. These noise parameter values allowed the IBL model to account for participant-to-participant variability in human data. Thus, overall, there was more reliance on recency and frequency of experiences in the IBL model.

Fig. 3 shows the proportion of attack-and-defend actions from human data and those from the IBL model. We observe that the IBL model is able to fit the human data with very low RMSDs (as can be seen in Table 1).

TABLE 1 IBL model with calibrated parameters and MSD values across Info and No-Info conditions.

Condition	d_H	d_A	σ_H	σ_A	$RMSD_H$	$RMSD_A$
Info	6.05	8.05	1.94	2.41	0.15	0.12
No-info	5.12	8.85	1.67	1.56	0.13	0.12

Note: The subscript "H" denotes the hacker and the subscript "A" denotes the analyst.

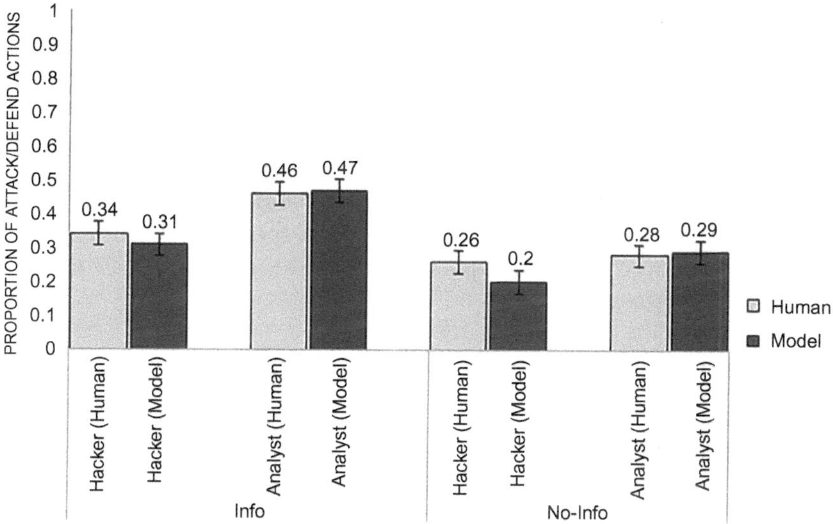

FIG. 3 The average proportion of attack-and-defend proportions across both Info and No-Info conditions from human participants and the IBL model. The error bars represent the 95% confidence interval.

8 Discussion

Cyber-attacks are becoming increasingly common, and they might cause major disruptions of work and the loss of important information. Therefore, it is important to investigate hackers' and analysts' behaviors and how these behaviors are influenced by interdependence information available in simulated cyber-attack situations. This chapter presented a laboratory experiment in which we varied interdependence information and observed its effects on human attack-and-defend proportions.

Our results show that the presence or absence of interdependence information made the proportion of attack-and-defend actions deviate significantly from their Nash equilibriums. The proportion of attack-and-defend actions were significantly higher when the interdependence information was present among players compared to when it was absent. One likely reason for this finding is that the reliance on recency was more in cases when the interdependence information was absent compared to when it was present. Perhaps, due to the reliance on recency, participants were more impacted by losses in the game. To reduce losses, participants tended to reduce their proportion of attack-and-defend actions in the absence of interdependence information.

We also developed a cognitive model based on Instance-Based Learning Theory (IBLT) to represent human participants' decision-making in simulated environments. We observed higher values for d and σ

parameters across the conditions suggesting excessive reliance on recency and frequency (high *d* values). We also observed a high trial-to-trial variability among hackers and analysts across both the conditions due to the high σ values. Furthermore, the trial-to-trial variability was particularly higher in case of the Info condition. One likely reason for this observation may be due to the availability of interdependence information about opponents' trial-to-trial actions and payoffs. Model participants seemed to perform similarly to human participants in the case of Info condition as well as No-Info condition implying that the model fits the human data with a better approximation (see Fig. 3).

Our findings do have real-world implications. First, we observed that the proportion of actions (attack/defend) was higher in the case of Info condition compared to No-Info condition, suggesting that availability of interdependence information regarding opponent's actions and payoffs leads to an increase in the proportion of attacks as well as defends. In the real world, analysts may obtain information about the possible actions of hackers by employing honeypots (Gary & Diane, 2011). Similarly, hackers may be able to obtain information about analyst actions and payoffs via insiders (Fernandes, Soares, Gomes, Freire, & Inácio, 2014). Overall, when the interdependence information is known, hackers and analysts are likely to attack and defend more, and these attacks and defend actions may prove to be costly. However, in case the interdependence information is absent, the attacks and defends would be less. Thus, there may be some value to protecting the leakage of interdependence information, which might help organizations save costs of defending against cyber-attacks. Also, due to the lack of interdependence information, the hackers may also wage fewer cyber-attacks on networks. Second, a model based on IBLT was able to closely mimic the actions of hackers and analysts in the presence or absence of interdependence information. Such IBL models may be useful in deriving predictions in scenarios where the interdependence information may be partially known to one of the parties (hackers or analysts). Also, such IBL models may enable the penetration testing of networks against analyst cyber-security actions.

There are a number of ideas as part of future research. First, there may be a number of decision aids (like intrusion detection systems or honeypots) that may be available to analysts for their daily decisions. It would be interesting to study the role of such decision aids as part of our future work both via human experiments and cognitive models. Second, we would also like to study the role of presence or absence of interdependence information of the efficacy of the patching process. For example, if a defender patches the network, then it is reasonable to assume that the payoffs for any subsequent attack made by the attacker will favor the defender if the network responds consistently to the defenders' last actions. However, if the network has technological inefficiencies, it might not respond consistently to the defenders' last actions. In addition, the payoff information may change due to inaccuracies in the network. Thus, the presence or absence of information related to inaccuracies of the network and the payoffs and their influence on decision making of hackers and analysts may be studied as part of our future research.

9 Conclusion

In the real world, hackers and analysts may know very little about the actions and payoffs of their opponents. Thus, it is likely that the proportion of attack-and-defend actions taking place in the real world are much less than what it would be if hackers and analysts could know each other's actions. This lack of information about the opponent's actions may make both roles perform close to their Nash optimal proportions. However, analysts could get to know about their opponent's actions to a certain extent via tools like honeypots. In such cases, analysts may not be able to perform in an optimal manner compared to what they do when they do not have such tools available at their end.

Acknowledgment

This research was supported by the Department of Science and Technology, Government of India award (National mission on interdisciplinary cyber physical system (NM-ICPS) implementation mechanisms—Technology innovation hubs (TIHs) IITM/DST/VD/288) to Dr. Prem Felix Siril (PI), Dr. Varun Dutt, Dr. Arnav Bhavsar, Dr. Anil K. Sao, Dr. Aditya Nigam, Dr. Gopi Srikanth Reddy, Dr. Srikant Srinivasan, Dr. Dileep AD, and Dr. Satyajit Thakor. Also, we are grateful to the Indian Institute of Technology, Mandi, for providing the necessary computational resources for this project.

References

Aggarwal, P., & Dutt, V. (2020). The role of information about opponent's actions and intrusion-detection alerts on cyber decisions in cyber security games. *Cyber Security: A Peer-Reviewed Journal, 3*(4), 363–378.

Alpcan, T., & Başar, T. (2010). *Network security: A decision and game-theoretic approach.* Cambridge University Press.

Anderson, J., & Lebiere, C. (2014). *The atomic components of thought.* Hoboken: Taylor and Francis.

Anderson, J. R., Reder, L. M., & Lebiere, C. (1996). Working memory: Activation limitations on retrieval. *Cognitive Psychology, 30*(3), 221–256.

Arora, A., & Dutt, V. (2013). Cyber security: Evaluating the effects of attack strategy and base rate through instance based learning. In *12th International conference on cognitive modeling. Ottawa, Canada.*

Camerer, C. F. (2011). *Behavioral game theory: Experiments in strategic interaction.* Princeton University Press.

Digital Guardian, 2020. Insider vs. Outsider Data Security Threats: What's the Greater Risk? Retrieved from: https://digitalguardian.com/blog/insider-outsider-data-security-threats, September 18, 2020.
Dutt, V., Ahn, Y. S., & Gonzalez, C. (2013). Cyber situation awareness: Modeling detection of cyber attacks with instance-based learning theory. *Human Factors, 55*(3), 605–618.
Dutt, V., & Gonzalez, C. (2012). The role of inertia in modeling decisions from experience with instance-based learning. *Frontiers in Psychology, 3, no. xx*, 177.
Dutt, V., & Kaur, A. (2013). Cyber security: Testing the effects of attack strategy similarity and experience on cyber attack detection. *International Journal of Trust Management in Computing and Communications, 1*(3–4), 261–273.
Dutt, V., Moisan, F., & Gonzalez, C. (2016). Role of intrusion-detection systems in cyber-attack detection. In *Advances in human factors in cybersecurity* (pp. 97–109). Cham: Springer.
Fernandes, D. A., Soares, L. F., Gomes, J. V., Freire, M. M., & Inácio, P. R. (2014). A quick perspective on the current state in cybersecurity. In *Emerging trends in ICT security* (pp. 423–442). Morgan Kaufmann.
Forbes, 2014. "Your-cybersecurity-don't-count-on-the-government" [online]. Available: http://www.forbes.com/sites/marcwebertobias/2014/05/12/your-cybersecurity-dont-count-on-the-government/#2715e4857a0b7f13779679dc. (Accessed [11-September-2015).
Gary, K., & Diane, G. (2011). Analysis of attacks using a honeypot. In *International cybercrime, security and digital forensics conference*.
Gonzalez, C., Ben-Asher, N., Martin, J. M., & Dutt, V. (2015). A cognitive model of dynamic cooperation with varied interdependency information. *Cognitive Science, 39*(3), 457–495.
Gonzalez, C., & Dutt, V. (2011). Instance-based learning: Integrating sampling and repeated decisions from experience. *Psychological Review, 118*(4), 523–530.
Gonzalez, C., & Dutt, V. (2012). "Refuting data aggregation arguments and how the IBL model stands criticism": A reply to Hills and Hertwig. *Psychological Review, 119*(4), 893–898.
Gonzalez, C., Lerch, J. F., & Lebiere, C. (2003). Instance-based learning in dynamic decision making. *Cognitive Science, 27*(4), 591–635.
Katakwar, H., Aggarwal, P., Maqbool, Z., & Dutt, V. (2020). Influence of network size on adversarial decisions in a deception game involving honeypots. *Frontiers in Psychology, 11*, 2385.
Lejarraga, T., Dutt, V., & Gonzalez, C. (2012). Instance-based learning: A general model of repeated binary choice. *Journal of Behavioral Decision Making, 25*(2), 143–153.
Maqbool, Z., Makhijani, N., Pammi, V. C., & Dutt, V. (2017). Effects of motivation: Rewarding hackers for undetected attacks cause analysts to perform poorly. *Human Factors, 59*(3), 420–431.
Maqbool, Z., Pammi, V. C., & Dutt, V. (2016, June). Cybersecurity: Effect of information availability in security games. In *2016 International conference on cyber situational awareness, data analytics and assessment (CyberSA)* (pp. 1–5). IEEE.

Index

Note: Page numbers followed by *f* indicate figures, *t* indicate tables, and *np* indicate footnotes.

K

L

Q

R

S

T

Kolibri 360 GmbH
Ferdinand-Jühlke-Straße 7
99095 Erfurt, Deutschland
produktsicherheit@kolibri360.de